W9-CYU-379

Studies in Consciousness / Russell Targ Editions

S ome of the twentieth century's best texts on the scientific study of consciousness are out of print, hard to find, and unknown to most readers; yet they are still of great importance. Their insights into human consciousness and its dynamics are still valuable and vital. Hampton Roads Publishing Company—in partnership with physicist and consciousness research pioneer Russell Targ—is proud to bring some of these texts back into print, introducing classics in the fields of science and consciousness studies to a new generation of readers. Upcoming titles in the *Studies in Consciousness* series will cover such perennially exciting topics as telepathy, astral projection, the after-death survival of consciousness, psychic abilities, long-distance hypnosis, and more.

BOOKS IN THE STUDIES IN CONSCIOUSNESS SERIES

An Experiment with Time by J. W. Dunne

Mental Radio by Upton Sinclair

Human Personality and Its Survival of Bodily Death by F. W. H. Myers

Mind to Mind by René Warcollier

Experiments in Mental Suggestion by L. L. Vasiliev

Mind at Large edited by Charles T. Tart, Harold E. Puthoff, and Russell Targ

Dream Telepathy by Montague Ullman, M.D., and Stanley Krippner, Ph.D., with Alan Vaughan

Distant Mental Influence by William Braud, Ph.D.

Thoughts Through Space by Sir Hubert Wilkins and Harold M. Sherman

The Future and Beyond by H. F. Saltmarsh

Mind-Reach by Russell Targ and Harold Puthoff

The Secret Vaults of Time by Stephan A. Schwartz

RUSSELL TARG EDITIONS

UFOs and the National Security State by Richard M. Dolan

The Heart of the Internet by Jacques Vallee, Ph.D.

STUDIES IN ⊂⊙ CONSCIOUSNESS

Russell   Targ   Editions

The Secret Vaults of Time

Psychic Archeology and the Quest for Man's Beginnings

Stephan A. Schwartz

HAMPTON ROADS
PUBLISHING COMPANY, INC.

Originally published in 1978 by Grosset & Dunlap
Original copyright © 1978, 2001 by Stephan A. Schwartz

Cover design by Bookwrights Design copyright © 2005
Cover photos © Digital Stock, adapted by Mayapriya Long. All rights reserved.

All rights reserved, including the right to reproduce this
work in any form whatsoever, without permission
in writing from the publisher, except for brief passages
in connection with a review.

Hampton Roads Publishing Company, Inc.
1125 Stoney Ridge Road
Charlottesville, VA 22902

434-296-2772
fax: 434-296-5096
e-mail: hrpc@hrpub.com
www.hrpub.com

If you are unable to order this book from your local
bookseller, you may order directly from the publisher.
Call 1-800-766-8009, toll-free.

Library of Congress Catalog Card Number: 77-71742
ISBN 1-57174-431-2
10 9 8 7 6 5 4 3 2 1
Printed on acid-free paper in Canada

For

my mother

B.W.S.

and in memory

of

my father

A.L.S.

Contents

INTRODUCTION TO *THE ENGINEERING OF PSI* TRILOGY iii

PREFACE vii

ACKNOWLEDGEMENTS xi

I. THE GLASTONBURY SCRIPTS: FREDERICK BLIGH BOND 1

II. THE EYES WHICH SEE EVERYTHING: OSSOWIECKI AND
PONIATOWSKI 57

III. THE SCOTTISH GENERAL AND THE RUSSIAN RODWALKERS:
SCOTT ELLIOT AND PLUZHNIKOV 108

IV. THE TRANSITIONAL MAN: J. NORMAN EMERSON 136

V. THE CANADIAN AND THE SEER'S SON:
EMERSON AND CAYCE 165

VI. CHILDREN OF THE CHANGE: GARRAD AND REID 198

VII. INTO THE AMERICAN MAINSTREAM: WEIANT, SWANTON,
LONG, AND THE MEXICO CITY AAA MEETING 222

VIII. KUHN, CONTEXT, AND REVOLUTION: THOMAS KUHN 247

IX. PREJUDICE, PAIN, AND PARADIGM 263

X. COMPLEXITIES, PROMISE, AND IMPLICATIONS 290

APPENDIX I: PREFIELDWORK TESTING 317

APPENDIX II: FIELDWORK PROTOCOL 326

SOURCES AND NOTES 331

BIBLIOGRAPHY 353

INDEX 367

INTRODUCTION TO THE ENGINEERING OF PSI TRILOGY

The three books in this series, *The Secret Vaults of Time, The Alexandria Project,* and *Mind Rover: Explorations with Remote Viewing* comprise the bulk of my experimental research, as well as the intellectual lineage from whence it springs. Each volume is complete within itself. However, they are also a trilogy. Each explores the practical application of Remote Viewing, a mysterious human ability to know things in spite of being shielded from that knowledge by time or space, or both.

Volume One—*The Secret Vaults of Time:* Tells the stories of researchers from around the world who over a century prior to my involvement successfully located and reconstructed archaeological sites using Remote Viewing. These are the documented stories of researchers and Remote Viewers, the sites they found, and the reconstructions they described. It is from this lineage that my own work in archaeological Remote Viewing sprang.

Volume Two—*The Alexandria Project:* In the Fall of 1995, a combined French-Egyptian archaeological team working in the Eastern Harbor of Alexandria, Egypt announced they had discovered the palace of Cleopatra, the remains of the Lighthouse of Pharos, and many other sites. It created the next dayís headlines, and ongoing media coverage. The work is fascinating, important, and worthy of the attention it has received. What it is not, for the most part though is original discovery. Sixteen years earlier, in 1979, The Mobius Society had surveyed the harbor, and discovered the same sites, and many of the same artifacts.

What followed, and what is happening now, provides a case study in how science and the media deal with things they can not explain, do not like for ideological reasons, yet which they can not make go away, because they are authentic, and because strong popular interest will not be denied. *The Alexandria Project* is the true story of how researchers from five universities and organizations put the claims of Remote Viewing to the ultimate test: Was it is possible under rigorously controlled conditions for some part of human consciousness to direct us to lost chapters and places in history? Were the sites where the Remote Viewers marked them on their maps? Were the artifacts they described at those sites actually there?

A second test was also carried out: comparing the relative accuracy of information derived from traditional remote sensing, in this case side-scan sonar, and proton precession magnetometer, with information provided by Remote Viewing. This book, and the papers, and film record that accompany it, have lain dormant as a kind of time capsule, awaiting some independent confirming or disproving event. That event, in the form of the Franco-Egyptian expeditionís reports has now happened, and it is possible in this newly revised edition to look back on both the skeptics and the research they criticized and make a determination as to who was correct. In both the generalities and specifics the second expedition has confirmed the claims of the first.

Volume Three—*Mind Rover: Explorations with Remote Viewing:* The final volume of the series is comprised of the actual research papers I presented at various scientific conferences over almost 20 years. This work, carried out on both land and sea, in the United States, Egypt, Jamaica, and the Bahamas, with the help of dozens of specialists from universities and institutions around the world, represents a unique body of exploratory science. Many of these reports include comparisons between Remote Viewing and a variety of electronic remote sensing technologies, and demonstrate that Remote Viewing has proven capable of producing results when other approaches have failed.

Together the three books argue for the replacement of the materialist/ physicalist view of the world, offering in its stead, a world view in which all consciousness is inter-connected and interdependent. Here are some of the defining parameters of the materialist world view:

The Physicalist/Materialist Model

1.)   The mind is solely the result of physiologic processes;
2.)   Each consciousness is a discreet entity;
3.)   No communication is possible except through the defined physiologic senses;
4.)   Consciousness dwells entirely within the time/space continuum.

This way of looking at the world is till the dominant perspective, but it is being challenged by an interdependent paradigm that is emerging. Some of its principal hallmarks are:

The Interdependent Inter-connected Model

1)   Only certain aspects of the mind are the result of physiologic processes;
2)   Consciousness is causal, and physical reality is its manifestation;
3)   All consciousnesses, regardless of their physical manifestations, are part of a network of life which they both inform and influence and are informed and influenced by: there is a passage back and forth between the individual and the collective;
4)   Some aspects of consciousness are not limited by the time/space continuum.

The most important thing to remember is that *The Engineering of Psi* is but one facet in a growing body of rigorously conducted research in a wide variety of disciplines, from medicine to physics, all pointing in the same direction.

PREFACE

This book began, as I suspect many do, as a personal quest; initially it had nothing to do with either the psychic, or archaeology. It started with a fascination for what I then thought of as science; a monolithic whole (as I saw it originally) which increasingly dominated our culture and determined the paths which it chose to tread.

Research into the literature of many disciplines quickly presented me with the paradox that science was indeed a whole— and yet was not. Sciences, nothing more than the products of human consciousness, often disagreed (sometimes violently) on specific points, but under all there was most definitely a monolith—the general agreement under which all sciences were conducted—this was the almost universal concurrence as to what constituted reality. But a background in history told me that reality changes, literally. Today, assumptions as to what things are differ radically from those held in the time of Newton, when man was considered to be the result of a Special Act of Creation and fossils were frozen fairies. These shifts in thinking are not limited only to an understanding of evolution and the nature of fossilization.

If this were true—that reality changes—then, in order to see what new reality was potentially emerging I needed to look at science's leading edge. From this vantage point it was no great leap to an interest in the psychic since research in this area is unquestionably on the far edge, and the psychic, after all, is a cause concerned with nothing but human consciousness and its relationship with reality. The difficulty here lay in the fact that even academic parapsychologists, those researchers under whose special purview the psychic fell, were not sure it existed. Consequently, their experiments were primarily oriented toward small, easily replicable psychic events such as card calling, which amply demonstrated by statistical analysis the presence of psi phenomena but did little to integrate psi into the scientific mainstream.

My starting point here was to operate on the assumption that psi did actually exist, a premise which I felt was fully tenable in light of the statistical research already done and which, in any case, was at least as good a working stance as the supposition that psi did not exist. This position, although I suppose it is controversial in one sense, is precisely the same philosophical posture taken by researchers whose field of interest is gravity, light, or electricity; no one knows what these forces really are, but for centuries that has not stopped scientists from studying their effects and defining reality from the conclusions they draw from their work.

This working hypothesis (that psi exists, therefore we must find what we can learn about ourselves from applying practically the information it has to offer) still left the problem of how to devise an experiment which allowed, at least to the limits of present calibration, for no alternative explanation to the resolution of a problem but through the intervention of psi-phenomena. What was required was an experimental protocol which was universally agreed to be triple-blind; that is, the answer to the challenge posed by the experiment could not be ascertained from any known source (thus eliminating questions of telepathy or cheating).

Such a protocol would obviously have to be inter-disciplinary, since no single experimenter or even a group within a single scientific discipline could possibly command the spectrum of information necessary to assure experimental rigor. But it would also have to focus primarily in a single discipline, in terms of the attempted achievements, if the results were to have application to mainstream orthodox science. This perspective dictated the creation of a team of researchers (from several disciplines) and psychics, but still left the experimental focus in question.

At first I thought of astronomy. To ask a psychic to locate a previously unknown black hole seemed to me the height of elegance. There could be no question of the target being there or not—objective instrumentation would determine that—nor could there be any vulnerability to the criticism that the psychic had somehow done sleight-of-hand, or heard the answer on his grandmother's knee. No magician, however artful, can put a celestial body in the sky. But there were considerations that ruled against astronomy.

To begin with, it is a discipline whose practice requires extraordinarily expensive equipment; as a result those instruments that do exist are always in demand for more orthodox experiments. I thought it unlikely that I could convince an astronomer to give up his limited time on what, to most, must surely seem a fool's chase. Even more than that, by the time I had arrived at the design of a triple-blind protocol I had also arrived at the conviction that the experiment should say more than something about a single individual's capabilities—although admittedly success would, by extension, say something about us all. To explore the full potential of the idea the experiment should also address the nature of man's condition; give some insight into his past that might possibly guide his future. At the very least the results, if validated, should help to resolve closed and ancient puzzles which did not admit to solution through orthodox channels. For this, anthropological archaeology seemed ideally suited.

The triple-blind conditions could be maintained. It is after all just as "clean" an experiment if the psychic first locates a previously unknown site; then describes the surface geography and sub-surface geology; then locates and identifies artifacts within the location and finally, describes their positions, as it is to locate a hidden body in space. And if these conditions are met, then the reconstruction of what the objects meant to the people who built and used them must be undeniably compelling; it is at least as good as a reconstruction contrived *after* objects are discovered.

To my surprise, as I began what has become a multi-year search through archaeological literature using such primary sources as field notes, site reports, and personal interviews, I found that the particular questions psychic archaeology was peculiarly well-fitted to answer were the very ones over which archaeologists as a group were agonizing, and which they felt no orthodox techniques would ever be able to solve. But something far more exciting and intriguing emerged from my search through little-used files, long-unread reports, and dusty, attic-stored notes: I was by no means the first person down this path. Beneath the uniformity of orthodoxy flowed another stream, in many countries, over many years. Others before me had not only considered the use of psychic data in archaeology, but also had put these considerations into practice, and succeeded!

The first seven chapters of this book are the results of my quest, the work of many individuals whose psychic research is largely unknown (or for reasons of self-protection and career security never published), although the actual finds they made, and they themselves, may be famous. For three-quarters of a century psychic archaeology has been a reality.

But having finished, I realized I had not gone far enough. I was still trapped at the level of reporting phenomena which, however interesting, are still relatively meaningless unless given context, first within the history of archaeology, and then within science as a whole. The desire to provide this was the impetus for the last three chapters but, again, as I was writing something else emerged.

Psychic archaeology, and the strange anomalies it produces by implication (where has this information been that the psychics pick up and how do they get it?), presents a radically different view of humankind from the materialist (proof oriented), theological (faith oriented) dichotomy that currently rules most of Western thought. It speaks to both worlds, and provides something each is lacking.

Psychic archaeology is ultimately important not because it deals with the psychic or archaeology, nor because it answers questions thought to be unsolvable (which it does), nor because it is one facet of a coming shift in scientific reality as profound as that which came after Copernicus or Darwin (which it is). It is important because of what it has to offer us, to you and me, in resolution of our timeless quest to know ourselves. As Nietzsche says, "It would seem as though we had before us, as a reward for all our toil, a country still undiscovered, the horizons of which no one has seen, a beyond to every country and every refuge of the ideal that man has ever known, a world so overflowing with beauty, strangeness, doubt, terror and divinity, that both our curiosity and our lust of possession are frantic with eagerness."

STEPHAN A. SCHWARTZ

ACKNOWLEDGMENTS

ONE MUST REALLY write a book to realize why these pages are amongst the most important in an author's life; it is his chance to pay back in some small way the people who finally made it possible for him to put his name on a finished work.

In my case I must begin by thanking all those whose names are scattered throughout the text. The interviews they granted, and sometimes this meant literally dozens of hours generously taken from schedules already overfilled; the often long-forgotten site reports they rescued from obscure attic files; and the correspondence resurrected only after long searches make up the primary foundation upon which the structure of the book is laid. The list is a long one and, since the names will be more meaningful as they emerge in their proper places as the book develops, I will only say here that I give them all my heartfelt thanks for their unstinting support and cooperation.

But just as a film has people both before and behind the cameras, so this book profits by the assistance of many whose names do not appear. I want to begin by thanking my typists, Diane Ellison, Nancy Mathern, and Nadine Woods, each of whom labored uncomplainingly first through handwritten manuscript and then draft after draft of changes. Also I thank Dane Clark who transcribed several hundred hours of taped interviews, and who even checked every name and book title to see that each was accurately denominated and spelled. His careful attention to details such as these made my task immeasurably easier.

To Christopher Bird and Andrew Nowina-Sapinski I owe a special debt. Bird not only supplied me with many technical papers from Eastern Europe and the Soviet Union previously unknown in the West, but also took time out from his own research and writing to provide me with accurate translations. Nowina-Sapinski cheerfully undertook the translation, from Polish into English, not only of the Poniatowski manuscript and the Poniatowski-Ossowiecki experiments in prehistory—a number of which were still in Poniatowski's original handwriting—and a wide selection of old newspaper clippings and affadavits, but also opened the door for me to a Polish culture long since vanished. His contributions and those of Bird greatly enrich my finished efforts.

Each person quoted or mentioned in the book who was still alive read the section of the manuscript in which he or she was mentioned to assure accuracy, yet a further act of generosity. In addition, I wish to thank Professor William Rathje, of the Department of Anthropology of the University of Arizona, who read the entire manuscript, for his suggestions and commentary. Furthermore, I would like to single out his departmental chairman, Professor Raymond H. Thompson, whose suggestions, introductions, and encouragement meant a great deal to me as I was doing my research.

To Peter Tompkins, I would also like to extend my thanks. His careful page-by-page commentary is responsible for much of the polish the work contains.

Recognition is due also to the entire staff of the Library of the University of

Arizona. Time after time these men and women helped me track down obscure reports, maps, and books. I regret that I did not keep a list of their names, for they truly played a significant role. At the Arizona State Museum Library of Anthropology I also received every assistance and, for this, I would like to particularly single out the cooperation of librarian Dauphne Smith who not only found the things for which I was searching, but also suggested papers and reports that she felt would contribute to the depth of my research. The library staffs of the Association for Research and Enlightenment, and the Philosophical Research Society are also due thanks for providing rare books from their collections.

Every book, I suspect, has one or more individuals who function as something of a cross between fairy godparent and patron; and this book is no exception. In my case the role was filled by four people. Jerry Lane and Laurie Allen read every word of the manuscript, and their suggestions helped smooth out many rough passages. But this was far from constituting their total support. Laurie Allen opened the shelves of The Book Stop in Tucson, one of the most marvellous used book stores it has ever been my good fortune to discover. From her stock I was loaned dozens of books, including many unavailable from any library, and my bibliography in large measure is a reflection of this bounty. Equally important, was the support of Dorothy and Frank Adams. It was in their house that this book was written; it was their friendship and encouragement that helped over many difficult times, and it was their careful proofing which made the task of my editor a little easier. I would also like to thank Fred and Ruth Cole who loaned a virtual stranger their mountain cabin for the critical last phases of the rewriting.

To my editor Diana Price I would like to give a very tremendous thanks. She shared my vision of what this book should be and say and in every way fought to see that the final work represented that original vision.

Finally to my wife Catherine and daughter Catherine I pay a special tribute. They not only tolerated my strange hours and frequent absences but provided a cradle of love and support in which the book could live and grow.

The
Secret
Vaults
of
Time

UNITED KINGDOM

I

THE GLASTONBURY SCRIPTS

Frederick Bligh Bond

"Wee laid down seventy and two, but they builded longer, and he who followed made new schemes. Ye must use your talents. Piece by piece ye shall rebuild it ... *Benedicte. Whyttinge, nuper Abbas* [Blessings. Whitting, the late Abbott]." These words were written on June 16, 1908, as other words, some thirty-two times before, had been psychically produced while two Victorian gentlemen sat in Frederick Bligh Bond's architectural office in the Star Life Building in Bristol, England.

For over a year now Bond had been consumed with a desire to go back into history and learn the past of Glastonbury Abbey—one of the most famous Christian shrines in Great Britain. His chosen vehicle for this movement in time was automatic writing, a psychic technique in which while the hand writes, the mind is allowed to go blank or to focus on something other than the writing, in hopes that either the unconscious or some paranormal impulse will assert itself.

At each session Bond and his long-time friend, Captain John Allen Bartlett, would seat themselves in Bond's office (both men needed to be present), put between them a sheaf of foolscap, and, with Bond's right hand laid lightly over Captain Bartlett's, which held a pencil, begin to talk casually to each other.

Sometimes in the midst of their conversation their joined hands would begin to trace across the paper, but the two men ignored the movement, keeping carefully to their conversation, or Bond would read to his friend from one of the lighter popular works of the times. As a page was filled, they would quietly take it up, turn it over, and place it face down on a second pile of paper—never looking at what had been written.

Frederick Bligh Bond, as he looked after losing his post at Glastonbury. It was during his editorship of *Psychic Science* that he began his friendship with Sir Arthur Conan Doyle.

Only when the cold in the unheated room or a sense of "flagging energy" touched them would they stop. Only then would they examine the results of what became known as a "sitting."

Their questions, addressed to the air and interspersed with conversation or reading aloud, were almost always about the Abbey, and their interest was far from casual. In 1908 Bond, an ecclesiastical architect and member of the Somerset Archaeological and Natural History Society, had been placed in charge of operations to excavate the old Abbey, which was then in ruins.

It was unusual for a professional man in Bond's field to use the psychic as a working tool, more unusual still to get a self-described "conservative" old retired Indian Army officer to help, and most unusual of all to assume he could conduct an archaeological excavation under the direction of a psychic source that identified itself as a band of long-dead monks who claimed to have lived in and built the Abbey.

Both men had developed an interest in the paranormal over many years, and after discussing the possibilities of getting information about Glastonbury through some type of psychic technique, they determined, almost jokingly, to put their proposition to the test. Their chosen approach, after rejecting mediumistic seances, was automatic writing.

What happened during that first session, held on a November afternoon in 1907, changed Bond's life. At that initial sitting, after asking the question, "Can you tell us anything of Glastonbury?" he first received words telling him that all information survived, and then, as Bond tells it: "JA's hand moved and began to trace a line, ultimately making a drawing which on inspection looked like a recumbent cross, but which when examined proved to be a fairly correct outline of the main features of the Abbey church, traced by a single continuous line . . . down the middle of the plan were written the words Gulielmus Monachus (William the monk).

At his very first sitting, Bond received information on the legendary Edgar Chapel. This crude but accurate drawing was his first contact with the Company of Avalon. Note the double-lined section at the left; this is the famous chapel.

Let's see it!

"Next followed what appeared to be an elaborate plan of the great enclosure of the Abbey Church, with a sketch of a central tower, with square pinnacle top, a west front or gabled façade, with two peaked turrets and a large arched light or window between them" as well as other highly specialized architectural details.

Both Bond and Bartlett could wait no longer. Taking up the paper, they looked at what their hands had done. It was the kind of drawing a blindfolded man might produce—shaky but definitely recognizable.

After the immediate excitement had passed, Bond's elation turned to doubt. There was a major error. How could the east end be that large? It did not conform to anything he had read in the literature on Glastonbury. Perhaps their automatic drawing meant nothing; perhaps it was a subconscious fantasy, produced to meet their intense desire for a psychic answer. But perhaps the drawing was correct, and what Bond knew about Glastonbury was wrong. There seemed only one way to find out. They resolved to try again.

Bond asked the source, "Please give us a more careful drawing of the chapel sketched just now at the east end of the great church."

Again, after a few moments, Bartlett's hand began to move, and a new outline of the chapel was produced. Again the line at the east end showed the unknown structure. It was drawn double for emphasis. Below this was written in cramped characters, difficult to decipher: *"Capella St. Edgar. Abbas Beere fecit hanc capellam St.*

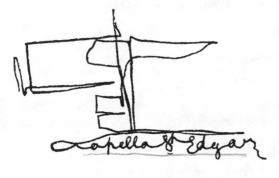

The first outline was so at variance with everything else Bond knew about Glastonbury, that he asked for verification. It came along with the words, Chapel of St. Edgar.

Edgar . . . martyri et hic edificavit vel fecit voltam . . . fecit voltam petriam quod voratur quadripartus sed Abbot Whiting . . . destruix et . . . et restoravit . . . eam cum nov . . . multipart . . . nescimus eam quod vocator."

When they were finally able to determine what the tortured writing said, the answer was even more disturbing than the fact that they were receiving it.

The scrawl in Latin told of a chapel, named for King Edgar, built by Abbot Beere in the first part of the sixteenth century, added to, along with other building projects, by his successor, an Abbot Whiting, in whose reign it was destroyed. A chapel whose existence, although alluded to in a few ancient manuscripts and vaguely mentioned in local lore, had never seriously been considered.

The information was fascinating—but Bond had to admit to himself that much of what is received psychically *is* fascinating, and that a good tale is not necessarily a true one. Nothing short of digging in the Abbey grounds would resolve the question of accuracy. The following summer Bond got this chance, as director of excavations. Using the directions provided by the monks, Bond discovered the Edgar Chapel, and much more, all of it of great significance. In the process this strange, touchy Victorian became the father of psychic archaeology and initiated an entirely new approach to locating, reconstructing, and understanding the past.

Why was Bond so absorbed with the Abbey and its past? The answer is obvious in light of Glastonbury's history; indeed, it would have been surprising if the ruined shrine had not captivated him. Bond was from the area, and for years had steeped himself in medieval architecture as well as the lifestyles of the monks who were the Gothic Age's source of both religion and learning. And of all medieval abbeys, one stood preeminent: Glastonbury.

It is by most accounts the oldest church in England, with a Christian history dating back perhaps to A.D. 47, but almost certainly to A.D. 166, when an earlier structure was named for St. Mary. As with most major European religious centers, however, its real history begins long before the coming of Christ. All that can be said with any certainty is that there were pre-Christian religious sites in the area, that the Celts had an early Christian church essentially independent of Rome on this ground, and that the Great Abbey was only the last in a long succession of shrines.

Legend has it that Joseph of Arimathea, who is reputed to have introduced Christianity into Britain, brought the blood of Jesus to Glastonbury and founded that first Celtic church; but such legends are not easily proved.

There is also an intriguing story that Glastonbury was the Isle of Avalon, that Arthur was buried there, and that Camelot was nearby. In several of his "sittings" Bond's dead monks claimed that this myth was correct, and described a black marble tomb, which Bond did locate. Whether it was Arthur's grave is still unclear but it is certain that Bond's psychic sources led him to an unusual burial place of a very important, but unidentified, person.

Aside from legend and a few scattered records from the past, what is supported by historic record begins with King Ine of Wessex, who built the first portions of the "modern" Glastonbury Abbey in the eighth century. The next one hundred and fifty years are vague but the Abbey was refounded—under Benedictine, not Celtic rules—in about 943 by a priest known today as St. Dunstan. From then until the sixteenth century the Abbey prospered more than any other in England. Indeed, to some its religious significance was comparable to that of Rome; and three kings, Edmund, Edgar, and Edmund Ironside, vouchsafed its importance by choosing to be buried there.

How firm is this history?

In 1184 the Abbey and its buildings were almost totally destroyed by fire, but were rebuilt almost immediately by the first Plantagenet, Henry II, who also returned to Glastonbury at least some of the bones of St. Dunstan. These relics had been confiscated by a jealous Archbishop of Canterbury after St. Dunstan's death in 988, foreshadowing a competition that would exist from that time on.

to Canterbury they went . . .

From Henry II's rebuilding, Glastonbury had an unbroken four-hundred-year history of expansion and prestige. Thousands of pilgrims came to its doors, and merchants from all over England, and even from the Continent, traded with the skilled monks and artisans who manufactured everything from ale to clocks to gloves. From these sources, and great land holdings, came enormous wealth, which made it possible for a succession of Abbots to maintain an almost continuous building and refurbishing program until the Abbey was virtually a town in itself—as well as an ecclesiastical seaport. Some reports list as many as one hundred buildings.

from 1066 there shd be good records

It is easy to see why this church, with its altars weighted with gold and jewels, its library, workshops, vineyards, fields, and university, stood far above common—even noble—medieval life. But within little more than a decade its glory ended.

When Henry VIII refused to recognize the Pope's authority Glastonbury's day was over and its Benedictine tradition doomed. It had sufficient power to be one of the last Catholic churches suppressed, but within a very short time after its closing—with the encouragement of the Protestant regime—every building but the Abbot's kitchen had been pulled down and the stones taken away to be used for roads, barns, and farmhouses. By the reign of Elizabeth I, except for legend, nothing but the Abbot's kitchen, a few ruined walls, and some half-buried foundations remained of what had once been the most important church in England.

That is how Glastonbury remained until Bond sought to reconstruct the past with psychic archaeology. Then, for a few years, what had been called Britain's "most precious relic" once again became a center of activity.

Despite his unorthodox methods Bond was the perfect researcher into the Abbey's past. His lineage, for instance, was exquisitely correct. The family was of the lesser gentry; his father was a cleric and headmaster of a "public school"; his uncles, cousins, and brothers were clergymen, admirals, physicians, and schoolmasters. They were tied not to the mercantile involvements of the great cities but to the countries and the rural social hierarchy that lies at the root of England's culture.

Frederick Bligh Bond was born on June 30, 1864, in Upper Kennet County, at Master's House where his father was headmaster of the Marlborough Royal Free Grammar School. His early years seem one of the more pleasant periods of Bond's life. There were always playmates to be found among his father's schoolboys, and the sense of privileged ease that comes from being the son of the senior family at the place.

In 1876, however, all of this changed; his father, the Reverend Frederick Hookey Bond, a "classicist," gave up "on principle" the security of Marlborough's headmastership rather than take part in altering the school's curriculum to accommodate the progressivism and more practical orientation being demanded by the business classes. Taking his family, which eventually included seven sons and four daughters, he moved to Bath, where he settled into a considerably diminished position at Bath College and faced financial difficulties for the rest of his life.

According to William W. Kenawell, Bond's biographer, Bond was from the first a clever, almost overglib person and his mother's favorite child. For all the charm, however, he was considered moody and "different." He himself seems to have felt very early some sense of apartness and, possibly because of this self-perception, he turned while still a child to the intuitive and imaginative. In Bond's words: "Always ailing and sickly, I shrank from contact with the more robust personalities of my young associates, preferring a dream life in which imagination afforded me always a strange solace and a sense of companionship . . . but about the age of fourteen, the glory faded and I was shot out, unreconciled, into a world I neither understood nor took any critical interest in. I shall not attempt to describe a suffering which can never be put into words."

Although there is no evidence other than his own statement, it sounds as if Bond had some psychic awareness as a young boy that, with the onset of puberty, was suddenly shut off. The phenomenon of the psychically sensitive child is by no means unusual, and today it is coming under increasing scrutiny by parapsychologists.

The only other aspect of Bond's early life that seems relevant to his later work with psychic archaeology was his exposure in adolescence to a book entitled *The Night Side of Nature,* written by a Mrs. Catherine Crowe. Using the style of a Gothic novel, the book attempted to compile all manner of psychic material into a format that the author apparently thought would force "official" science to take notice. Science did not, but Bond did and, by his own admission, "practically knew [it] by heart when I was fifteen or so."

In spite of his early interest in the intuitional and occult, and despite, perhaps, his psychic awareness, Bond seemed destined to take his place in a comfortable establishment niche such as generations of Bonds before him had occupied. Although he never went beyond Bath College (essentially a high school), Bond was quick in learning whatever subjects held his interest. From late adolescence until death this meant architectural drawing, the Middle Ages, and ecclesiastical architecture.

Probably because of his disinterest in formal education and lack of family funds, Bond learned architecture by apprenticing himself to established architects, first Sir Arthur Blomfield, A.R.A. (Associate of the Royal Academy), and then, two years later, Charles F. Hanson, F.R.I.B.A. (Fellow of the Royal Institute of British Architects) of Bristol. And since Hanson's reputation rested on his work with the Gothic churches of England, it is clear that from the beginning, Bond pursued his special interests.

His work from the beginning was good, if specialized, and he steadily progressed from pupil to full partner to independent architect with his own practice. In his own eyes his most important accomplishment of this period was not simply becoming an architect, but receiving official recognition as such. In 1897, Bond passed the examinations of the Royal Institute of British Architects and was admitted as a fellow. (Perhaps because he had few claims to academic or professional respectability Bond sought titles all his life. Once he had become a fellow, F.R.I.B.A. always appeared after his name.)

Almost as significant as his admission to the Royal Institute was Bond's acceptance as a member of the Somerset Archaeological and Natural History Society. This was one of those peculiarly English amateur organizations that have provided much of the basis for what is known scientifically about England and its past. As Bond's biographer Kenawell notes, "Many of these amateur scholars laid foundations upon which later professional experts built their structures and claimed as their own." (One such amateur, Augustus Henry Lane-Fox Pitt-Rivers, is the father of archaeology as it is practiced in most of the world today.)

The Somerset Archaeological and Natural History Society was considered one of the best in the British Isles. Founded in 1849, by the turn of the century, it has almost eight hundred members and several branch or affiliated groups, as well as a headquarters and museum in Taunton Castle. When Bond joined in 1903 he almost immediately began to take an extremely active role. He published in the society's *Proceedings* that same year, and was soon giving lectures with lantern slides, commentaries, and walking tours for the annual meetings. Within a very short time he was one of the most prominent members and was accepted as he never had been before. His words were listened to, his counsel given heavy weight. Because he was liked and respected, this county society of gentlemen became for Bond a kind of psychological home.

With this security, his research, which had always centered around churches, especially their interiors, became even more ambitious. Soon his work was appearing in journals of the Cambridge Antiquarian Society, St. Paul's Ecclesiological Society, and the Royal Institute of British Architects. In 1909 this "orthodox" portion of his life reached a pinnacle with the publication of *Roodscreens and Roodlofts*. It was the definitive work on the subject and, as Kenawell points out, it established Bond as "one of the foremost authorities on medieval church structure and ornamentation in England." Although he had little formal education, and lacked the "right" university and city connections in intensely connection-conscious English academic and ecclesiastical circles, Bond had become what he had sought to be—a recognized authority and acknowledged scholar.

His engaging manner and the status achieved through his research brought another reward: clients for his architectural practice in Bristol. Since Bond had no private means, this was important. Soon he was able to set up on his own and hire a small staff and, after he became involved with the Abbey, to open a second office in Glastonbury. Every indication pointed to a minor but satisfying career.

On the surface, in fact, he had only one problem. In 1894 Bond had married Mary Louis Mills. It was a disaster from the start. Divorce, for a man of Bond's background, was impossible, and so after four years they separated, with Bond taking the only child, a daughter. At first the separation seemed merely an example of that tolerable but peculiarly Victorian hell in which there is neither marriage nor divorce, but Mrs. Bond's extreme antagonism ended up playing a major role in Bond's life and research.

The successes in his career and the failure in his personal life had become secondary by the time *Roodscreens* was published in 1909, for in 1907, when he was researching his book, Bond had entered into the real work of his life: association with Glastonbury Abbey and a lifelong psychic search.

At the Somerset Archaeological Society's annual meeting in June of 1907 the Reverend F. W. Weaver announced, "Your committee notes with great satisfaction that through the instrumentality of the Bishop of Bath and Wells and Mr. Ernest Jardine, Glastonbury Abbey has recently been secured."

The newly elected president, Mr. A. Fownes Somerville, emphasized the importance of returning the Abbey to the church, rather than having it auctioned off, by saying, "June 6th, 1907 will be hereafter a red-letter day in our Calendar." He then went on to give perhaps the most revealing commentary recorded on how the society felt about the Abbey, a statement that does much to explain the organization's position in the coming years of controversy.

"The stern historian analyzes and discards the tradition and legendary lore; he cannot see below them; for him there is no sun except the light of documentary evidence. But most of us have an innate, it may be child-like, faith in that hidden landscape, and we shall continue to believe, in spite of our stern historian's warnings and rebukes, that in Glastonbury we have a link with the earliest history of the British church of today with the church of the first century."

At this meeting it was stated officially that nothing had been settled as to who would do the excavating or how the project would be administered, but behind the careful remarks arrangements were being worked out. Bond, who was well into his research on *Roodscreens*, was, as one of the society's leading members, privy to these maneuverings and was actively lobbying to be named excavation director. After all, he reasoned with considerable justification, who was better equipped to represent the society and oversee the archaeological digging?

The society was either persuaded by Bond's arguments or was already inclined in that direction; at any rate, Bond was named director of excavations. (In the later difficult years the exact name and nature of even the title would be disputed.)

Bond's interest in Glastonbury, while intense, was at this time limited to the archaeological, architectural, and medieval. There is little doubt that he saw Glastonbury as providing the capstone to a satisfying establishment career in ecclesiastical architecture. His original intentions were not controversial, and understandable ambition, not the desire to try out a new psychic model for research, was what elated him at his appointment. Unfortunately ambition also blinded him to what should have been a clear danger signal on the road of organizational politics. He ignored the way the Abbey power structure was being arranged. Perhaps he saw what was happening, decided he could do nothing about it, and resolved to let the results he was sure he could produce speak for him. If this last is true, it may explain why he finally did turn to the psychic.

Essentially, Bond found himself involved in a dual line of command, with the Abbey belonging to still a third party, the diocese, which meant the Bishop of Bath and Wells. (By a strange twist, although Bath and Wells owned the shrine, the deed was held not by the diocese but by the Archbishop of Canterbury.) Two committees—one appointed by the Bishop of Bath and Wells, and known as the Trustees of the Abbey; and the other appointed from the Somerset Archaeological and Natural History Society, with

Bond as its director of excavations—would jointly administer the work. But Bond was not the only one to receive an appointment. After much gentlemanly jockeying for status and power, and following consultations with the powerful National Society of Antiquaries (of which Bond was not a member), it was agreed that another architect, W. D. Caroe, would be assigned the rather vague task of preserving the ruins on a coequal status with Bond. Caroe was one of England's preeminent professionals in this field and a university man with strong church, academic, and social connections.

In so complicated a situation, there were bound to be problems about authority, but Bond, ignored, either through naiveté or for lack of interest critical developments in organizational politics and concentrated on the work. Although there was no money to organize a systematic digging schedule, he began within weeks of his appointment, in the summer of 1907, to personally work on the Abbey grounds, and to read even further into the literature of Glastonbury's history and such minimal past research as had already been done.

Exactly why Bond turned to psychic sources remains one of his secrets; he never gave his reasons and no one else seems to know. It is probable, however, that he had three reasons: First, he realized that if he was not to be overshadowed by Caroe he would have to produce results, and quickly; secondly, his research indicated that there were more questions than answers about Glastonbury and that orthodoxly selected excavations, especially with the penurious budgets he had to face, would not be very efficient at producing impressive results; and finally, he had always been interested in the psychic and perhaps felt paranormal information would make such efficiency and success possible. In any case, a few months after his appointment, and after working on the grounds through the summer, Bond sat down with Bartlett in his office at the Star Life Building and decided to try for a psychic contact.

It was cold and damp outside with fading gray light coming through the windows of Bond's small inner office. Both men, despite attempts at humor, were nervous and ill at ease; only the most fanatical believer would find it easy to sit and talk to the air, and it was more uncomfortable still for two grown Englishmen to sit across the table from each other holding hands.

Bond was perhaps the more nervous of the two and he moved about restlessly, dressed in a dark rather formal suit, occasionally pushing his absolutely round dark wire glasses up on his prominent nose.

After several false starts they settled down at the plain wooden table across from each other; Bond steadied the paper between them with his left hand, made a few final adjustments to the position of his right hand, which was laid lightly on top of Bartlett's, and posed the first question in as normal a tone as he could manage. The response was contact with a flow of information that would eventually overpower all else in his life. He had begun his conversations with the Company of Avalon, the "Watchers" from the other side.

First came their strange, stilted words in a mixture of English and Latin, then the drawings of the Edgar Chapel. Four days later, at a second sitting, they announced, "Monks anxious to communicate . . . they want you to know about the Abbey. They say the times are ripe for the glory to return . . . they have been endeavoring to reproduce things in your minds."

Bond had been chosen to help those who "spoke" to accomplish their wish—which, after his summer's work, had become his wish also. Through his excavations, Glastonbury could be restored as a place of power and significance, a place where all religions could meet in peace.

For the rest of his life the Company of Avalon, whoever or whatever it might be, would guide and counsel Bond until the incorporeal world became, as it had been with him in childhood, as real as his outward life.

To Bond, it was not a matter of retreating into fantasy and being unable to separate the two realities. It was more simply that the spectral world of the monks turned out to be a fascinating place to be, even if it could only be reached through a tracing pencil. There he encountered fully formed personalities, ranging from Abbots, to knights, to medieval publicans and even a farmer from the nearby town of Glastonbury; not only did he receive correct testable information, but the world of the Watchers met certain of his emotional needs.

From their perspective, the world was not chaos or a random succession of events, as modern man seemed to believe. In the monks' view there was an underlying spiritual plan, one that satisfied Bond's own desire for order and reason. But perhaps most important of all, Bond felt he was needed, that he could help the Monks by revealing their long-forgotten work and the message for all ages they said it carried; that he could be as special to them as they were for him.

Long afternoons passed as the monks used Bartlett and Bond as channels of communication, coming through to them one by one and then signing off, the handwriting often changing as the personalities changed. Although their only voice was the scratching of pencil lead on paper, they nonetheless opened to the two men vistas of another world, as well as providing the direct information about excavations at the Abbey.

Eventually over a dozen personalities gave messages and it is not difficult to understand why Bond was so entranced, and why he pursued with such singlemindedness a technique that seemed to open wide the gate to history and offer a reconstruction of the past.

How could anyone, whether steeped in medieval lore, as Bond was, or not, resist Ambrosius the Cellarer, who relayed a tale of the Abbot's Alehouse? Or Peter Lightfoot, the clockmaker with his story of how several stonemasons on holiday caused a clock to be built and jealousy to be ended? Or Johannes Bryant and his description of holy days when the Abbey was filled with pilgrims?

But for all his fascination, Bond was a good researcher and he realized that only excavation would determine how much of his information was accurate. Apparently the monks agreed because on more than one occasion they told Bond, "We worked in our day; ye must work in yours."

The problems though lay not in their world but in Bond's: Raising money to put the psychic guidance to the test was difficult. Through private subscriptions it was finally accomplished, and a full digging season was scheduled as soon as the spring weather cleared in 1908.

By that time, Bond had developed a plan based on information obtained in the fifteen sittings he and Bartlett had conducted at Glastonbury since November, 1907. The plan had two objectives.

First and most obviously he had to determine whether there was in fact a "significant structure at the east end of the great church," as the drawings from the first session had indicated. If such evidence were found, he reasoned, it could only mean that the legendary Edgar Chapel was not pure legend. Second, if the words received from the monk named Johannes Bryant on April 20, 1908,

The condition of the best-preserved portions of the Abbey at the time Bond began his excavations.

were true, Bond should find the remains of two large and, until now, unsuspected towers at the west end.

In May the rains broke and Bond set his crew to work on the Abbey grounds, digging what appeared to be a random series of trenches with little if any logic behind their placement. To Bond, however, they were far from random, and by the time that first long season ended eight months later he had both major discoveries to report and proof of what his psychic source could produce.

It must have been hard on Bond when he presented his first paper to the Somerset Society. Although he could talk about and take pride in his truly significant discoveries, he had determined not to reveal his source of direction and was therefore in the position of someone with a delicious secret that cannot be shared. He could tell only half the story. It was, nevertheless, quite a story.

Describing his search for the Edgar Chapel, Bond reported that he had been "rewarded by finding the edge of a massive crosswall running north and south for a length of thirty-one feet." This finding and the results of further digging in 1909 confirmed what Bond described as a "strong conviction" that a chapel would be found on

This is all that remained of this enormous Abbey. Most of what Bond worked with was below ground, usually mere footings.

that spot. But he could not explain that the conviction was based on information gained half a year earlier from the Watchers.

Thanks to Johannes Bryant's directions, Bond's results regarding the towers were equally impressive. Although it would take another year's digging to outline the foundations fully, Bond felt justified in stating that his initial work strongly supported "the theory that the aisles were terminated with two massive towers . . . flanking the great west gable of the nave."

Bond officially could say only that other monastic churches had such towers and that historic precedence guided his search. The impression he left was that good research plus good luck had accounted for this important find. Bond had prepared himself well but luck played a very small role. During the April 20 sitting he had been told exactly where to look. "Ye shall find proof of ye goodly towers at ye west end."

It was an extraordinary irony. Bond had worked out an entirely new archaeological method, one capable of locating sites with ease, thus saving thousands of pounds and hours of fruitless labor, and yet he could not talk about it. He felt that to do so, especially since he was a researcher with no academic credentials, might well bury the solid archaeological achievements he had made in an essentially irrelevant controversy over the nature of his source. In light of what happened when he did reveal the psychic component of his work, this decision was a wise one, although it meant he had to present most of his findings in a distorted manner.

Nowhere is this more evident than in the case of the skeleton, which illustrates a second benefit to be derived from psychic archaeology: the explanation of what an event means, or why it occurred.

Because of the constraints under which he was forced to operate, Bond was obliged to present the discovery of this skeleton, actually

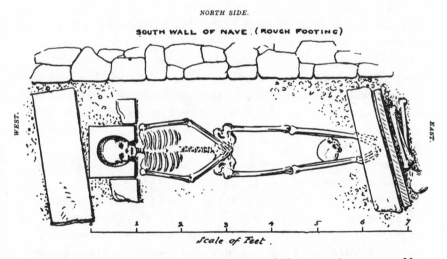

The skeleton Bond discovered during the 1908 excavation season. No orthodox archeological method could have explained what the burial meant, but the Watchers gave Bond a full reconstruction and explanation, down to the man's name and to whom the skull at his feet belonged.

one of the most fascinating human stories to come out of his work at Glastonbury, almost as an aside. He had been digging along the south side of the nave in his search for the towers when, at a "depth of three feet," he came across "a very curious interment," the skeleton of an elderly man who must have been close to six feet nine inches when alive, buried not in a graveyard but casually along the wall. Even more peculiarly, the skeleton had no coffin, though there was a complicated kind of stone cradle for its head and between the legs at the feet lay the skull and the bones of yet another body.

Here the problem was not locating the object, since Bond had stumbled onto this peculiar burial quite by accident, but interpreting it. Although there was no coffin, the burial could hardly have been either an accident or an attempt to hide some crime—the carefully carved stone cradle for the head ruled out these hypotheses. And how to explain the skull between the leg bones of the elderly giant? A traditional archaeologist would be hard pressed to come up with a plausible explanation. Bond recognized this fact and, in his report, did not even try to explain it. He knew what had happened, though, because on September 19, 1908, shortly after he had found the bones, he was given the following reconstruction of events:

"Radulphus Cencellarius, who slew Eawulf in fair fight, did nevertheless suffer by his foeman's seaxe, which broke his bones asunder. He dying after many years, desired that they who loved him should bury him without the church where he was wont to feed the birds in his chair. The sunne did shine there, as he loved it, for his blood was cold. It is strange yet we know it is true. The head of Eawulf [who apparently had been slain and buried at the spot many years before] was [there]. As they digged around the body they knew not that the head of Eawulf fell, and so lay betwixt his feet. And thus ye have found it."

In answer to Bond's question about Radulphus and his fight with Eawulf, the monks went on to explain:

"Radulphus the Treasurer was Norman of the time of Turstinus [also known as Thurstan, first Norman Abbot of Glastonbury] —annose One Thousand and Eighty-Seven. Ralph was hee. Eorwulf [Eawulf] of Edgarley, on in years, was wroth because the soldiers of Turstinus did slay Saxon monks. Ralph the Norman and Treasurer of Turstinus slew him."

It was tragic, both for Bond and for science. If the purpose of archaeology extended beyond locating and accumulating to reconstructing and explaining the past, as the discipline claimed, then here was a unique chance to do so. The Watchers had presented new material, inaccessible to archaeologists through any traditional method; material that provided lines of future research not only for archaeologists but for anthropologists and historians as well.

Abbot Thurstan is a known historical figure, and his outrages against the Saxon monks of Glastonbury, after the Benedictine monastic order was imposed on them by Rome and a Norman king in place of the earlier Irish order, is also part of the historic record. Now came this gift of insight into the puzzle of how the Saxon monks had reacted to the eleventh-century Norman takeover of the English church and their Abbey. But did the psychic reconstruction find any support in more orthodox sources? Was there any mention of Eawulf in the literature that has come down to us? What of Radulphus, the Norman treasurer whose skeleton had prompted the sitting?

Although Bond's channel was highly unusual, in light of its earlier accuracy, the information deserved at least as much attention as a hypothesis conceived by a living investigator, especially since there was an internal check. After he had been told the story and had gone back to examine the skeleton, Bond discovered, just as described, "the right forearm was fractured." Encouraged by this discovery, Bond next began a review of ancient church records, and other surviving documents of the ninth century. The reference to Radulphus proved fairly easy to verify but the matter of Eawulf was far more complex.

Bond's point of entry was "Yarl (i.e. Earl) of Edgarley" which was the title the Monks had ascribed to Eawulf during the September 16, 1908 sitting. Initially the search was barren of results, and anyone besides Bond probably would have let so obscure a matter drop. Bond though was nothing if not tenacious and through the years he worried about seeking a resolution to the problem. So great was his dedication to thoroughness, in fact, that in 1921, in a period which was virtually the lowest of his life, he was still concerned about the matter. It must have been some solace for him then when, while studying Book IV of the *Chronicle of Fabius Ethelwerd,* a work dated A.D. 866, he read: "During this year died Eawulf, nobleman of the province of Somerset" and that the body was buried "in the monastery at Glastingabyrig (i.e. Glastonbury)."

Unfortunately, there was not a way to justify acceptably in-

The Edgar Chapel, as Bond was able to outline it by 1909. There seems little question that this chapel was discovered entirely through psychic guidance.

troducing Radulphus or Eawulf as names in some way connected to the skeleton's excavation, let alone tell the whole story, and so Bond's report of 1908 is devoid of this reconstructed explanation, as every report would be until his last one in 1919. Each year there are only the meticulous descriptions of what Bond saw; nothing of what he knew.

In all, there were nine reports, with two supplements, one in 1916 and another, unpublished, in 1917; there were no reports for the years 1916 through 1918 because of World War I. Although Bond started the series with his most impressive find, the Edgar Chapel, it was by no means the end of his surprises. Along with the Western Towers and the skeleton, each year produced a special high point. In the third report (1910) it was the discovery of the Abbey clock and bell tower. The fourth report (1911) presented the Monks' Kitchen.

Altogether, Bond's reports on the discoveries made during the nine-year excavation stand as milestones along a road leading back from the meager ruins that remained of Glastonbury Abbey to comprehensive knowledge of men from the twilight of legend. In Bond's carefully phrased academic pages modern man first discovered: the Refectory, the Monks' Dormitory, the Cloisters, the Chapter House, a glass and pottery kiln, the Great North Porch of the Abbey, the Monks' Kitchen, Peter Lightfoot's clock and the bell tower, the Western Towers, the Skeleton. Five unknown chapels were discovered and described—the Chapel of the Holy Sepulchre, St. Dunstan's, St. Michael's, the Edgar Chapel, and what may be the most controversial find of all, the Loretto Chapel. In addition, the reports contain an extraordinary wealth of specific architectural detail, a sequence of dates beginning in pre-Normal times, and the descriptions of large numbers of artifacts, medallions, potsherds, stained-glass-window fragments, gilded wood, and similar objects recovered.

The reports also provide a revealing picture of Bond as a professional. In fulfillment of a research investigator's prime obligation to his fellows, Bond published his work promptly. Moreover, an evaluation of the content of these publications shows Bond to have been, at least in the areas that interested him, an extremely competent architect. Most important, unlike the majority of researchers, Bond did more than just report facts in an orderly way; he communicated. His writing was clear, and his eye for a phrase worth respecting. It is doubtful that the members of the Somerset Archaeological and Natural History Society were ever bored by anything Bligh Bond said or wrote. Antagonized or exasperated perhaps, but not bored.

What is less obvious is that no traditional archaeologist or architect could have matched Bond's achievements; they simply would not have had his unique expertise. The recovery of Glastonbury Abbey's past could only have been accomplished by someone possessed of extensive skills in both disciplines and an encyclopedic knowledge of the historic literature and legend surrounding the Abbey and English history. The self-taught Bond had that unique collection of qualifications.

The reports are also impressive for their painstakingly detailed content. In excavations as complex and drawn out as those at Glastonbury the temptation is always present to skip over tedious details not essential to the main issue. Collateral details are not usually glamorous. The time and effort needed to define the condition of a fragment of twelfth-century plinth can be so much more excitingly employed in searching for another unknown chapel. Bond, however, never let quality slip.

In the middle of his fourth report, at a time when he was working on the massive excavations of the Refectory, the kiln, and the Monks' Kitchen, Bond took time to draw and describe groining details of the Refectory Subvault: "The respond stands upon a square chamfered base, and the shaft, which is simply a section of XII Century work, with keeled 4½ in. roll, and hollow, is 3 ft. 1 in. in height. This work is obviously a piece prepared for some other purpose (probably for inclusion in a clustered shaft of the XII Century Church) and made to do duty here." Dry, but of great value and interest to another professional.

Although Bond is often accused of arrogance, even by Kenawell, his only biographer, and there seems little doubt that in his private life he did show the touchy assertiveness of a man who feels inferior, in his professional writing, however, he displayed a remarkable humility. When he reached the limits of his knowledge, he neither skipped over an issue with generalities nor attempted to downplay its importance. Again and again, even when the point was tangential to his major theses, Bond searched out an expert in the necessary field and then, with full attribution, reported his views within the body of his own report.

A good example of this professional openness is found in the sixth report (1913). Bond had discovered what he thought was an eggstone (a large boulder that was carved at some point in the unrecorded past and that was in some way associated with pre-Christian religious ceremonies). The issue could have been dismissed in a sentence by stating, "A large sandstone boulder was found, situated in sandstone, that bore signs of primitive working by man, and that may well be an eggstone." Simple enough and adequate; Bond, after all, was working on a Christian shrine of immense proportions, not minor primitive pre-Christian relics.

Instead, he first discussed the history of eggstones, referring his readers to appropriate literature, then the etymology of the word, and finally the geology of the stone and its surroundings. Because he felt he lacked sufficient geologic background to perform this last service himself, Bond enlisted the aid of W. A. E. Ussher, F.G.S., who provided the information in exhaustive detail. The entire excursion into eggstones takes ten pages (with photographs), and to this day is a significant contribution to the literature on the subject.

Some of this, especially in cases directly involving the Abbey, may have been scholastic gamesmanship. One of the oldest ploys in research literature is to get some well-known expert to say what one wants to say but doesn't have the facts or the reputation to say. Certainly Bond was no innocent. But he was so consistent about using men from other disciplines that, on balance, it seems justifiable to conclude that his primary motive was to make the record as complete as possible. And he was truly willing to look anywhere for guidance; he never imperiously turned away from any source. In the

"Egg-Stone," Glastonbury Abbey.

course of the reports, he discussed his conversations with gardeners in the area, local workmen in his crew, and even a woman in the county poorhouse. For instance, Bond's open-mindedness not only put Bond in the forefront of the burgeoning subdiscipline of scientific ethnography, it also paid off with solid archaeological results. The case of the tunnels is a good example.

Bond seems always to have been interested in secret passageways and as early as June 16, 1908, he received psychic information on the subject:

"Digge east beyond the beds of feathered grasses. There was a passage to the east doore in ye walle to the streete. There was a lodging where now is the great howse." He was also told that only part of the tunnel still remained.

As there is seldom any specific literature referring to secret tunnels (or they would not be secret), Bond's probable course was simply to look in the general direction the monks gave and hope to stumble across these subterranean ways.

There was another way, though, and Bond took it. He began interviewing the old people of Glastonbury to see if any of them remembered hearing about or having been in a tunnel. His efforts were rewarded and in the eighth report (1915), he was able to expound at length on three tunnels located by following ethnographic material—which he reports on—and psychic information—which he does not..

Such finds may be of only secondary importance in comparison to lost chapels, but they do much to give a feel for what life was like at a medieval abbey where visitors, whether for reasons of physical safety or political secrecy, wished to pass unnoticed.

Bond's open way of listening paid another dividend; it made others feel he was approachable with old documents, that they would not be ridiculed if the material turned out to be worthless. In at least two instances such unsolicited information was to play the critical role of providing historic checks supporting both the messages from the Watchers and the actual excavations.

The first example of this openness to unknown literature appears in the third report (1910), and again in the supplement of 1916. In both, Bond described receiving confirming evidence of the Edgar Chapel, St. Dunstan's Chapel, and the Loretto Chapel. The special impact of this collection of manuscripts and prints, and Bond's excitement about it, are understandable since he was shown this previously unknown material by a Colonel William Long of Clevedon almost a year *after* he had discovered the Edgar Chapel. In one of the drawings the Edgar Chapel is clearly shown.

In the sixth report (1913), Bond was able to present still further support for his conviction that the Edgar Chapel existed, and for his theory about its size and shape. As Bond related this event, "In the summer of 1910 there came to light [an] MS Diary of John Cannon (Schoolmaster, of Meare [Mere], b. 1684)." The diary also mentioned the Edgar Chapel, as well as St. Dunstan's Chapel, which Bond had just excavated. There was even a reference to the Loretto Chapel, as yet undiscovered, but which Bond believed existed because the Watchers told him that it did.

These source documents are worth examining, not only because they support Bond's work, but also because such material disproves the most serious allegation against Bond—that he had all the information the monks gave him in his subconscious mind, and that the sittings were simply a gimmick he thought up, or if he was sincere, a delusion to which he was subject. Beyond doubt the Long and Cannon documents and the descriptions they contained were completely unknown to Bond when he received information on these chapels from the Watchers. The documents are also important because they contain information crucial to a line of reasoning Bond pursued for the rest of his life.

In all the reports, starting with the first one, Bond is extremely meticulous about the measurements of the various buildings on the Abbey grounds, particularly the Abbey itself. This may be considered excellent archaeology, but the motive behind such fine calibration lies elsewhere. Like many medievalists, Bond had succumbed to an occupational hazard—Gematria, the theory that buildings carried within their measurements codes that reveal a secret interpretation of Scripture. This hypothesis predates the Christian period—the best examples being the Great Pyramid of Giza and Stonehenge—and recent research has indicated that this may be a valid line of study. In the first two decades of this century, however, the idea that building dimensions contained codes or hidden information of any kind seemed mainly the product of such questionable avenues of research as the Cabala and alchemy.

It is unclear whether Bond had an interest in Gematria before he began the automatic writing sessions with Bartlett, but there is no question that the Watchers introduced the subject early on and that it caught Bond's interest. In 1917, a year before the publication of *The Gate of Remembrance,* Bond brought out a book, *The Greek Cabala,* written in conjunction with the Reverend Thomas S. Lea, vicar of St. Austell.

It was this business of measurements that led Bond to take what were, by the standards of the times, extremely accurate measurements of all the buildings, or footings, at Glastonbury. Who could have done otherwise after being told, "Our Abbey was a message in stones. In ye foundations and ye distances be a mystery—the mystery of our Faith, which ye have forgotten."

In sum in any case, the question of measurements, and whether Edgar Chapel ended in an apse or a simple rectangle, became heated issues between Bond, the Society of Antiquaries, and the church. It is easy to see why Bond took succor in both the Long and Cannon historical material; they documented the Chapel's shape and the length of 580 feet that the monks had given him.

If there are any criticisms to level against Bond's reports they are that he did not always include as many drawings as some might like, and that their scale and orientation were not always consistent from report to report. He also had a tendency to assume knowledge his readers might not possess and, perhaps as a result, certain of his citations seem rather cryptic to the reader today. What kind of archaeologist, then, was Frederick Bligh Bond? Perhaps the best answer comes from the present director of excavations at Glastonbury, C. A. Ralegh Radford, a well-known English archaeologist:

"The interim reports which he published in the Somerset Archaeological Society are factual and precise and show that he was working scientifically with a technique as advanced as any at the time. . . . In brief, I would say that . . . the errors in the . . . reports were such as inevitably come to light when earlier work is reexamined in the light of fuller knowledge and advanced technique."

Dr. Radford's comments pose an interesting question: What more could Bond have given us if he had had radiocarbon dating, thermoluminescence, and pollen analysis? The Monks' Kitchen, for instance, is a puzzle which obviously would benefit from the use of these techniques.

Bond was told in 1910 that next to the kitchen building was "ye mint garden" and that nearby was a stable for "four horses and room for ye guesten horses in ye same stable." Bond discovered signs of the kitchen during the digging season of 1910-1911, and reported on it in his fourth report. The work was only preliminary, however, and he never got a chance to follow it up. Here is a clear opportunity for a researcher today to test the psychic material with methods not available when Bond was at Glastonbury. Pollen analysis could determine whether mint was grown at this spot and also whether horses were stabled there, for their droppings and feed would have contained seeds that would still show up in laboratory tests. If the tests were positive, they would be a validation of the monks description of life at their Abbey.

Bond, then, clearly fulfills the first goal of his discipline. Almost all of what we know today of the physical structure of Glastonbury Abbey we know because of the work of this controversial man whose research was completed half a century ago—against a backdrop of penury, war, and administrative back-stabbing. In addition, his work—the part he told about in his official reports—began to recreate the life and flavor of this historic religious community, from its rude beginnings to its unequaled power, ending finally with its total destruction at the time of the Reformation. It is truly an extraordinary record, and one that involves three disciplines: archaeology, architecture, and anthropology.

The tragedy lies in what Bond did *not* say in those nine reports. Material began coming in to him, through his friend Bartlett, in 1907, and continued until about 1920 (on a sporadic basis after 1912). Bond came to the United States in 1923, but before he left, he began receiving the unsolicited scripts of a woman medium living near Winchester. Once he was in the States, he was also sent material by the psychic Philip Lloyd, beginning in July 1923. About a year later he began working on Glastonbury with the sensitive Hester Travers Smith; eventually a fifth sensitive would be involved.

In general, this psychic information from sources other than Captain Bartlett covers a period earlier than the Abbey's medieval era and it purports to tell the story from A.D. 47, when Joseph of Arimathea founded the first Christian shrine in the British Isles at Glastonbury, to 1200, when Hugh was Bishop of Lincoln. Published by Bond in a privately subscribed series called *The Glastonbury Scripts*, and a small volume entitled *The Company of Avalon*, published in 1924, it adds yet another challenge to the already formidable list of challenges Bond left behind.

Together these two volumes present a coherent, and largely internally consistent, tale that, Bond says, makes it possible to "lay out a fairly detailed plan of the buildings . . . from the first foundations to the final . . . work of Abbot Herlewin [abbot from 1101 to 1125?]," when *The Gate of Remembrance* picks up the tale and continues to the Abbey's destruction.

However, this later psychic material has one major problem: It comes after Bond has fallen from favor and is no longer doing work of any substance at the Abbey. These sittings have never been put to the test of the shovel—they have no context within a rigorous program of excavation.

The material covered by Bond's 1918 book, *The Gate of Remembrance,* is entirely free of this dilemma, and it not only makes archaeology more than a shadow science, it provides a model for the practice of psychic archaeology which is valid today. Indeed, Bond achieved the true goal of all the sciences that study man—the answer to the riddle that lies beyond location, beyond physical reconstruction, even beyond explanation; the answers to the question: Why?

The archaeologist must usually be satisfied with only the first two parts of the puzzle: location and physical reconstruction. He can surmise an explanation for why something was done only in the most rudimentary manner. By the very nature of what he works with, the orthodox archaeologist's explanations must dwell within the parameters of an arbitrary and incomplete collection of *things.* The anthropologist aspires to a somewhat higher state, and concerns himself little if at all with location or reconstruction. It is his desire to produce a cultural or behavioral explanation. Neither discipline, however, does more than attempt the next step, the subtle and most hidden part of the problem—the real reason why something was done.

We can say that the Magna Carta resulted from the effort of powerful forces to diffuse centralized power, thus laying the groundwork for democracy in modern times. A neat and tidy explanation. But anyone familiar with the dynamics of government knows that individuals rarely feel they are acting for the furtherance of historic trends—at least not until after the fact. Orthodox methods can no more discover the true reasons the individual barons acted than they can explain the true motivation of a President set on reshaping the way government will operate. It is this last, this deepest part of ourselves, known to anthropology as the *emic*—the real why—that has eluded the orthodox researcher no matter how strict his experimental protocol.

And it is this material that the Watchers, the Company of Avalon, lay before Bond like a banquet—highly detailed information to be validated, checked through the test of actual field excavations, so the burden now fell on those who would dispute his explanation.

The story of Rudolphus and Eawulf, referring to the skeleton Bond located in 1908, is one example of how the monks explained the "why" behind an event. The bell tower and the clock at Glastonbury is another. Bond was first led by the monks to locate the bell tower, which he then quite orthodoxly established as having been built relatively late in the Abbey's overall development. But the location of the bell tower was only a partial solution. To any other archaeologist (and to Bond overtly) several critical questions still remained unanswered: Why had the monks not put up a clock before—the need had been present one hundred years or more. Certainly they had the capability and the funds, so why this particular order of construction? What prompted the act? Traditional avenues of research, as usual, provided no answer, but, as usual, psychic information, this time in the words of Peter Lightfoot, who built the clock, could, and did. It was jealousy:

"Then, when they were building at Welles, we were jealous of our howse, and certain masons coming on holiday across the causeway which led straight across the marsh, did tell us we were lacking. They sedde our howse was over smalle for our community, and the choir thereof was not long enow for our processions and for the brethren to sitte at the service of the church—for we were three hundred and forty-seven in number . . . And moreover [our] tower was too lowe for beauty," Lightfoot recounted to Bond, and the masons' taunt hurts no less for being six hundred and fifty years old. Nor is the response any less predictable or human for all that no orthodox reconstruction could have assumed it as a reason.

Since Wells, where the masons came from, was "new and faire with carven stone, our Abbot was moved to beautify our howse. And so was our choir elonged and afterwards the towre was beautifyed with certain panelling." Lightfoot goes on, adding another point that would be a source of confusion to a traditionally constructed recreation of· events. Glastonbury had just undergone a massive refurbishing of its interior; so much is known. Why, then, build a clock—which was a considerable expense—or finish off so lavishly a tower that would be seen only by the monks? The answer again is the human desire to be as good as one's neighbor. "The gabell was finished like unto Welles, and the clock and certain belles did hang there."

Lightfoot's story is by no means the only answer Bond received as to why something was built. And this information is of great importance not only because Glastonbury was considered by some to be a second Rome, but also because its building history traces the rise and fall of religious strength and the development of religious thought throughout England's early history.

It is this importance, in fact, that causes confusion, for even after Glastonbury ceded preeminence to others—such as the Archbishopric of Canterbury and Westminister Abbey—it continued to grow. The monks even continued to build just prior to and after the Reformation, when Henry VIII split off England's church from papal authority.

Why keep building when you know you have no future? The answer is not to be found in literature, nor in discernible logic. But there was a reason and Bond was told what it was; and again, the truth was tied to human emotion, that elusive facet of man's past that archaeology and anthropology have so rarely been able to touch.

Just as the monks told Bond that the bell tower and its clock had been built because of jealousy, so they also told him that many of the chapels were apparently built for love. Not the ecstatic love of religious devotion, for that, they say, had cooled in both the brothers and the people, but the more human love simple men feel for the land and the place that has been their home.

Henry VIII paid a visit to the Abbey's guest house and the monks turned out their best effort. It was an effort so good that "in that . . . was death to our howse, for the Kinge did lust after our meats and wines and [cared] not to save us."

The King was lusting after more than a fat collection plate, or the altar jewels. Glastonbury was one of the most extensive economic combinations in the entire country, with vast land holdings, craft industries, manor houses, and herds of animals. It would make a sizable addition to Henry's purse.

The monks knew their assets, just as they knew King Henry as a man in constant need of new sources of money. They also knew that even before the Reformation his minister Thomas Cromwell, Earl of Essex, had begun suppressing England's abbeys and expropriating their wealth. Why, then, did they make this display, or extend the invitation; for, according to their account to Bond, the King had come on invitation. The obvious answer, that they simply wanted to indulge their natural urge to show off before their sovereign, is not plausible in light of what they knew about Henry's greed and Cromwell's counsel. Bond's automatic scripts provide a rational answer for this odd behavior:

At the death of Abbot Beere in 1525 the monks could not agree on a successor and deferred to Cardinal Thomas Wolsey, adviser to the King. In March of the same year he announced his decision—an elderly priest named Richard Whiting, who, according to both the testimony he gives Bond and the few surviving documents, was both unwell and diffident about accepting the post.

Because he had close contact with Wolsey, who was an intimate of the King, Whiting was advised that Cromwell's efforts and the King's shaky finances were endangering the abbeys. There were also growing problems with the Pope.

The Watchers explained that Whiting, apparently at Wolsey's urging, extended the invitation in hopes that a successful visit might put the King in a more accommodating mood toward the Abbey should it ever need his goodwill. It was a disastrous miscalculation.

Wolsey soon fell from favor and Henry's visit had a very different effect than the one planned. The monks told Bond that they knew, by the time he left, that Henry looked upon Glastonbury as a rich and defenseless prize.

After the break with Rome, when both religious and political loyalty flowed to the state, it was almost pathetically easy for Cromwell to devise charges against both the Abbey and Whiting. The Abbey was closed, its wealth and property appropriated by the King, and Whiting was brutally murdered. The day was over for the religious center that inspired William Blake to write "And Did Those Feet in Ancient Time," Britain's alternative national anthem.

But the faithful servants of Glastonbury's glory had a final say. Although the Catholic faith was weakened, and despite their uncertain future, they built as never before—out of love.

"Chappells, a many! Everywhere! Why cumbered they the ground when faith was dead and there was no longer any need [for them]? The purse was full; it must be spent, and so, when nor barn

A fairly typical example of Bond's automatic writing. This sample refers to the Edgar Chapel.

nor byre nor pent [corral or pen] called for it, it was yspent. Why should roysterers and evil men have it to spend? So we builded much . . . Chapels everywhere—no need of them."

And what they could not spend, so local legend says, they buried. Although Henry's agents confiscated a truly enormous list of assets, the Watchers told Bond there was more. They also gave him directions where to look. But, as with so much information he received, no real effort has been made to test the veracity of these statements. Bond once had a dowser named Tims, under the guise of searching for drains, go over the grounds. Tims indicated fourteen locations for buried treasure but none was ever investigated. And so, although other excavations demonstrated the accuracy of the Watchers' statements, their assertions about treasure remain improved. Along with so many other lines of inquiry, they stand as a challenge for the future.

Not all the chapels, however, were built to avoid giving over the Abbey's riches to "roysterers and evil men," two chapels, the Edgar and the one Bond called the Loretto Chapel, both almost unknown before Bond's work, are examples. Here again though, if nothing more than the evidence of excavation was available, all that could be said was that if such chapels did exist, they were the work of one abbot, Richard Beere.

Beere, the next to the last Abbot at Glastonbury, is a fascinating man who ruled the house for some twenty-five years. A close friend of Henry VII, who helped the Abbey as much as his son would later harm it, he is described by British scholar Geoffrey Ashe as a "curious figure." He was a man appointed by the King for delicate diplomatic missions, he was at home in the most advanced and sophisticated circles, and yet he seemed sincerely dedicated to esoteric Christian mysteries.

Why did Beere build the two chapels? By means of the psychic pen, he told Bond that the Edgar was built for fairly orthodox reasons of respect for a saintly king whose life had been tied to the Abbey and who chose to be buried there. But the Loretto Chapel? That is quite another story—one of highwaymen, an overweight abbot, a thorn bush, and a vow. Beere told it to Bond on June 13, 1911—almost five hundred years after the incident occurred.

"Wee were borne downe by rude men in foreign parts and the mule which bore me fell, for I was a grete and heavy man. And being like to fall downe a steepe place or be trampled by ye mules, I called on Oure Lady and shee heard me, soe that my cloke catching on a thorne I was prevented, and then said I: 'Lo! When I return I will build a chapel to our Lady of the Loretto, and soe [insistent] was I inn [my vowe] that the brethren were grieved, for yt was arranged in Chaiptre [meeting of the monks] that wee shold build a Chapel to oure Edgars before I went in ye shyppe. Therefore builded I hym first, for it was a public vowe: but mine own vowe I fulfilled afterer and soe all was well."

It is an explanation going far beyond what an artifact could tell, going far beyond a simple reconstruction of events.

Not all of the reconstructions and explanations Bond received related to buildings or were issues of such moment. Many were offered almost as asides, interspersed among digging directions, and they dealt primarily with how the monks felt about being monks and what life was like in the Abbey—the trivia that truly brings a culture to life. Of all the stories he was told, at least during the period in which Bond was actively digging at Glastonbury, perhaps the most human and endearing are those of Brother Johannes Bryant, who describes himself as a stonemason and curator of the Edgar Chapel. He was, by his own admission, never cut out to be a monk, but was instead a simple man who loved the Abbey with an unquestioning devotion that made him a key member of the Company of Avalon and central to the question of the nature of Bond's psychic source. But his love was not that of an intellectual ecclesiastic. Gulielmus, a fellow monk, told Bond:

". . . He ever loved the woods and pleasant places which lie without our house. It was good for he learnt in the temple of nature much that he would never hear in [choir]. His herte was of the country and he heard it calling without the walls. . . . He went a-fishing did Johannes, and tarried oft in the lanes to listen to the birds and to watch the shadows lengthening over all the woods of Mere.

"He loved them well, and many times no fish had he, for that he had forgot them . . . but we cared not, for he came with talk and pleasant converse, as nutbrown ale, and it was well.

"And because he was of nature his soul pure and he is of the company that doth watch and wait for the glories to be renewed."

It is Johannes who gave Bond information about such specific excavation projects as the Vault and the Edgar Chapel. And, because he proved to be accurate in instances where his words can be checked,

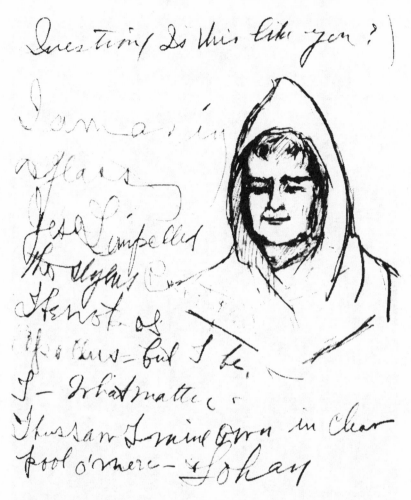

A drawing of the monk who served as Bond's main connection with the incorporeal world of the Watchers.

his reconstructions of events and his explanation of their meaning take on a special weight. Perhaps his greatest gift to Bond (and to us) is the glimpse he offers into what it felt like to be a monk in the most powerful abbey in Great Britain. His discussion gives great insight into the medieval mind and reveals why at least one man chose to enter Glastonbury's monastic life.

"We have sat in the grate gallery under the west window and watched the pylgrims when the sun went downe. It was in truth a brave sight, and one to move the soul of one there. The orgayne that

Bond's conjectural recreation of the Abbey's interior. It is easy to see why a medieval pilgrim, and the monks who served him, would all be in awe of such a structure.

did stande in the gallery did answer hym that spake on the great screene, and men were amazed not knowing which did answer which. Then did ye bellows blowe and ye . . . man who beat with his hands upon the manual [keys] did strike yet harder, and all did shout *Te Deums,* so that all ye town heard the noise of the shouting, and ye little orgaynes in ye chapels did join in triumph. Then ye belles did ringe and we thought hyt must have gone to ye gates of Heaven."

Here is the exaltation of medieval religion and the life it offered. Here is a living statement supportable by actual fieldwork. Here is the step that gives archaeology the power to describe the true substance of the past, not the shadow.

If all this is so—if Bond did locate sites, if he did reconstruct their usage, and if he did offer testable explanations of what they meant to the people of the time—why is his work almost unknown and frequently dismissed by those who do know of it? Why, instead of being incorporated into university courses, does this remarkable record languish in out-of-print books? Why, even if archaeologists reject Bond, isn't his work presented by parapsychologists as a beautiful demonstration of applied psychical research? What went wrong?

The obvious answer to these questions—that there was something wrong with Bond's professional standards—is incorrect. Bond's archaeology and architecture were, as his successor at Glastonbury, Ralegh Radford, makes clear, as good as any being done at the time. Frederick Bligh Bond's work is obscure not because of professional incompetence but because of the man's personality, his private life, his place in the academic community, and perhaps most important, his source of information—the Company of Avalon.

The fact is that Bond's work has never really been evaluated beyond a few superficial criticisms. Instead, issues wholly unrelated to archaeology, psychic or otherwise, have been allowed to cloud the record.

To begin with, Bond was both eccentric and paranoid, convinced that there were plots around him to deprive him of his just credit and success. Added to this mix was his poor political sense when dealing with others, especially others in positions of authority, and his misfortune in being juxtaposed with Dean Joseph Armitage Robinson and W. D. Caroe.

Kenawell is convinced that neither Robinson nor Caroe actually meant to hurt Bond, although the record of their actions he presents belies that conviction. It is clear, however, at least in hindsight, that Bond would have run into trouble at Glastonbury without ever receiving any psychic messages. As early as 1913 Caroe and his patron Robinson had formed an alliance, one of whose aims was the removal of Bond. Not out of spite, of course, but simply in the way that two "great" men deal with a "small" man who was not only offensive to them socially but who continually disrupted their orderly plan with new ideas and radical schemes. With his intense, almost obsessive interest in the ruined Abbey, Bond was a nuisance and something of an embarrassment. And so, impersonally and with no overt ill will, they went about having him deleted from Glastonbury affairs.

Through a tiresome and complicated series of maneuvers Bond lost his post as diocesan architect. Then the always troublesome dual line of command under which the Abbey was being excavated and restored was resolved in Caroe's favor. Robinson became president of the Somerset Archaeological and Natural History Society, and soon after the decision was made that the Society would cease its efforts at Glastonbury. Therefore, there was no need for a director of excavations and Bond was dismissed.

Bond, as might have been expected, did nothing to help his cause. His letters to various professional societies seeking redress against Caroe and others, his efforts to make everyone even remotely involved choose sides for or against him, and his often difficult manner courted trouble.

His attitude is best displayed by his insistence on discussing Gematria, the esoteric key to Scripture, which he felt was embodied in the physical dimensions of the Abbey. Anyone with Bond's awareness of church history should have known that secret teachings, cabalas, and the like could only enrage a modern Anglican priest, especially one with the worldly ambitions and academic aspirations of Dean Robinson. And yet in 1916 Bond delivered a lecture at the Society's annual meeting on "The Lady Chapel of Glastonbury Abbey: A Study of Measures and Proportions," before an audience that included the newly elected president.

This question of messages in the measurements of the Edgar Chapel was the source of an emotional debate when Bond was working at Glastonbury, and is a point of contention to this day. As late as 1939 the actual markings on the ground were changed and Bond's outline of key parts of the Abbey were obliterated to obfuscate his claim that the Edgar Chapel ended in a polygonal apse rather than a simple wall. In the official reports Bond explained his conclusions as to the shape of the Edgar Chapel as in the interest of accuracy, which in part it undoubtedly was. But what he did not say was that the apse was needed to get the measurement Gematria demanded of 580 feet, as well as to produce the necessary rhombus of two equilateral triangles.

It is unclear whether Robinson or Caroe realized Bond's real reasons for making these claims, but they may have. In any case, the issue, which at first seems a peculiar one for violent emotion, became a central point of contention very early in Bond's career at Glastonbury. Even Bond's being proved correct did not ease the tension. For years his opponents simply ignored the proof. It was never central to their real feelings anyway. Bond, not measurements, was what disturbed their tranquility.

To aggravate his professional difficulties, Bond's wife, after their legal separation, turned her considerable energies first to destroying her husband's architectural practice (the title, director of excavations, was honorific; there was no pay), his reputation, and finally his purse.

The story, like most Bond stories, is often contradictory and always complicated. There were seemingly endless lawsuits, all of which Bond won, none of which helped his reputation. There was a demeaning custody row over their only child; she wanted to be with her father and was eventually successful in that wish. Mrs. Bond charged to the National Vigilance Association that Bond was debasing the child by using her as a medium as well as involving her in other unnamed but obviously scandalous occult practices. It came to nothing, there being no substance to the charge, but Bond suffered further embarrassment because of it. Still, he would not divorce his wife. There was also a continuing financial drain created by a woman dedicated to using any weapon at hand. Bond paid all bills as long as he was able, but in 1915, he was driven to bankruptcy, the ultimate degradation of a Victorian gentleman. Even after bankruptcy and the loss of his architectural practice, Bond paid what he could from a small income based on his writing and lecturing. Perhaps the best way to summarize Mary Bond's effect is to quote Bond's biographer:

"Throughout the years the sheer amount of her activity and its results in the community, and farther, were truly remarkable. Though it is difficult from this point in time to be positive about it, it is very likely that her whispers everywhere played a large part in his finally losing his archaeological position at Glastonbury. Too much importance has perhaps been attached to the displeasure of his employers over his psychic concerns and not enough to the malice of

Mrs. Bond. Bond's attempts to cope with her seem to have been courageous and responsible on the whole, but in the end he did not have enough strength and psychological insight to deal with her adequately and so bring him a quiet mind."

It is easy to see why the local gentry of the Somerset Society, who had known the Bond family and Bond himself for generations, might be sympathetic, but that such men as Caroe and Dean Robinson, with larger horizons and the more fastidious natures of those one step up on the social ladder, would find Bond distasteful. The psychic did not so much cause Bond's downfall as provide both the excuse and the final provocation.

Bond ultimately was hopelessly outclassed in his battles with two major opponents, and lacked allies in the interlocked world of academia and the church. Further weakening his position was his paranoia and unfortunate manner of unpleasant humility coupled with aggressive arrogance, which only gave his opponents the leverage they sought. Further worsening the situation, his private affairs tainted his professional existence and even the source of his information carried disturbing implications for the theology of the religion he sought to serve. What was the church to do with a band of dead monks who not only gave directions for digging but presented a philosophy of soul survival, hidden knowledge, and other esoteric information that had been declared heretical even before the Church of England split from Rome?

In spite of all this, however, the truth remains that Bond formulated a psychic method of approaching the past that produced results and that was never really scrutinized. It is ironic that amid all the shouting and arguing, no evaluation of Bond's techniques was made. Perhaps now, after most of the participants are dead, taking with them their emotional ardor, it can finally be done.

The first difficulty in making such an evaluation lies in the way Bond chose to reveal his information. While it is understandable that he wanted people to concentrate on the results of his excavations, rather than the controversial source of his directions, an evaluation of his psychic methods would certainly be easier if he had more firmly established the order in which events took place. In *The Gate of Remembrance* he did publish a letter from Everard Fielding, secretary of the Society for Psychical Research, stating, "There is no question but that the writing about the Edgar Chapel preceded the discovery of it by many months. I was present at . . . the beginning of [Bartlett's] automatism ... and that was before you ever started your work at Glastonbury, and before you were even appointed to the work."

It would have been better if, in addition to letting some people in on the secret from the beginning, Bond had had the writings notarized and sent several copies, in dated envelopes, to men and women of unquestioned probity. Then a clear conclusion would be possible as to what was psychic and what was either good archaeology or simply good luck.

As it is, there are four classes of information in Bond's report. First is material such as the original outlines and description of the Edgar Chapel, which are almost unquestionably psychic in origin. This excavation is also the freest from the charge that Bond had simply run across the chapel in his extensive survey of old maps and documents. That is why the Long prints and papers were so important; they surfaced only after Bond had discovered the chapel and reported on it.

The second class is psychic data that seem to predate actual excavation, when the excavation itself is sufficiently inconclusive to be questionable. The bell tower is a classic example. There was no reason, other than the guidance the Watchers gave, for Bond to look where he did, but although something was found, it was so sketchy that whether it was a bell tower cannot be finally established.

The third class of information is comprised of sites about which there is some question over the order of psychic comment and excavation. The St. Michael Chapel is a good representative. The monks appear to have told Bond about the chapel before he discovered it. Certainly it strikes one as odd that he could walk out into what was by then an open field and, on the first try, locate the footings. But it could also be argued that a good ecclesiastical archaeologist would know that the existence of such a chapel was probable, for the field had been monks' cemetery and a graveyard chapel almost certainly would exist, according to custom. No conclusions can be drawn about the psychic component of Bond's archaeology in such cases.

The fourth and final category includes those discoveries in which there is complete confusion as to what is psychic and what is good judgment or luck. Such things as subvaults and altars fall into this category.

Further complicating the issue is the fact that even when Bond did elect to go public, he did not tell the whole story. In *The Gate of Remembrance* Bartlett—who, incidentally, is not identified by his true name, although this is supposed to be the full story—states that there were "'fifty communications dating from 7th November 1907, to 30th November 1911, and also some supplementary writings produced in 1912 and later." Of this fifty Bond includes, mostly in extracts, only twenty-five. He does, however give a sequential numbering (in Roman numerals, naturally) and the sitting dates.

From this we can infer that all but three of the sittings relating to his work as director of excavations were held by the first week of December 1908, and that of this 1907-1908 group the majority—fifteen—were held before the first digging year began, although not until several months after Bond either received or knew he was going to receive his appointment. This would seem to indicate that most of Bond's major discoveries were based on psychic data.

Bond himself felt sure enough of the psychic information that guided his archaeological decisions on the Edgar Chapel alone to advance what he called "the veridical passages in the automatic script, sixteen in number, referring to the Edgar Chapel and the east end of quire [sic]." Admittedly, this chapel and the east end are Bond's most impressive excavated discoveries, his fully supported claims going down to details of gilt and color. If there is a weakness it is that Bond did not seem particularly concerned with presenting in the same systematic way the reconstruction and *emic* material he received, although in the early part of the century in England this was not unusual. The modern systematic study of such material came much later, in the United States.

It can definitely be said, then, that Bond did indeed locate and define at least some sites on the basis of psychic information received before those excavations began, information that could not have been obtained from the literature available to him at the time.

He also did attempt to establish the order of events with Fielding's letter, and he did develop his hypotheses prior to digging and then test the Watchers' words by actual fieldwork; thus providing the first systematic model for psychic archaeology.

Perhaps the biggest tragedy is not that Bond's psychic archaeology never received a proper evaluation but what happened to the man, and to the mind that could have added much more to science than this already significant contribution. By 1921 Bond was reduced to cataloguing and cleaning, for ten pounds a month, material found in his earlier successful years—and happy to be doing these mundane tasks if they would let him stay near his beloved Abbey. But after everything that had happened, even this pathetic job could not last.

The cleaning was to take place in the one original building still standing on the Abbey grounds, the Abbot's Kitchen. Bond wanted a key so that he could come and go at his own schedule. He obtained one, but when Dean Robinson heard of it, he demanded the key's return; Bond was not to have such freedom at Glastonbury. The fight was degrading to both men and, of course, had nothing to do with a key. It was the final seedy chapter in their struggle for power, and Bond was doomed to lose. He returned the key.

The letter he received the following spring, in 1922, was almost an anticlimax, but it made everything official: "The Council of this Society [i.e., Somerset] informs you that they are no longer responsible for the excavations at Glastonbury Abbey and that the Excavation Committee appointed by them is therefore dissolved."

For the next four years Bond lived a life of frayed-cuff gentility, subsisting on lectures and the sales of his books. Even these activities were to suffer, as the Abbey trustees engaged in one last malicious exercise of their authority. They ordered that all books by Bond be

The kitchen where Bond worked for £ 10 a month as a laborer cleaning artifacts he had discovered as director of excavations.

removed from sale at the Abbey bookstore (even the innocuous *An Architectural Handbook to Glastonbury Abbey* was not exempt), an order that remains after fifty years.

When he became too successful as a lecturer, this avenue also was denied Bond. He found that invitations he had accepted were suddenly withdrawn. Although he was never able to determine who was behind these tactics, he finally came to understand that he had been listed as an undesirable by the Church of England and therefore anything he attempted in which the church had any influence could never succeed.

The only bright spot during this four-year period occurred when Bond was asked to become the editor of a new journal, *Psychic Science,* the official publication of the newly founded British College of Psychic Science. The position, like so many others in Bond's life, offered no pay, but at least the people were friendly toward him, including Sir Arthur Conan Doyle, who became the president of the group's Advisory Council in 1923.

By 1926 though it was clear that England held nothing further for Bond, so he planned a trip to America. He had recently acquired patrons, an American couple, and their offer to bring him to the United States seemed to present the opportunity for a new life.

Things began auspiciously. He was successful on a lecture tour that included several universities. There was an offer to join with a New York architect in his practice, and a chance to become involved in psychical research with the American Society for Psychical Research (ASPR) . . . Everything seemed to prosper. His architectural work was giving him a small but regular income; the ASPR asked him first to become its editor and then educational director; speaking engagements were offered all over the East Coast; and the negative aspects of his reputation had not crossed the Atlantic.

It was an illusion, however. Bond was only going through a charade of living and building a new career. He was still with his Abbey. In a sense, he had never left it, since he continued to talk with the Abbey monks through a series of local psychics in America. And when his patrons, who supported him for the rest of his life, paid for two visits to England, even though one trip's purpose was the unpleasant task of helping his daughter, who had remained in Britain, to get a divorce, as soon as the social requirements had been met, Bond traveled down to Glastonbury to see the Abbey and walk the grounds as an ordinary tourist passing through the turnstile.

The pull of the Abbey was so great, in fact, that when he returned to America from the second trip in September 1928, although he had taken out permanent visitor status, he seemed always to be looking back to England. Its influence coupled with Bond's personality problems made an impossible combination and, by 1935 he had either quarreled with or cut himself off from everyone in the States except his patrons and a few other supporters. He was seventy-two years old, broke, and homeless.

In 1935, again at his patrons' expense, he returned to England, this time for good, taking with him only one thing. In the best Bond tradition it too would cause him trouble. As Kenawell describes it, "Sometime during his first lecture tour in America he met 'The Most Reverend William H. Francis,' self-styled primate in a vague Episcopal order called The Old Catholic Church in America." Their friendship resulted in Bond's ordination as a priest and his assumption of the title "The Right Reverend Monsignor Bond, Vicar General of The Old Catholic Church in America and Prior of St. Dunstan's Abbey."

Bond arrived in England in good spirits, although at first there was some question about where he would live since neither his daughter nor his surviving brothers and sisters were able or willing to extend an invitation for more than a brief visit. Finally an arrangement was made in North Wales, and there, except for an occasional excursion to the Abbey or a trip to his daughter's home or to see others in his family, he stayed until his death, at eighty-two, in Cottage Hospital, Dolgelly.

These final years were bitter and pinched, Bond's only relief being the death of his wife in 1938. Even his funeral brought complications. When Bond realized he was dying, he insisted that his ordination in the Old Catholic Church entitled him to the same recognition as an Anglican clergyman. Here again, the Church of England was to have the final word. He was buried with Church of England rites at Llanelltyd Church in North Wales.

And so it was over. All that remained were books rapidly going out of print and the magnificent Abbey he had reconstructed, where his work received the minimum recognition possible.

Bond's greatest legacy to us, however, is not the shape of Glastonbury's physical mass. That may be fascinating and substantive, but it is essentially a written chapter. Footnotes have been added by subsequent researchers, and although several major finds described

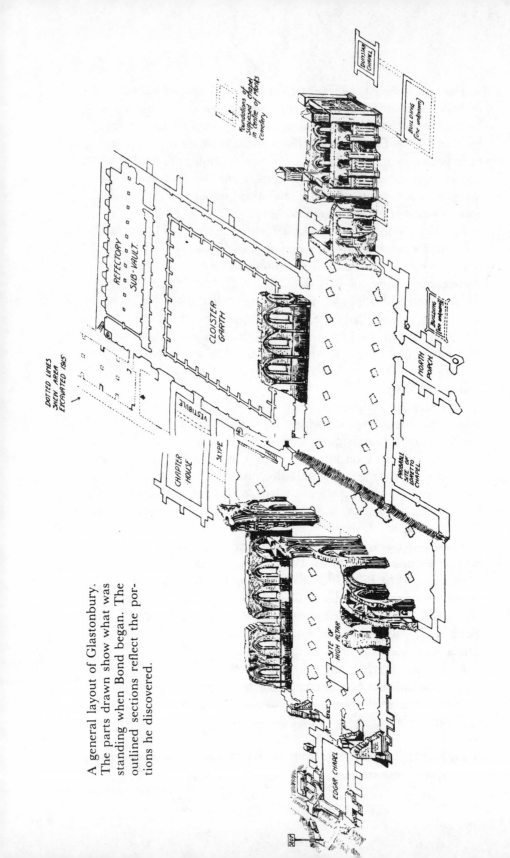

A general layout of Glastonbury. The parts drawn show what was standing when Bond began. The outlined sections reflect the portions he discovered.

DOTTED LINES SHEW AREA EXCAVATED 1915

Foundations of supposed chapel in centre of cemetery

DUNSTAN CHAPEL

BUILDING (use unknown)

REFECTORY SUB-VAULT

CLOISTER GARTH

VESTIBULE

CHAPTER HOUSE

SLYPE

NORTH PORCH

BUILDING (use unknown)

PROBABLE SITE OF LORETTO CHAPEL

SITE OF HIGH ALTAR

EDGAR CHAPEL

by the monks have yet to be excavated, the main script is finished. Bond's major legacy is method, and the research he did to understand the psychic tools with which he worked. It is this that has meaning for science today, and that extends beyond the borders of a single discipline or a single project. That is why Bond is not only the father of psychic archaeology but one of the seminal thinkers in parapsychology. It is a contribution even less understood than his work in archaeology.

Bond wrote in *The Gate of Remembrance*, "Intuition must bring all her results to the bar of Reason for provisional acceptance, and when this test is passed then the matter becomes ripe for further research.

"From the depths of the subconscious mind her power has evoked these images. Let us [now] analyse the facts, such as they are, which bear upon the case, and in light of the intuitive results see whether an argument may be built up which will be capable of supporting weight.

"They are not to be accepted with credulity, but are subjects for critical analysis . . . assisted by every useful means of normal research and exploration.

"Prove all things and hold fast that which is good.

"In this lies the true utility of the method."

It is the primary statement on how science can create a practical application for psychic material.

But formulating a method is only half the problem; the other half lies in developing the tools necessary to put the new method into successful operation. Here again, Bond was far in advance of his time and he is still in the forefront of parapsychology.

Once psychic information is divested—as Bond wanted it to be—of its quasi-religious overtones and the syntax and language so often associated with such data sources are relegated to their proper secondary position, the question becomes quite orthodox: Can an instrument be developed to pick up data in this frequency range, and is the information good—is it fact or fiction?

The psychic explorer is no longer a lunatic on the occult fringe but someone comparable to a submarine detection technician aboard a destroyer. Both are searching for a target that operates in another medium and moves at a speed differing from theirs, the one in water, the other in the incorporeal world of the *psyche*. Both have as their objective something that can only be "seen" through the inter-positioning of an instrument, either sonar or the human sensitive. Both must contend with other objects and energies that cause "static" and cloud the picture: fish and thermal barriers, or other psychic informants such as discarnate personalities who intrude, and negative emotions, which decrease a sensitive's ability. And both rely on some kind of screening mechanism to focus on what they seek: either an electronic computer or the human personality of the sensitive.

Viewed from this perspective, the psychical researcher faces the same problems as any other scientist: how to make the tools that provide the data the method demands as sophisticated, efficient, and accurate as possible.

Bond obviously understood this and, from the first, geared his work not so much toward getting information for a single dig as toward developing a model for psychic research and a method for working with the model's instrument—the sensitive. It was because he took this broader view that he rejected the most popular psychic approach of his time—the séance—in favor of automatic writing.

He says he objected to the loss of free will and self-control inherent in the trance state. To be a medium through which the dead speak is, by definition, to lose control and the ability to observe objectively what is going on—a position no conscientious scientist would allow himself to fall into during an experiment.

What Bond was looking for was a psychic technology that would allow a sufficiently relaxed state to let the unconscious mind surface, but permit the conscious mind to be fully aware of its surroundings and exercise its will. Automatic writing, in which both he and Bartlett were channels, seemed to answer his research design (interestingly, neither Bond nor Bartlett seemed capable of doing automatic writing when alone).

Having worked out his technique, Bond immediately encountered what a scientist in any other field would call instrumentation problems. From the very first sitting, in answer to his first question about information on Glastonbury, came the reply, "I cannot find a monk yet." As if no one was home to take his call.

He also felt drained from the session and was told at the second session two nights later that "the material influences were at fault;" that is, Bartlett and he weren't working properly as receivers. In the third session he encountered the obverse of the problem when his source said, "I think I am wrong in some things. Other influences cross my own . . . why do they want to talk Latin? Why can't they talk English? . . . Benedicte Johannes. . . . It is difficult to talk the Latin tongue."

In three sittings he had as many problems: one with the sender, one with the receiver, and one with the transmission path—the three main hurdles to applying psychic data in science.

Bond dealt first with the second problem, difficulties on the receiving end. He discovered that the cherished assumption that the researcher could simply function as a decision maker and an unaffected observer did not hold true in psychical investigations. The researcher was a factor in the problem, and his mental state, ideals, expectations, and physical well-being had to be considered.

When either he or Bartlett was not feeling well or was under tension, the sittings did not go well. When they had high expectations, supported by all the preparation they could do at their end—reading available documents, trying to work the problem out as best they could—the sitting results improved. Consequently, there are gaps in the regular rhythm of sitting dates. When Bond was feeling especially pressed about his marital problems, for instance, things did not go well and so he suspended the sittings. For the rest of his life, whenever he tried to contact a psychic source, he attempted to be rested and in good physical condition, and to ask questions whose answers would have a real meaning for his work, not just settle his curiosity. It is a lesson that other people working in this area have had to learn.

It is also a radical departure from tradition, and one that to this day is capable of provoking argument. Even now, despite increasingly persuasive evidence to the contrary, there are still parapsychologists who believe they are only observers.

The other two problems, the source or sources and the transmission lines, were not so easy to resolve because Bond had less leverage to bring to bear. Still, he was able to achieve both considerable success for himself and outline guidance for researchers who would follow him. He discovered, for instance, that there seem to be laws governing psychic contact that may be as "hard" as those of chemistry and that certainly compare in rigor to anything in the social sciences.

Because his contact with his sources was on the mental level,

Bond found intent and ideals particularly important, for it was these factors that seemed to attract informants and determine the quality of their information. Bond was interested in a specific period, the medieval, and it was his hope that his work would make the Abbey once again a center for religious thought and activity, a place where all points of view could meet in peace. This seemed to attract sources with the same aims. Although the Abbey's history long predated the medieval period, almost all the best material Bond received and almost all the "personalities" who contacted him came from this time. It was only after he had left the Abbey work and his horizons broadened that information concerning earlier times came through; but perhaps because he was no longer actively working on the Abbey, the quality does not seem as high or the syntax as lucid. It is almost as if common areas of intense interest made the contact possible, and following the research up in the material world affected its quality.

This feeling is strengthened by the fact that when Bond was excavating, he soon learned that the Watchers had certain information they wanted to get across; and no matter what direction he wanted the sitting to take, they almost always appear to have advanced their own points before they would address themselves to Bond's questions. He learned to let them have their say and then to put forward his queries. Frequently, however, he found that it was almost as ift gyey were reading his mind, for if his interest in something was especially high, the question would be answered before it was asked.

He also discovered that even when a successful contact was made during a sitting, he could not command a certain answer because the monk who knew was not available. "Beere, Abbot, is not with us now. He has a work to perform," Bond was told in one instance.

Listening to Bond describe what happened, and his reactions, reminds one of a person making unannounced telephone calls to a strange land, dialing a number almost at random, and getting people who have either just wakened from a deep sleep or who are paying only partial attention to what is being said while they work at something else.

The most revealing information Bond provided on the nature and dynamics of psychic sources came from his exchanges with and about Brother Johannes Bryant. Johannes appeared as early as the second sitting and almost immediately seemed to become the dominant personality among Bond's sources. At first glance he seems to have functioned as a kind of "spirit guide" in much the same way that such famous psychics as Arthur Ford and Laurie Piper had guides. But this would be a wrong interpretation. Instead of being a guide, Johannes seems to have been the osmotic membrane through which the information passed:

"Johannes mystified and bewildered by its beauty gave [Glastonbury Abbey] his heart as one gives his heart to a beloved mistress; and so, being earthbound by that love, his spirit clings in dreams to the vanished vision his spirit eyes even still see.

"Even as of old he wandered by the mere and saw the sunset shining on her far-off towers, and now in dreams the earth-love part of him strives to picture the vanished glories, and led by the masonry of love, he knows that ye also love what he has loved, and so he strives to glimpses of his dreams.

"Those others [other monks], the great and simple, are passed and gone to other fields, and they remember not save when the love of Johannes compels their mind to some memory. . . .

"Then through his soul do they dimly speak, and Johannes, who understands not, is the link that binds you to them.

"Learn and understand."

This was the description Bond received during the fortieth sitting in 1911.

If Bond learns from and through Johannes, however, the help is not one-sided. Bond does Johannes a service, and this not only provides the most poignant story in all the scripts, but also gives a deep insight into the largely unknown nature of psychic channels. Johannes, at the time of his first contacts with Bond, is described by another monk, who says of himself, "My punishment is past," as being "in pain." Apparently the fixation on the Abbey that makes him a conduit for Bond is a form of obsession that traps his awareness in a time period existing now only in his memory. It is as if Johannes were self-condemned to a dream that, while it began comfortably enough, has become an agonizing maze from which he can find no exit. Six months later, after many exchanges with Bond, Johannes speaks of his role and there seems to be a change in his awareness.

"Why cling I to that which is not? It is I, and it is not I, butt parte of me which dwelleth in the past and is bound to that which my carnal soul loved and called 'home,' . . . Yet, I Johannes, amm of

52

The Abbey proper as Bond was able to determine it.
This plan does not include the many outbuildings.

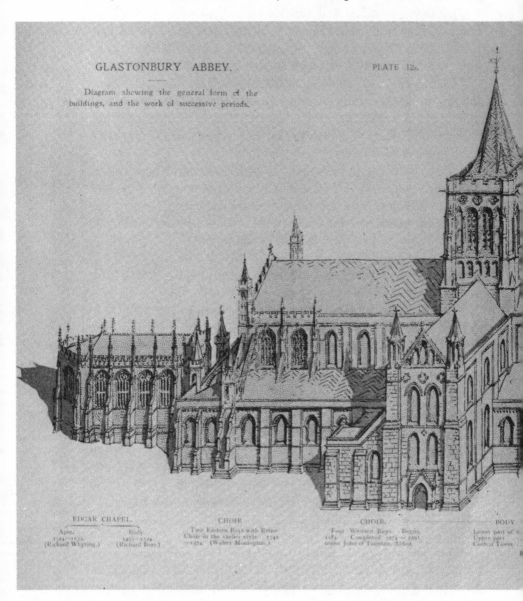

GLASTONBURY ABBEY.

PLATE 12r.

Diagram shewing the general form of the
buildings, and the work of successive periods.

EDGAR CHAPEL. CHOIR. CHOIR. BODY.

H. (NAVE AND TRANSEPTS)   WEST END OF NAVE.   GALILEE PORCH   ST. MARY'S CHAPEL

1184–1189. (Ralph Fitz-Stephen.)   Lower part, 1274–91 (John of   1274–1291.   1184–1187. (Ralph Fitz-Stephen
1234–1263. (Michael Ambresbury.)   Taunton, Abbot.)   (John of Taunton.)   (On site of Early British Church.)
1303–1322. (Geoffrey Fromond.   Upper part, with the vaulting,
    1303–35 (Adam de Sodbury.)
lted by the same Abbot.)   Western Towers: the same date.

many partes, and ye better parte doeth other things . . . only that
parte which remembereth clingeth like memory to what is seeth
yet.''

The liberation seems to come to maturation, for in August of the
following year another monk speaks again for Johannes, saying,
''Johannes [is] now very far away; far, in that the force is weake . . .
but the weakness here is strength gathered for other duties.'' It is as
if his conversations with Bond have freed Johannes of his obsession—as psychoanalysis relieves a living patient.

None of this, however, is unique in psychical literature. What
gives its beauty the possibility of truth is Bond's archaeology.
Where we can check the monks, they are accurate. When there are
gaps in time between questions on the same point, they can pick up
a thread as if no break had taken place; even the most accomplished
liar would have trouble keeping minor details straight and internally consistent over a period of years. When Bond seems unsure of
quite what to ask or how to phrase it, they demonstrate the ability to
''see'' what he is doing and to answer the question he needs to ask
but may not know. Even the strange mixture of Latin and English is
reassuring, since that is exactly the way a medieval monk would
speak; to this day there is a patois called by those who use it monastery Latin.

Strangely, the one question Bond never really answered is what
or who his sources actually were. Were they the personalities of
dead monks, as would seem the case? Or was there, as Bond became
increasingly convinced, a kind of universal species memory bank in
which informatior is stored in the ''computer banks'' in a personality format? Does each of us contribute to this bank, and if so, do
we do it not so much through our intellectual faculties as through
our emotional experiences? Were the dead monks not so much incorporeal personalities as personalized thought forms in a species
memory bank whose circuitry was activated by Bond's interests and
need to know? The answer eluded Bond; it eludes us still.

Although Bond was unable to resolve this final mystery, he did
successfully overcome the three main hurdles to the practical application of psychic data: difficulties with the sender, the receiver, and
the transmission line. And his answers to these challenges not only
improved the quality of his own archaeological work but pointed the
way to the use of the psychic as a research tool for any discipline.
For perhaps the first time in history the flaws that plagued psychic

information were overcome and combined with orthodox science to solve problems that neither alone could have handled as well.

One last point remains. Bond left archaeologists several further tests of the accuracy of his source, and a final challenge to his detractors—the Loretto Chapel. Within the automatic scripts are to be found its exact location, references to its Italianate decoration and style (an influence brought back by Beere from his Italian embassy for Henry VII), the name of its architect, and the Italian church after which it was modeled.

Bond began the search for this chapel—which was totally unknown through orthodox sources except for a brief mention by the medieval writer Leland—as early as 1911. In 1917, when he first published *The Gate of Remembrance,* although he thought he had located its foundations, he was still not certain. In 1919, however, after a lapse of some five years in his excavations, Bond was able to have a new cutting made, following directions received on December 4, 1916. In this session, the Watchers had admonished him, "Ye did not go far enough beyond the bank . . ." in the earlier searches, and that he would have to dig a "full five feet. . . ." Now, in 1919, although his director's position was already shaky, he was determined to present this final proof, to demonstrate that his book, published two years earlier, was solidly based no matter what controversy surrounded it.

Within two hours he was able to do so. Masonry was encountered just where he expected it to be, and by the next day "there was revealed the south-west angle of a building, in the form of a solidly built foundation of rough stonework." Eventually Bond outlined the entire chapel, finishing this work in August, 1921. Everything appeared just as the Company of Avalon had said it would. Bond even had time to begin a search for architectural debris to test the psychic writings on the important issue of the building's decorations. He soon discovered, however, that the destruction of the Loretto Chapel had been unusually thorough, and before he could dig further, his days at Glastonbury were over.

The dismissal did not stop him from adding a new section to *The Gate of Remembrance,* and with the fourth edition there was appended "a record of the finding of the Loretto Chapel." With Bond's usual clear directions, plus the excavations he was able to carry out, it would be an easy matter for a later archaeological team to follow on and finish Bond's explorations.

But this task, like so many other promising possibilities in the psychic scripts produced by Bond and Bartlett, has gone unattempted for more than half a century. This eccentric Englishman is waiting to make one final statement, and to supply, posthumously, further proofs of the reality of psychic archaeology.

II

THE EYES WHICH SEE EVERYTHING

Ossowiecki and Poniatowski

In 1952, seven years after the end of the most devastating of the many wars her country had known, Zofia Swida Ossowiecki was walking home to her apartment at No. 1 Adama Pluga in Warsaw. She hardly noticed the seedy man in the shabby topcoat. He looked no worse than many other middle-aged Poles and his distressed appearance was completely consistent with a Warsaw still more ruins than city.

"You are Mrs. Ossowiecki?"

"Yes, I am Mrs. Ossowiecki."

"Your husband was the clairvoyant?"

"Yes."

"Do you know that he made some prehistoric experiments?"

"Yes . . . but they were lost."

"They are not lost. . . . Would you like to get them back?"

"Of course, but who are—"

"It doesn't matter. All that matters is whether you want to own them again. It must be done in secret . . . or no one will ever have them."

Zofia Ossowiecki never knew the man's name nor how he came to possess the research done by her husband Stefan with his friend Professor Stanislaw Poniatowski. He was right about one thing, though: All that mattered was getting back the unpublished manuscript on their work. Alone, unable to contact others for fear the man would carry out his threat, and with only limited resources to meet his demands, she still accomplished her task. This act of preservation was a last statement of love for the extraordinary man who had been her husband.

Stefan Ossowiecki had come into the world as he died, at a time

STEFAN OSSOWIECKI
This is one of the few pictures known to exist of the Polish clairvoyant, and the last ever taken of him; probably no more than three weeks before he was murdered by the Nazis in August 1944. In light of his extraordinary abilities in dealing with animals, it is fitting that his last picture shows him holding a small bird whose life he had just saved.

MARIAN SWIDA

when the very survival of Poland was in question. In 1877 his homeland did not even exist as a nation, and there were few prospects that it ever would again. For over a hundred years it had been partitioned between Austria, Germany, and Russia, and many of its people dispersed, among them the aristocratic family from which Ossowiecki came. His father had been born, reared, and married in Moscow. He started and ran what soon became a successful chemical factory. But though his profits might be Russian, he was not. He chose a Polish wife, tried to keep that language alive in his home, and taught his son Stefan, who was also born in Moscow, to think of himself as a Pole. This training would have an enormous effect on the development of psychic archaeology and ultimately would be the cause of Stefan's becoming a legend in Poland during World War II.

But if the Ossowieckis thought of themselves as Polish, this did not stop them from moving in the best circles of Russian society, and this training would also prove important. Perhaps it was this social access, in combination with his inner sense of being an outsider, that helped Ossowiecki avoid the arrogance hanging like a fog over the Russian Czarist court; he developed, instead, into a man totally at ease in any situation, with any type of person.

There is some evidence that his grandmother, and possibly his brother, had psychic capabilities. But these must have been slight, and the family interest in such things only casual, because Ossowiecki's obvious talents were neither noticed nor, when brought to the family's attention, understood. What they saw was "a normal lively child" popular with his playmates. They did not realize that from a very early age their son saw flickering auric bands of color around people and was surprised, when he questioned his mother about them, to discover that she thought something was wrong with his eyes. She took him to an eye doctor. The duly prescribed drops "irritated my eyes but did not diminish my ability," Ossowiecki recalled years later.

Stefan Ossowiecki was born into one of the most opulent societies in the world—the upper class in Russia before the Revolution. Although Polish, his family had considerable stature within this group, and the means, as this picture of his family's home in Moscow makes clear, to meet its standards.

By the time he was fourteen, and a member of the elite Russian Cadet Corps in 1891, his psychic talents were developing as fast as his body. At eighteen, when he was selected for the St. Petersburg Technological Institute (situated in what is now Leningrad) to train in his father's profession of chemical engineering, his capabilities were frightening. Particularly awesome was their major manifestation, which may be history's most unusual example of telekinesis (the ability to move things with the mind alone). Research being done in Russia and the United States today with such sensitives as Nina Kulagina and Alla Vinogradova concentrates on moving matchsticks and other small lightweight objects. On one occasion during this period of his life, after a dinner party, Ossowiecki was stripped, wrapped in a straitjacket, and laid on the floor of an empty ballroom. As he lay there supine, three husky servants moved, with difficulty, a large marble statue into place at the other end of the room. As witnesses stood around him, Ossowiecki moved

Ossowiecki as he looked in about 1895 when, as a cadet at the St. Petersburg Technological Institute, he first met "my master," the Jewish mystic Wrobel.

MARIAN SWIDA

what amounted to several hundred pounds of dead weight about eight feet toward him across the floor. Those present were stunned, and even Ossowiecki himself was a little shaken.

He also discovered that he could psychically "read" his examination questions at a distance, although they lay covered on a table. Ever generous with his friends, he was willing to share.

"Ah, how many of my colleagues I helped with this ability! The most entertaining part was the reaction of the professor of higher mathematics who was a decided positivist. He noticed that I had this gift. . . . It upset him because it didn't fit into his world view. But what upset him even more, I think, was that he couldn't punish me for it."

For all their power, however, Ossowiecki's psychic skills were erratic and, by his own admission, would have been "wasted" on little more than parlor shows if the Russian railway schedules had been different. In 1898, when he was twenty-one years old and in his fourth year at the St. Petersburg Institute, one of his final assignments was to study the making of paper. Oddly, the subject held a real interest for him, sufficient, in fact, for him to write another Pole living in Russia, Antoni Stulginski, the director of a paper factory, and ask if he might visit his plant in Dobruz. Because no train ran direct from St. Petersburg to Dobruz, Ossowiecki found himself in the railway station of the town of Homel with several hours at his disposal.

In answer to his query, "What is there to see in Homel?" the stationmaster had only two suggestions: a statue and a visit to "a locally famous person to whom the greatest dignitaries of Russia came for advice." Ossowiecki took this last as typical civic hyperbole, but was intrigued enough to ask further questions, thus learning that the man's name was Wrobel, and that he was "an old Jewish yoga who had spent all of his life in India studying secret knowledge" before he "returned to the city of his birth at a very advanced age to 'die at home.'"

Surprisingly, Ossowiecki had no real interest in meeting this man. Despite his youth, he had no burning questions about life he wanted answered, nor did he feel the need for guidance. He knew who he was, how he fitted into the social scheme, and what he wanted to do. It never occurred to him to ask about his unusual abilities. But "for lack of anything to do," he decided to go into Homel's suburbs in search of the house where the mystic lived. He took the matter more seriously when he got there and opened the door to the small wooden house. The stationmaster had not been exaggerating; there in the foyer were "two very famous personages" who had come for consultations.

Looking at Ossowiecki, they saw a rather handsome six-foot-tall blue-eyed young man with light-colored hair. From his clothes and manner it was obvious that he was a gentleman and so they spoke to him, learned he was just passing through, and "gallantly allowed me to go ahead of them. I shall never forget that moment."

"I enter. On the bed there rests an ancient man. Head of a patriarch. Subtly chiseled Semitic features. Spirituality manifesting itself through the eyes. Long, white beard."

Despite his psychic abilities, Ossowiecki in those days did not believe in clairvoyance, and though impressed by Wrobel's appearance, he admitted his whole posture "was that of skepticism." He was completely unprepared for what the old man did.

"Upon seeing me—even before I had a chance to introduce myself—he extended a skinny arm toward me . . . looked me straight in the eye and said, simply, 'Your name is Stefan.'"

Then he proceeded to describe "my past and future in minute detail. You must realize this was before the war [World War I and the Russian Revolution of 1917], in a completely different political system. Then much of what he said seemed improbable . . . it created a feeling of some dreadful nightmare. And yet, he saw all that happened later: the fall of then unquestioned powers, the oceans of blood, heroism, sacrifice of human lives, terror, fear and separation. He sketched for me the whole skeleton of political games, nationalistic battles, and victory. He talked about all this which I later encountered, took part in, and was shocked by. All that affected the fate of my life . . . foreseeing even my inner reactions which I myself could not have then predicted . . . Without any information forthcoming from myself he said then, 'You see auras around people.' I asked naively, 'What is an aura?'"

By the time he had to leave to catch his train, "in an old man, demolished by an ascetic life," Ossowiecki had found his first and only real teacher.

To complete his degree in chemical engineering Ossowiecki had to put in a period of practical work. There was never any question of what he would do. He moved to Dobruz and began assisting Stulginski in his paper factory. But, as would be true all his life, his outward work was entirely secondary. His real motivation was spiritual and sometime in 1899 he apprenticed himself without reserve to the dying Jew.

"I would spend every free moment with him, not to mention all Sundays, . . . which passed almost unnoticed in long . . . discussions . . . and exercises . . . the concentration of thought and will . . . which bridged the way from the conscious to the superconscious state."

Under Wrobel's tutelage Ossowiecki learned the yoga skills of visualization and concentration on a single object. It was a torturous experience but exhilarating at the same time and gradually "I conquered time and space" and finally reached a stage when "myself as myself seemed not to exist. I understood, then, that I had crossed the border of average human consciousness. My visions were 'superconscious.'"

His teacher explained to him that the superconscious was very different from the subconscious, which "is a state in which our will does not participate. Our dreams, for instance, are a manifestation of the subconscious and sometimes they have an almost prophetic character. . . . Superconsciousness, however, is a higher level of organization. It requires the necessary presence of a new element—the element of spirit, of trained will. For superconsciousness the barriers of time and space do not exist. . . ."

From Wrobel Ossowiecki also learned the arts of psychometry, the use of objects to focus the superconscious mind so that the object becomes a kind of psychic bannister allowing the sensitive to move along a single thread in time's skein.

For two months, and possibly longer, the man he called "my Master" worked with Ossowiecki. Although the actual time was short, it had an almost unworldly intensity. More than that—it was effective: A wild powerful native talent was transformed into a coherent highly disciplined skill whose limits were never fully explored. It had been for Ossowiecki the major spiritual experience of his life, and what he became under Wrobel's guidance was to make his experiments in psychic archaeology, four decades later, unequaled in the history of this method of searching man's past.

Surprisingly, however, Ossowiecki did not become outwardly esoteric or mystical and for many years only practiced his skills intermittently. He moved back to Moscow, probably shortly after the beginning of the new century, and took his place as a regular member of the city's society. Before he was thirty he had established the external pattern he would follow all his life. He loved good food (indeed, soon became stout as a result), good wine, lovely women, creative artists in any field, but particularly writers, and parties where there was lively conversation. He knew everyone. Though the capital was in St. Petersburg and he was mostly in Moscow, he became close to the Czar, the Imperial family, men like Tolstoy, as well as unknown artists, dukes, even workmen in the chemical industry.

And, if prevailed upon, he would exhibit his psychic abilities, particularly his telekinetic strength, moving all manner of things—including a grandfather clock. The feats were so unprecedented that "while I moved these objects, there were many cases of common hysteria . . . even among men." Such reactions are not hard to understand when one pictures the scene: a relaxed gathering of Moscow society, replete with food, a little tight on vodka and wine, standing around Ossowiecki, the men in white tie or dress uniform, the women in sumptuous gowns and jewels, all of them watching an event that contradicted everything they had ever learned of nature and reality.

For all the public acclaim these performances brought him, Ossowiecki himself in later years had little interest in such feats. He talked little of these early paranormal demonstrations, and much else of this period is also blank. One can only speculate, for instance, that since he was close to the imperial family he must at least have met the most famous psychic ever to come out of the Russian Empire, Grigori Efimovich, known to the world as Rasputin. But Ossowiecki is silent. All that is certain is that his father died in 1915 and that Stefan, at the age of thirty-eight, became the owner and principal director of what had developed into a very large chemical works. He was now a genuinely rich man, but strangely his entire Russian period appears like an interlude—a pleasant space between his time with Wrobel and what was to come. The revolution that ended the Romanov dynasty freed Ossowiecki of the encumbrance of wealth. Nothing was left to tie him to Russia. Although outwardly everything had gone to pieces and worse was yet to come, to Ossowiecki, whose real life was entirely within, this was the second great awakening.

"A tremendous breakthrough in my *psyche,* a breakthrough which increased my spiritual awareness and consequently my powers. . . . I was arrested in Moscow toward the end of the year 1918 [the reasons probably being his wealth, ownership of an industry, and friendship with the Czar and his court]. This brought with it a long term in prison with the prospect of execution by firing squad.''

Wrobel had told him this would happen, but not the final outcome, perhaps because such things are not predestined but affected by the consciousness and will of the participants. Thus the experience and the possibility of death were both frighteningly real to Ossowiecki:

"The forced isolation brought the opportunity to think through many things; to ponder the broadness of spiritual horizons; to deepen one's spiritual personality in order to gain solace on the threshold of death.

"Death passed me by but what did remain was the faith, the understanding of higher, immortal life for which, unconsciously perhaps, all yearn. Even those without faith. It was then that I began to fully value this gift given me by the creator and I understood that by utilization of it I could help others.''

Sometime in 1919, at the age of forty-two, almost penniless after confiscation of all his Russian holdings, having nothing but his profession, his mother, and his sister Victoria, his surviving friends, and most important, his psychic gifts, Stefan Ossowiecki left Moscow to return to a Poland he knew only as a visitor, whose language he spoke ungrammatically and with a heavy Russian accent. Poland, which itself had been barely restored to nationhood, for the first time since 1795. Left behind in Russia, along with all his possessions, were his telekinetic powers; he had transformed them while in prison to the almost unequaled powers of clairvoyance that were soon to make him the first scientifically studied traveler in time.

By 1920, thanks to his personality, friends, and training, Ossowiecki had found his place in Warsaw, and although he would never again be wealthy, he had enough to live comfortably. For the rest of his life money was of secondary interest. And even when he had it, he seemed little interested in keeping it. Always an easy touch, if he had funds, he would invest in a company to help a friend or just lend the money outright. Usually the company failed or the loan went unrepaid.

What he specifically would not do was charge for his psychic skills, which were now so much in evidence. Wrobel had warned him of this years before. At the time the words had been largely meaningless; there was no temptation, for his family was rich. To make money from psychic talents would have been unnecessary and, worse, unseemly. During his stay in prison when he was faced with poverty, however, this appears to have been one of the issues Ossowiecki considered. When all had been lost, the temptation was suddenly there, particularly since he now had developed such extraordinary clairvoyant abilities. But after his release, even in his first weeks in Warsaw, he did not succumb. He would say later, "I am of the opinion that the domain of the spirit cannot be mixed with monetary considerations because the pure, abstract value would be lessened. . . . I have a vocation which gives me work and assures my existence. I strictly separate these two areas of myself and professional engineer Ossowiecki, the man of business, has nothing in common with Ossowiecki, the spiritual seer."

Just how strongly he felt about this issue is revealed in a story told by Marian Swida, the man who became his stepson. "In 1938, when things were becoming unsettled and finances were not so good, Eugene Rothschild of the Vienna line, the family of bankers, financiers, came to him . . . secretly, very privately, in a private airplane . . . a rare thing in those days. He asked Ossowiecki to make an experiment, to see if he could locate some family papers. I do not know all the details . . . only that Ossowiecki said that there were millions and millions involved. Rothschild asked only, 'Was it stolen?' Ossowiecki said 'yes.' He next asked, 'By whom?' Ossowiecki replied, 'By a daughter of your butler and they are in London.' He then said where they were in London, where the papers were. They checked . . . everything was found, just as Ossowiecki had said! A check arrived, signed by this Rothschild, to my stepfather, to write himself any amount. The money would have been helpful at that time. But he would not do such a thing as this. He returned the check saying money was not the purpose for which he performed such experiences."

But if Ossowiecki was uninterested in using his gift to make money, he was most interested in using it to help others. He scheduled his business life so that he had time between engineering consulting assignments to meet with those whose problems might be eased by his insight. Always using an object as a psychic focal point, he would answer queries and offer guidance. Although he would not take money, there was one thing he would accept, often just to get grateful people to give up on the matter of payment. He liked a souvenir, a memento of the experience, particularly the guide object he had used.

These personal consultations though were not enough. Also as a result of his prison experience Ossowiecki felt a need to be scientifically tested—not because he had any doubts about his ability to show off for scientists, but because he hoped that in testing him these researchers would alter their view of the world. Consequently, in Warsaw and in Paris, to which he traveled every year throughout the 1920s, and in Germany, which he also visited frequently, he let it be known that he would entertain any request seriously presented to him by a responsible scientist. This sentiment went out through what had become a truly astonishing network of friends, a list of virtually all the leading minds in the arts, sciences, and aristocracy. (Only politicians were not well represented; Ossowiecki was never politically oriented, perhaps because Wrobel had described in detail how ephemeral European political fortunes would be during the psychic's lifetime—and how irrelevant in any case.) The response was immediate. As early as 1920 he met Professor Gustave Geley, one of the great French pioneer parapsychologists. This meeting was soon followed by others such as (a partial list): Nobel Prize-winning physiologist Professor Dr. Charles R. Richet, also of France; Theodore Besterman, Lord Charles Hope, Miss A. Reutiner, E. J. Dingwall, all of Great Britain; Baron Dr. von Schrenck-Notzing and Dr. M. Gravier of Poland.

There were many experiments; indeed, the studies conducted with Ossowiecki, if collected between covers without commentary, would make a substantial volume. Basically, however, they fell into the two major categories of research being done in Europe at that time: astral projection, the movement of one's consciousness to a point different from the body's point of view; and reading messages secreted in sealed envelopes, metal tubes, and boxes. What all this determined was that Ossowiecki was not in any way a medium (a spiritualist), nor was he ever fraudulent. Most important of all, it was found that Ossowiecki, unlike most other psychics, had absolute control over his talents and could turn them on and off at will, as well as provide a uniformly high standard of performance. This was (and is) so rarely encountered that the researchers studying him had no explanation as to why he could do this and other psychics could not. Ossowiecki, however, gave them an answer.

He made it clear to anyone who would listen that he was not so much a psychic as a man deeply committed to a spiritual pilgrimage in his inner life. Because of this, he happened to be psychic. He explained that there was a difference between being a man seeking enlightenment, for whom the psychic is a kind of byproduct, and being a person seeking psychic ability as an end in itself. He himself was obsessed not with the psychic but with the true nature of consciousness, the human will, and the relationship these had to man's destiny. Unfortunately, such words apparently had little impact on those who heard them, and it would have been a violation of his principles for Ossowiecki ever to push his philosophy on anyone else. It would also have been out of character.

Ossowiecki had built the public persona of a slightly eccentric gentleman who enjoyed the best that was offered in a Warsaw rapidly becoming one of Europe's most sophisticated cities—"the Paris of Eastern Europe." He wore his father's large gold pocket watch, but never remembered to wind it. And he could (and more than once did) invite a prince for dinner, forget he had done so, show up uninvited at a friend's, under the mistaken assumption that they expected him, be happily invited in anyway, only to receive a call from a second friend asking where he was, go there, and just be sitting down to table when his maid called saying the prince had gotten tired of waiting and had gone off before the master could be located.

And when he did do an experiment, more often than not it was after dinner dishes had been cleared. He did not like people to stare, or treat him with awe, and if they did so, he would stop and tell them to talk or play the piano. Nor, except with very few people, could he be dragged into a discussion about spiritual philosophy—a popular topic among the upper classes of the day. But if he appeared a "hail fellow," he still always made sure to provide serious researchers with what they needed: results. And these were such that in the book *Our Sixth Sense,* which the Nobel laureate Professor Richet would write about his years as a parapsychologist, after describing many other psychics, he would say, "It would seem . . . that there would be nothing more to add, or rather it would be impossible to find better.

"Well! There is better.

"If any doubt concerning the sixth sense remains . . . this doubt will be dissipated by the sum total of the experiments made by Geley, by myself, and by others, with Stefan Ossowiecki . . . the most positive of all psychics."

Surprisingly, however, although he was studied by every leading researcher in the field (indeed, no psychic has ever been studied so thoroughly and by so eminent a group of men and women as Ossowiecki), only a few brief and superficial experiments into the past were ever tried in this period from 1920 to 1935.

The reason for this lapse is both simple and revealing of how psychic researchers approach the psychic. Clairvoyance was then explained as some form of telepathy and telepathy was considered nothing more than an unknown form of radio waves. Strange, admittedly, but to science, the explanation least harmful to its world view. Similarly, psychometry was seen as only a variation of telepathy in which, for a short space of time, the thought waves of a person were impressed upon an object in some way, much as an iron

For some fifteen years, during the 1920s and '30s, Ossowiecki participated in every psychic experiment and test that the best European scientists could contrive. In this, one of the more fantastic conducted by the Nobel laureate Professor Charles Richet, a partial drawing of Marshal Pilsudski was placed in a metal tube, which was then welded shut. Presented with this challenge, Ossowiecki correctly identified the person as Pilsudski, named the prominent features, and described the man's character. When the tube was cut open, Ossowiecki was shown to be completely accurate.

bar can be magnetized. The sensitive was just that; he read the object by being sensitive to its thought *impregnations*. Obviously, with this theoretical base, experiments involving actual movement in time, which would have been devastating to science's assumptions, were not often attempted. And the few experiments carried out in contacting the past seemed to support this theoretical position. If the length of time was greater than a few decades, the information always seemed to come from spiritualist sources.

Survivalism, which until the 1940s was the third great interest of psychic research, was not paranormal time travel. It involved an entirely different issue—whether the personality survived corporeal death. Thus it was not exceptional to find parapsychologists who believed on the one hand that it was impossible for a living man to travel backward in time beyond, say, a normal life expectancy (traveling forward involved living people and thus could be explained as telepathy) while, on the other hand, believing with equal conviction that there was personality survival after death. It was for them an apples and oranges proposition.

Ossowiecki, since he was primarily a seeker of nonintellectual supersensible knowledge as opposed to being a kind of paranormal technician, did not agree with this. He "knew" that while the theory of survivalism was valid, the accepted explanation for telepathy was not. Time and space were not the limitations science thought them to be. But he could not think of an experimental protocol that would make this point, one that could withstand the charge of the thought-wave explanation. In 1935, however, he found he didn't have to think up an experiment; it was presented to him already set up, and its results would set in motion the last and greatest research epoch of his life.

A wealthy Hungarian by the name of Dionizy Jonky had died in 1927, leaving among his effects a small package measuring 7 by 4.5 by 4 centimeters. It weighted 59.5 grams (just a little over 2 ounces) and was about as big as a man's fist. To add to the mystery, its contents had obviously been carefully sealed against observation. The package was wrapped in fabric, sewn tight, tied in cruciform fashion with string, and the seams and ends of the string were secured with sealing wax.

Only when Mr. Jonky's will was opened did the purpose of this odd little bundle come to light. The Hungarian had always been interested in the psychic, and he too questioned whether there might not be more to paranormal awareness than the wave theory. Consequently, he had prepared an experiment. To test the wave explanation, he stipulated that the package should remain sealed and that no psychic attempt to determine its contents should be carried out until eight years after his death. By then, any putative wave energy should have dissipated. And if the information source was a dead person, presumably the kind of psychic to discover its contents would be a medium. In any case, Jonky said in his will that no one had seen what he had put inside the package, and after eight years any psychic who wished to could try to divine what the box held, as long as his answers were carefully witnessed and recorded. Those charged with seeing that these terms were carried out followed them scrupulously; not until 1935 did the research begin.

Ossowiecki was approached last on the matter, after "the best of Europe's clairvoyants" had already tried. A committee of researchers came to him that January and he, of course, consistent with his policy, immediately accepted. He was told nothing of the package's background, not even Jonky's name. The question: Can you determine what is inside this? was his only information. Fifty people, mostly scientists, were present when he made his attempt, beginning the proceedings by looking at fourteen photographs arranged on a table and selecting the one of Mr. Jonky, "a man whom I had never met," he would say later.

Six individuals then stationed themselves next to Ossowiecki, and never moved beyond arm's distance of his side until he was through. Everything he said was "taken down and later signed by the whole committee."

"After ten minutes of concentration I stilled my consciousness and moved to the realm of the superconscious. I began to speak." The following are his exact words.

"Interesting and convoluted story. I see the owner of this object. He's long dead now. He was successful in life; he had his own house. It was a man of advanced years, with a white beard . . . worried. A man of large spiritual seances, read, wrote, traveled a lot. This package found its way into the hands of another older gentleman who looks like Professor Gravier [Ossowiecki's friend]. "But you had it!" Ossowiecki said, addressing Professor Gravier, who was one of the committee. "I see it in your place on top of a wardrobe with some boots.

"There are some pieces here . . . several . . . two . . . three . . .

more. They are minerals of some sort. Stone . . . metal. Color is gray-brown, something like lime and iron ores . . . volcanic minerals. All this was once in the hands of some young girl.

"There is something here that pulls me to other worlds . . . to another planet. . . . It's a great world . . . it has no similarities to ours. It hurtles with dizzying speed through endless space. Much fire . . . it collides with another body . . . cosmic catastrophe. Tears away . . . tears apart. Showers into small fragments. They fly . . . they speed, they fall in many places on earth. These are fragments of a meteorite.

"This experiment was thought up especially for me. Perhaps it was at one time intended as two experiments because it draws me in two directions. The elderly gentleman had some sugar ready and nearby was a box of minerals. In one sugar . . . in the other meteorites. At first, I clearly felt meteorites but also the presence of sugar. I can't say anything more."

Ossowiecki had been asked only for a prediction of what was in the package; he had given that, plus a re-creation of the events leading up to its preparation, and what had taken place up to the time he was first shown the target object. If he was right, the theory espoused by Richet and others was demolished. Without further delay the seals and wrappings were broken.

The report concludes by saying, "All this from beginning to end was verified. Indeed, the package contained meteorites and the paper in which it was wrapped was originally confectionery wrap and still had traces of powdered sugar and a Hungarian word meaning bonbons.

"Mr. Ossowiecki recreated then not only the nucleus of the package but also its surroundings.

"Because of the prolonged time since the death of Mr. Jonky until the experiment took place, the notion that clairvoyance is a sort of telepathy must be abandoned." Twenty psychics of various nationalities had tried to solve the riddle but only the Polish seer "was able to unlock this secret with 100 percent success."

The implications of this were not lost on those concerned with studying the past. Within months Ossowiecki was approached by a committee of historians and other scholars and asked to perform an experiment that was to take him the final step toward psychic archaeology. Their question: Could Ossowiecki determine what Nicholas Copernicus (1473-1543), the most famous Pole in history,

In his first essay into the past, Ossowiecki was asked to select the true portrait of Nicholas Copernicus (1473-1543). Presented by historians with a collection of portraits, all of which were then thought to be genuine, he chose the one that later proved to be the only portrait taken from life. More than that, Ossowiecki told the researchers that Copernicus had a sister, then unknown, and this, too, proved to be accurate.

looked like? Surprisingly, up until 1935 no one knew the lineaments of the remarkable Pole. There were a number of portraits identified as Copernicus, but they did not agree.

Ossowiecki asked that the available paintings and drawings be brought to him and he made a selection. The scholars would have been happy with just that but, using the portrait that he said was the correct one, and that he further indicated had been painted by Copernicus himself, he went on to re-create the life of this physician, administrator, and mathematician—the man whose work on astronomy, *De Revolutionibus,* had caused one of the major scientific revolutions in history. It was all taken down and the scholars left to check Ossowiecki's words. Every detail that could be traced proved to be absolutely accurate, including the fact that the scientist had had a sister.

This validated psychic venture, back almost five hundred years in time, did more than solve a specific problem about Copernicus. It established that Ossowiecki truly was a psychic time traveler; thus his claim that time and space did not exist in the superconscious state had to be taken seriously. Here was a phenomenon very different from the Bond experiments, which the Poles may well have known about since Bond's books had been in print for some years and Bond had known the parapsychologists, particularly the English ones, who had studied Ossowiecki, Bond had gotten his information from a source apparently other than himself, much as a man who tunes into a radio station. Ossowiecki, on the other hand, actually seemed personally to go back in time, even as his physical body sat in a twentieth-century room.

He told his friends that while he did not go into a trance, when he moved into his superconscious mind, this became his primary level of awareness and the room, even his own body, faded to a kind of shadow state. When the shift occurred, it was as if he were looking at a movie running in reverse; that is, from most recent to more distant events of the past. When he chose, he could freeze this reverse action. As that happened, it was as if he were suddenly in an airplane, first seeing a sort of broad overview, and then going lower until he was approximately at a point of view where his eye would normally be if he were physically standing on the spot he had chosen in the past. At this stage the action would begin again, only now going forward as it should. From then on, it was as if Ossowiecki were some kind of science fiction spy-eye under long-distance control. He could go anywhere he wanted and see anything in the scene he desired; all the while a part of his mind knew that he was sitting in his study or someone's house in Warsaw.

At first the potential Ossowiecki offered was too awesome; no one quite knew what to make of this opportunity, the parapsychologists least of all. Finally, Witold Balcer, an old friend who was not a psychic researcher but an engineer like Ossowiecki, became interested in the idea. It was he, and not one of the many scholars around Ossowiecki, who was to make the first rudimentary usage—excepting the relatively modern and limited experiment on Copernicus—of the seer's skills in psychic archaeology. Balcer himself, though, did not really expect success. He had talked with Ossowiecki about the matter, and he knew from the psychic about the commentaries of Dr. Rudolf Steiner, the Austrian philosopher, scientist, and founder of the General Anthroposophical Society in Switzerland. Steiner had spoken at length, but only in a general philosophic way, about past cultures based on knowledge that he said derived from supersensible awareness. Because of these conversations Balcer was willing to accept the idea that some sort of generalized record along the lines of racial memory might exist. But that was quite another thing from psychic time travel, in which the sensitive actually viewed an event that had taken place thousands of years in the past in the same detail as if he were physically present. Balcer believed—but he still wasn't sure. Nor, for this reason, was he entirely clear about how to conduct the experiment.

He finally settled on a kind of hybrid protocol. He decided to have a friend prepare another of those sealed boxes that litter the psychic research landscape of the period. Balcer was to be aware in an overall way of the content, but was not to have seen it, nor have any specific details on the matter. In this way, even if Ossowiecki could not provide a reconstruction, if he could successfully name the artifact and describe it, the event would still qualify as a responsible double-blind experiment.

On the appointed night, Thursday, February, 14, 1935, Ossowiecki and a small group of friends journeyed out to the Saska Kepa suburb of Warsaw on the east bank of the Vistula River for a dinner at No. 21 Obroncow Street. Ossowiecki had not felt well that day and wasn't sure he could do the experiment, but finally, at 11 P.M., as dinner was being finished, he said he would give it a try. He took the paper-wrapped box in his right hand, cautioned his friends to continue their conversation, and while they went through the motions of casual chatter, for about fifteen minutes Ossowiecki appeared lost in thought. Suddenly he began to speak. What transpired was carefully recorded by Balcer:

"I see a metal foil box. Its surface is reflective. The inside is brownish, wrapped in paper and cotton. Something like wood or stone. Something petrified. This is something very old and it originated several thousand years ago. Before the birth of Christ. This was unearthed by some scientific expedition. I see people in white pith helmets directing this excavation. Around sand and rocks. This is in some hot country. This object was a portion of some bigger object and served or was connected with some cult or religious procedure . . . wedding . . . or funeral. Yes, this was connected with a funeral. But what is it in fact? It's some figure . . . or idol. I don't understand. I see some fires, like torches, some strange people who bow in front of this or are praying. What is it? This object has some fibers, knots, in places as though it were covered with strips of fabric." Ossowiecki seemed to retreat into himself, then said distinctly, "I can see now what this is . . . it is a petrified human foot!"

When he had regained his normal consciousness he found a roomful of people visibly straining to unwrap the box. Balcer took it from the psychic's hands and unpacked it. "Revealed to our eyes," said Balcer "was a shiny metal box and inside it, carefully wrapped in cotton and white tissue, a mummy foot. Rather small, most likely female, brownish in color. In some places tendons were visible and toe bones as well as traces of the fabric bandage in which it had once been wrapped." Suddenly the room erupted in conversation. Balcer explained that he had known only that it was part of a mummy that a friend of a friend had recovered in Egypt around 1927 almost eight years earlier.

Ossowiecki had been accurate again on every point, but was so curious that he wished "once more to lift at least the corner of the mysterious veil which covered a happening thousands of years old." Another experiment with the mummy's foot was arranged for a little over a month later, March 18, 1935. Again, Balcer kept his careful record:

"Mr. Ossowiecki felt very good that evening. He took the box (unwrapped, of course) with the mummified foot and after just a few minutes crossed the border of normal seeing but did not seem removed from us. His face assumed the look of highest concentration and appeared inspired, as though illuminated by some inner light. I had never seen him look quite like this. I was under the impression that his face really emanated light of some kind.

"After a while he began to talk. . . .

"'I'm now entering this far, unknown world. Ah! This is so long ago. Thousands of years divide us. I see clearly this woman, her whole life. She is olive-skinned, young, pretty. She has a slightly hooked nose, a pleasant facial expression. She's dressed in a white . . . translucent . . . long garment. On her ankles and wrists she wears gold bracelets. Around the neck decorations of gold and silver. Hair . . . plaited . . . into small black braids under a high tiara culminating in a square shape. She is a daughter of a high dignitary . . . as though a prince . . . but not pharaoh. She lives in a huge stone palace. The courtyard of this palace is landscaped with trees and bushes. In the center there is a stone pool with a fountain. The princess has a husband. I can see him too. He is slender . . . wears black sandals made as though from wood. His garments are white fabric.

He has no tiara on his head . . . and braids fall down on both sides of his face. Precisely . . . they are not . . . braids, but a great many smaller braids. On his forehead a head band either soft or rigid . . . golden.

"'She dies during childbirth . . . in great suffering. The child died also. I see now . . . they are taking her body out on a stretcher. The stretcher is gilded. They are carrying her to some house by the river. That's where embalming takes place. Ah! This is interesting. They are taking out the intestines . . . taking the brain out, using some long implements . . . through the nose. They rub the body with oils and continuously sprinkle it with some powder. Now I see the funeral. Many people. Some people dressed in gray-black garb are walking in rows of six. They jump every so often. The husband and father are not present at the funeral . . . they do not walk behind the coffin. They remain at home . . . they are kneeling on the floor with heads down, covered with a shroud.

"'I see an opening in a rock and a corridor that leads deep inside. Further there are steps leading down. At the end of the corridor there is a fair-sized chamber hewn out of the rock. That's where they place the coffin with the corpse. Nearby they arrange various objects of daily use . . . and food . . . on terra cotta plates. Rice, wine, berries, resins . . . something else. All these things were brought in leather sacks.

"'I see a black urn. Something is being burned there. That's the entrails of the embalmed person. They are being burned here on the spot . . . next to her. All the group around the urn pray. Husband and father are not here. They stayed at the house. All in the tomb now pray and cry. I see some women who are hired mourners. They weep. They wear long veils and black garments. Now everyone files out, their arms crossed on their chests. I see the entrance of the chamber walled over and again another wall erected at the turn of the corridor . . . and yet a third one at the end . . . near the exit. The exit is blocked with huge rocks and filled with soil.

"'All this is happening quite far from the big pyramid and the Sphinx . . . near some mountain on which I observe carved bas-reliefs. . . .'

"With those words we interrupted the experiment," concluded Balcer. He had noticed that the "glow" he had seen on Osso-wiecki's face at the beginning of the session had faded and the man seemed slumped in exhaustion as if caught and mesmerized by what he was seeing. It took more than fifteen minutes for the psychic to become fully and normally conscious, and then only after drinking a glass of wine. Even after he was again alert no one said very much for a while. It was obvious to everyone present, and particularly to Ossowiecki and Balcer, that they had touched something far beyond their ability to explore. To each observer, the experience, even though shared secondhand, was so immediate and profound that frivolous conversation was no more appropriate than it would have been in a cathedral—or a forest.

Ossowiecki, although interested in doing experiments about the past, knew virtually nothing about ancient history, let alone the epochs of prehistory; Balcer's position was much the same. It was decided that they needed to consult with a specialist, but neither man knew an archaeologist or ethnologist who was also interested in the psychic. The problem was shelved for the moment, however, since in prewar Warsaw those who could do so left for the country or to travel when the warm weather began. Social life ended sometime in March or early April, not to resume again until September. It was agreed between Ossowiecki and Balcer that they would try to find a prehistorian in the hope of making further experiments during the coming fall season.

As if in preparation for this, sometime during the spring or summer of 1935 Ossowiecki ended what had been his major business connection for almost fifteen years. As a result he was no longer compelled to travel large portions of the year as he had done since the 1920s. But this was only one of several seemingly unrelated events that, taken together, marked the end of an epoch in the psychic's life. Before the year was over his great friend Professor Richet would be dead in Paris at the age of eighty-five. The other parapsychologists who had been studying Ossowiecki since 1920 turned to other work, retired, or also died. Their research, which had involved classic tests to see whether his psychic talents were genuine, had established that they were, and so was finished. It was the end of another fifteen-year cycle and, with one or two minor exceptions, Ossowiecki would never again perform an experiment that did not have a practical purpose.

Exactly how Ossowiecki met the man who would become the scientist half of the best psychic archaeological team in history is now unknown. Perhaps Balcer introduced them. It is clear, however, that they knew each other by March 1936.

Stanislaw Poniatowski was fifty-two, seven years younger than the psychic, and since this was to be his major work for the remainder of his life, it was almost as if he had come to Warsaw for just these experiments. After eighteen years as a professor of ethnology at Wolna Wszechnica (University) he had suddenly accepted the same post at the University of Warsaw in 1934.

From a collateral branch of one of Poland's oldest and noblest families, he was named for Prince Joseph Poniatowski, one of Napoleon's most brilliant generals. Stanislaw Poniatowski had added luster to an already illustrious family name. (Strangely, it had been Prince Poniatowski's statue that Ossowiecki had gone to see in Homel the day he met Wrobel.) He was considered by many to be the most eminent ethnologist in Poland at that time, and was acknowledged as the founder of the cultural-historical school of Polish ethnology. But as important as his impeccable scholarship was his interest in psychic archaeology. "The possibility of a clairvoyant look at a prehistoric time [had] appeared to me as highly enticing . . . as an ethnologist I was very interested in discovering if and what links exist between the most primitive and uncivilized contemporary societies and their counterparts in the oldest prehistoric cultures." Forty years later this study by analogy is still a major anthropological approach.

Nor had Poniatowski's interest been merely idle. He had been actively searching for "an appropriate clairvoyant," and had meticulously worked out the experimental protocol he would follow should his search be successful, concluding that the correct approach was multiple experiments by a psychic-archaeological *team.* His plan called for one major clairvoyant to serve as a baseline "and somewhere upward of ten other clairvoyants" as cross-checks. Indeed, Poniatowski's work is the model of how psychic archaeology should be practiced. Contained within his method were checks against any bias a psychic might have, as well as consideration for the limitations of his source. All the complexities that plague practical psychic research were fully thought through and compensated for, and amazingly, this had all been done before Poniatowski had completed even one experiment with Ossowiecki or anybody else.

Poniatowski immediately recognized that he had finally found "an appropriate clairvoyant" in Ossowiecki. The latter was equally enthusiastic, very possibly because he had been expecting such a meeting and knew what the consequences could be. In 1937, a year after they began to work together, Ossowiecki told Ms. Zawadzka of the journal *Goniec Warszawski* that Wrobel to that date had described his life "to the smallest fragment." Thus, while he may not have known the specific researcher's name, he had anticipated this final research.

At their first meeting it was agreed that the work should begin immediately. The first session was set for April 23, and on that appointed Thursday evening, a small group called at Ossowiecki's flat, at number 32 Polna Street. The scene was described by Ms. Zawadzka.

"Entrance. A hall or a waiting room. Attention focuses first on a white statue of Christ standing in the window. The figure's outstretched hand is raised in benediction.

"On the wall are lithographs of the Creator awakening the world from chaos. High gothic chairs complement the interior. . . . In the large, light study a collection of memorabilia personal and inherited from the family. . . . Several important paintings and many photographs signed by famous people. There are two huge, potted fig trees and among them, flying freely, several birds. Always open cages are their nests. They go in and out at will and often sleep in the branches of the trees. The gray nightingale is so tame that he comes over at the sound of Mr. Ossowiecki's voice. When the man is silent, working at his desk, the bird sits on the back rest of the chair and sings.

"The most interesting room, however, is the small salon—the favorite place of the clairvoyant. It is here that most of his experiments take place. It is truly his home.

"There is a wide day bed covered with a Persian rug; low Eastern coffee tables and exotic paraphernalia [a recreation of Wrobel's study?]. Each fragment displayed on the walls is connected with some experiment of the clairvoyant. Over the door a gold statuette of the Chinese deity Cian-Fu; next, a Japanese Christian Madonna; a silk tapestry in a Japanese frame; drawings by the anthroposophist [Rudolf] Steiner; a fragment of the fabric with which the face of Dabrowka* was covered in the coffin—a gift from Bishop Lanbnitz; a letter from the current Pope when he was still a nuncio; an authentic letter from General [Tadeusz Andrzej] Kosciuszko [who

---

*The wife of the first king of Poland and a major influence in the Polish conversion to Christianity.

fought in the American Revolutionary War] to General Dabrowski; a fragment from the original manuscript of *Quo Vadis* [by Nobel Laureate Henryk Sienkiewicz]. . . . In a word, a small museum." This was the description of the launch site for sixteen ventures along the uncharted river of time.

Included in the group, along with Poniatowski and Balcer, were Michael Kamienski, professor of astronomy and director of the Astronomical Observatory, University of Warsaw, whose work in celestial mechanics had brought him world fame; Jan Lukasiewicz, one of the most emminent mathematicians and logicians of the century; Stefan Manczarski, world class geophysicist; and Witold Henser, one of Poland's leading archaeologists. They were not there by chance or out of curiosity. It was a basic premise of Poniatowski's approach to psychic archaeology that the research should always be interdisciplinary.

Consequently, for the entire two-part sequence of thirty-three experiments men and women of world stature were present; each was invited to provide commentary on the results of questions Poniatowski planned to put to Ossowiecki in matters concerning their specialties. This list, in fact, is one of the more remarkable things about this unique research project. At a time when Dr. J. B. Rhine and his wife Dr. Louisa Rhine in the United States were finding it difficult to get American scientists to take their parapsychological research seriously, Professor Poniatowski established an advisory and support group seventy-five percent of whose membership is still to be found listed in encyclopedias. This is no absolute judgment on the character and quality of the experiments, certainly, but it is doubtful whether any other research program in the twentieth century—except perhaps the Manhattan Project—could make a similar claim.

At about 9:30 in the evening, after an early dinner, the group moved to Ossowiecki's salon. Once settled, Poniatowski explained the procedure he would follow. Mostly this was for Ossowiecki's benefit, since he was totally ignorant of what would be happening. The professor said that the psychic would be told nothing about the guide object, would not even see it until it was handed to him. Nor would he be told what culture had produced it or where it had been found. All the experiments would be conducted in exactly the same way and every attempt would be made to keep conditions as uniform as possible. Ossowiecki was told that certain controls had been built into the research series, but their exact nature was not explained. He was asked not to read any material dealing with archaeology or ethnology. He was then asked what conditions he wished the observers to meet. He said only that he preferred that they maintain a light conversation on some subject other than the experiment or the object, and that they not stare at him or do anything to make him feel conspicuous. Those present agreed to comply with this request insofar as they were able. Poniatowski then asked if Ossowiecki could take questions while giving the reading. Ossowiecki said he thought so, he would try. Finally it was agreed that Balcer, Kamienski, Kosieradzki, and particularly Stefan Radlinski would help Poniatowski maintain a complete and accurate record of what took place, and that each person present would agree to its correctness before leaving the room. (In fact, Poniatowski's reports are works of art. Artifacts are listed according to museum number, culture, and site, as well as description. The names of all present are given, as well as the location and the exact time the artifact was handed to Ossowiecki. There are notations as to how long it took him to begin and how long he spoke, what gestures he used, how he felt, and even the expressions on his face. Often there is a postexperiment commentary consisting of statements made by observers, and a final wrap-up by Poniatowski in which he reviews Ossowiecki's major statements and their validity, as well as any controls invoked.)

By the time the discussion ended, it was 10:15 and Poniatowski took from his pocket and handed to Ossowiecki a small flint tool he knew to be about ten thousand years old. After almost twenty minutes of concentration, Ossowiecki began with the words:

"Thick, thick forest . . . such a strange forest, black leaves, such dark color . . . vast distances . . . yonder there are places where there is no forest, clearings, and on them mushroomlike squat houses made of twigs, smeared with clay. I see them well in this moment."

He then went on to give a detailed description of a microlithic culture, its people, and customs, answering questions, all put to him by Poniatowski, although some seem to have been suggested by other observers. Fifty-seven minutes later he stopped, complaining of tiredness and "a weight in my head." After he recovered, Ossowiecki offered a few other remembered impressions and then Poniatowski solemnly went around the room asking for comments; only two major thoughts emerged and they were held in common.

First, there was general agreement that Ossowiecki's vision was accurate, at least to the degree that reconstructions are ever testable. Second, the use of questions was considered a major breakthrough of great importance to the project.

This business of questions seems so obvious an approach today that it is easy to overlook the fact that Poniatowski's scheme of asking for elaborations on an unclear point, or for a shift in focus to something he considered more significant, was unprecedented. Previously Ossowiecki and clairvoyants like him were given a question or questions at the very beginning of a session. They then responded until they felt the subject was exhausted. (This was essentially how Balcer had conducted the mummy's foot experiments.) Now Poniatowski had proved that it was possible to exert some control and to be selective. He had been handed a wondrous new telescope that he had learned could be pointed into the past and focused at will. Poniatowski would later discover that too many questions tended to cause confusion, but once he had found the balance, the question-and-answer format became a major tool.

Two weeks later, on May 7, the same group met again at Ossowiecki's apartment, and after discussing the first session, a second was begun, this time using a stone club. The first artifact had been one from Poniatowski's personal collection. After the second session almost every guide object was to come from the museum named for and founded by Professor Erazm Majewski.

This was yet another part of Poniatowski's careful plan: using diverse objects from a uniform and academically accepted source. Stone tools, or lithics, as they are more properly known, comprise a subtle and complex branch of archaeology. It is difficult for even a general archaeologist to tell whether these stones are naturally formed or man-altered, and that task is an easy one compared with dating and assigning these artifacts to their proper culture. Even today, forty years after Poniatowski's research, when archaeology is far more sophisticated, only a few men and women are acknowledged to be experts by the discipline as a whole. Poniatowski was not such a person, so he believed it was important that the artifacts to be used be precertified. In this way he felt he had a reasonable, traditionally developed baseline against which to measure Ossowiecki's reconstructions.

Having done two successful experiments, thus establishing both the validity of the concept and the usefulness of his experimental protocol, Poniatowski next resolved to attempt something so awesome that even four decades later it sounds preposterous.

On July 24, 1933, Dr. Berckhemer, the chief curator of the Natural History Museum in Wurttemberg, Germany, received a call from an acquaintance, a Mr. Sigrist, who owned a gravel pit near the town of Steinheim on the Mur. It was not the first time he had called, and Berckhemer anticipated the message as soon as he heard the name: Sigrist, he felt sure, had discovered something in his pit that he thought the scholar might find interesting. All the same, he was surprised at the agitation in the man's voice, until he heard what he had to say: an almost complete human skull had been found very deep in the pit!

It was in this manner that Steinheim man entered history. Two years later, in 1935, an English dentist who was an an amateur archaeologist, Dr. Alvan Theophilus Marston, was to discover a more fragmented but similar skull situated in the midst of some six hundred extremely primitive stone tools near the town of Swanscombe in England. The lithics at Swanscombe were identified as Acheulean; the gravel beds and the remains of two extinct animals found in proximity to the Steinheim skull dated it to the same epoch, the third interglacial period. As German writer Herbert Wendt notes in his book *In Search of Adam,* the two skulls looked like nothing so much as those of "a female chimpanzee." Yet their cranial capacity and certain other features established that they were unquestionably *Homo sapiens,* approximately twice as old as Neanderthal man and yet more modern in appearance. To this day they remain both anomalous and among the oldest remains of modern man discovered in Europe. At the time they were found, within the academic circles that concern themselves with such things, they were a sensation.

To Poniatowski, the challenge was irresistible. He was interested in very primitive man, the people of the Pleistocene Age. Now with two discoveries that confirmed each other he had some idea of what an anomalous variation that man must have looked like. He also had a stone club from the Majewski Museum certified by lithic specialists as having come from one of the best Acheulean sites, Abbeville in France. With this to validate what Ossowiecki might say, he wondered if contact with a past so distant could be achieved. Could a psychic travel back in time almost three hundred thousand years?

While reaching back in time, Ossowiecki frequently added to his commentary by making rough drawings of what he saw. In this instance, on the basis of nothing more than a primitive scraper (which to any one but a trained archaeologist would look like a plain rock), Ossowiecki astounded the researchers by accurately drawing Chancelade man. Excavation has confirmed that, just as the clairvoyant described and drew, this member of our ancestral tree had a wide nose, a hydrocephalic-like head, and a short, runty build.

The third experiment on May 21 was designed to discover just that:

"God how far. . . . They are very curious people . . . small people . . . not big heads and huge in the back," Ossowiecki said showing the shape of the head with his hands. "Hair matted . . . falling down, it hangs [on the sides of the cheeks—Poniatowski]. Very different types. Height 150-160 cm. [approx. 4 feet 9 inches to 5 feet 2 inches] . . . they are terrifically muscular . . . nude . . . skin . . . dark . . . like light chocolate . . . so dark. Women very well developed, fat. . . . Hair long, falling down . . . chins forward inclined, thin beards, eyes black, dark brown. . . . Noses very broad . . . hips thick, forehead low, eyes wide open . . . dark eyes. Women very ugly . . . not pretty. Ears protrude some . . .''

The words poured out, a rambling near-monologue by a man whose speech was alternately enthusiastic and quizzical, like a person telling a long and complicated anecdote. There was no trance, nothing occult. Ossowiecki was obviously alert; he asked for paper and pencil with which to draw pictures of animals, houses, and people of the past, even stopping the narrative to give instructions to his maid when she came into the room. And yet, those present had the inescapable feeling that it was they who were unreal and dream-like and this band of small dark naked ancestors that was solid and alive. Ossowiecki, surrounded by the accoutrements of a twentieth-century European parlor, gave the unmistakable impression of easily and casually looking on a scene *three hundred thousand years old*. His eyes were open and seemed to be moving back and forth as he followed this scene, actually seeing it. Indeed, very possibly on this occasion, and certainly on others, Poniatowski suddenly but quietly put a screen in front of Ossowiecki and the psychic complained that "his vision had been blocked."

Even stranger was the meaning of the words; as Poniatowski heard them, he knew that they made sense. More important, Ossowiecki's description passed another of his tests. He had never seen psychic archaeology as a new and independent science; rather, he believed it offered a new *technique*—one that would add to and complement more traditional approaches. Consequently, it was his conviction that a primary test of psychic data would be whether they supported solidly based orthodox observations on the same culture. And that was exactly what he was hearing. Of course, much of the psychic reconstruction was uncheckable. No artifact or bone, for instance, could reveal the skin color of early *Homo sapiens*, nor could Ossowiecki's statement that the women were wearing some kind of feather ankle bracelets be proved or refuted. But where orthodox data did exist, Ossowiecki's words were in agreement—the shape of the skull, for example. Ossowiecki's description would make even better sense to Poniatowski after he had done some further research on his own. The description of this early form of man, however, was just a start. Experiment three had much more to reveal.

To begin with, almost with his first words of the evening Ossowiecki had said, "This stone has a double history." He then suddenly began talking about another people: "Their skin is whiter beyond comparison. Their heads are different, not elongated here, broader jaws, broader. Black hair, brown hair's already here. Many people. Faces nicer, less savage. The noses not so broad . . . different people. The people are white . . . they are whiter than the others . . . they are more the color of brick, a bit, just a bit . . . they are dressed . . . have some fabric? . . . something—I don't understand."

A third culture emerged during the course of the evening, a people Ossowiecki merely described as "smaller." By the time he had finished, the stone's entire history of association with man was laid out for Poniatowski and the others. At various times the rock in an unaltered state had served as part of a hearth ring, a portion of the doorway to a primitive stone house, and, after being worked by man, a weapon.

In trying to sort the material out Poniatowski inadvertently arranged both an unplanned test of the information's accuracy and an experimental control. Seeking to clarify the various cultures, he asked Ossowiecki, in reference to the geography of the site, "Is there water?" The psychic, after looking around, said, "There is water on the right side. Flows fast, it's a stream." Later, during the description of what appears to be the most modern epoch, Poniatowski asked, "Is the locale the same as before?" To this Ossowiecki answered, "The locale is the same. I recognize it. But it is greatly changed. . . . Now I'll tell you something interesting . . . I see water, waves. Is this possible? I see the ocean . . . I'm closer to Belgium [millennia before such a nation existed, of course] . . . France. . . . Now I recognize this map . . . I see the peninsula of Normandy. . . . Need to go . . . I'm departing . . . countryside . . . forest . . . mountains . . . mass of water . . . huge waves, powerful, massive. . . ."

Upon returning to the present time two hours and ten minutes after he had begun (this was the longest session he would ever do), Ossowiecki was told by his friend that experiment three had contained much that seemed accurate, although it was confusing sometimes as to which time went with which people. But Poniatowski said there had also been a major error. The prehistoric settlement whose people had used the stone as a weapon could not have been near the ocean, for the site where the stone had been found was in the central portion of southern France. Ossowiecki listened carefully but "with complete conviction . . . that [the site] had to be on the shores of France between Belgium and Normandy." The following day Poniatowski checked. He was chagrined to learn that Abbeville was not in the French interior; it was on an "estuary of the River Somme to La Manche canal," easily within sight of the ocean. Ossowiecki had been correct! Now Poniatowski began to realize that he had been handed the keys to time's cabinet.

There would be nine experiments that first year, the greatest number in one year during the prewar period, and, just as Poniatowski had planned, the format never varied. Indeed, the only inconsistency in the entire sequence of sixteen sessions was the fact that in 1936 experiments were held in the spring and summer. Thereafter Poniatowski and Ossowiecki worked together only in the fall, when Warsaw's social life was at its height. It was a wonderfully gentle time and Andrew Norwina-Sapinski, who translated the Poniatowski research for the author in 1976, remembered it well:

"Prior to World War II the social pulse of Warsaw at the level Ossowiecki and Poniatowski occupied beat most strongly in the late hours. I remember my parents going out and receiving visitors until midnight . . . on week nights. Shops were open and people promenaded down main street and sat in Kawiarnia's . . . a kind of elegant super-coffee house. Warsaw smelled of fresh fruit and flowers because the shops that sold them displayed much of their merchandise outside their door front. In the early hours, while people still walked and talked, the streets were washed clean by water trucks and then their black-wet surfaces mirrored the neon lights. It was a lovely city. Full of life."

Only two things rippled this placid surface. In the fall of 1937 Ossowiecki became concerned enough about what he perceived as Europe's coming instability to give a series of lectures at universities in Warsaw, Poznan, and Lwow. He called these talks "Psychic Crisis and the Future of Mankind," but exactly what he said has been lost. It is known, however, that he decided to publish the talks in a somewhat different form, seeing this second book as the culmination of an earlier volume, *The World of My Spirit and Visions of the Future,* which had been published in 1933 from an "inception of pain" with the purpose of showing people "the way to overcome spiritual suffering." Particularly it reflected Ossowiecki's concern with technology run rampant without moral control.

The second and more immediately frightening event took place on

October 21, 1938, during experiment fourteen, the first one of that year.

The reading, for a culture identified only as Mezynska, had begun routinely enough at 10:07 P.M. at Ossowiecki's apartment, and had described for some time an extremely ancient and primitive people whose settlement the psychic had correctly located between Spain and the Italian peninsula.

Suddenly as the session seemed to be coming to a close, Ossowiecki became very agitated. "I cannot get out of here," he said, and then, almost against his will, began to describe a couple of these pre-*Homo sapiens* making love.

"I see him. They sit now. He makes advances to her. Takes her breasts and pulls to himself . . . he moves around her to and fro. She jumps and sits next to him. He begins . . . there are no kisses. With

After correctly locating the excavation site in question, Ossowiecki complied with Poniatowski's request for a description of a Paleolithic hut by not only drawing it, but describing it as being made of light saplings overlaid with skins, and having a single entrance and a smoke hole at the top. Later research by archaeologists who knew nothing of Ossowiecki or Poniatowski supported this, and Dr. Francois Bordes, the world's foremost authority on this prehistoric period, would later describe such huts in almost exactly the words the psychic had used—down to noting, as Ossowiecki had done, that the structures would have been "warm and comfortable."

Ossowiecki often provided Poniatowski with detailed descriptions of the animals whose presence was important to early man. As so often happened with what he said, the pictures and words later proved to be extremely accurate. Particularly exciting was the fact that the clairvoyant invariably correctly located the site from which the guide object had been taken. Indeed, on more than one occasion when Poniatowski thought Ossowiecki wrong, subsequent research would show that the information given to Poniatowski was in error, and that the sensitive was correct.

her hand he covers his face. She strokes his neck, back . . . he lies down. . . . She sits on him equestrian style. She sits still, looking into his eyes. Now she jumps up, jumps up . . . . She traces with her nose along his nose, around his face. He embraces her with his whole strength . . . she is active, not he . . . she jumps up again, so . . . she is washing his member with water, covers it with green grass . . . now she goes to fetch water . . . fabulous elasticity of body. . . . There are no normal movements, they are like monkeys." Again Ossowiecki cried out, *"I cannot return."* This time both he and the observers became uneasy.

Unable to stop, the psychic continued, describing a zebralike animal. "Long ears. Several there are . . . they graze near the house. Long tails, the end like thick gray brush. Rear white and from it toward the front darker. From spine white bands go to white belly. Huge ears, horns I do not see. . . ."

Talking about the animals seemed to calm and release him. Then the man of the vision abruptly came into view again. "God how wild it is here . . . savage . . . they are like monkeys. The man is so hairy . . . all are so hairy. Men's bodies are brownish, dark like fur. Hair on foreheads grows from bridge of nose." With the man's reentry some connection seemed reestablished for Ossowiecki. He yelled, "I can't get out of here!" and with those words dashed the stone-tool guide object to the floor. Even after he had returned to the world of the living, he still seemed to linger in the past and for another fifteen minutes felt as if the two realities had become confused and that he was alternating between them.

From the beginning the psychic and the scientist had realized that these experiences with prehistory were taking a toll on Ossowiecki. All too often he would end them saying he felt tired or had a headache. He also felt that doing such work "took phosphorus from my body." These were problems, however, they both were prepared to accept. But it apparently had never occurred to them that there might be far more dangerous factors involved in what Ossowiecki was doing. What had caused his entrapment they never knew. Possibly with a normal healthy man's curiosity he had become too interested in watching a couple make love hundreds of millennia ago. Or perhaps the explanation lay not in Ossowiecki but in the people he was watching, whose psychic energy generated a kind of whirlpool into which he was dragged. (What other event in human life is as intensely and singularly focused as a sexual act and its climax?) Such explanations can, of course, be dismissed with amused condescension as occult pseudoscience, but Ossowiecki did appear to be trapped and was observed in this state by men of unquestioned probity and considerable attainment in science.

Whatever the reason for Ossowiecki's seeming inability to shift his point of view, it had frightened everyone present. The complacency with which the session had begun was shattered; all now knew that psychic archaeology was not a simple process and that they were far from understanding it. One practical solution, however, emerged from the experience: Poniatowski concluded that *his* attitudes and state of mind had an effect on the experimental dynamics (something scientists rarely admit). He realized that he too had become entrapped because of his fascination with what Ossowiecki was describing, and in doing so, had relinquished some of his control over what was happening. In his superconscious state it was clear that Ossowiecki was vulnerable, pulled along by currents he only partially ruled. Poniatowski concluded that he himself would have to stay "awake" and avoid also being seduced into too intense a concentration. After the fourteenth experiment whenever Ossowiecki seemed to become too exclusively focused on a single scene, Poniatowski would draw him away by asking questions about some other topic. There were a few tense moments, but never again did the incident that had marred the fourteenth session repeat itself. Tiredness became an ever-increasing problem, but the remainder of the thirty-three experiments went off without tripping another psychic deadfall.

By February 24, 1939, fifteen psychometric contacts with eleven different cultures had been accomplished (since several guide objects were from very obscure Polish sites it is possible that the number is greater or smaller by perhaps one or two). The final (sixteenth) session in February was a distinct variation from the regimen. There were two good reasons for this early-spring variance in scheduling.

Poniatowski wanted to round off this first cycle of research with Ossowiecki by getting a reading on the relatively recent Magdalenian culture (approx. 15,000—10,000 B.C.) of the Stone Age. Then, in accordance with his plan, he wanted to try the same selection of artifacts with different sensitives, in this way assuring a psychic cross-check as well as an orthodox one. Unfortunately, he was able to find only one other clairvoyant who seemed to have the necessary skills, a professional commercial psychic he identified only as Mrs. S. Ch. After three sessions, however, he concluded that further work with her was useless. Although she had a twenty-year reputation as a gifted clairvoyant, he found her descriptions "muddy and devoid of originality." He even discreetly implied that she had read up on prehistory and filled in what she could not get psychically by regurgitating what she remembered of this cramming. The experience taught him a powerful lesson: " . . . professional [i.e. commercial] sensitives . . . can easily lead an uncritical researcher to error."

Ossowiecki, for his part, wanted to end this first sequence of experiments and be free for an extended period for another reason: He intended to marry again.

In 1922 Ossowiecki had married Alietta de la Carrière (presumably of French origin), but the marriage was not a happy one. They separated a few years later and were divorced in 1930. For nine years Ossowiecki had been a bachelor, and although he loved beautiful women, he also loved a close family life; he wanted to remarry.

He had met Zofia Skibinski and her family some years before, probably while he was legally separated and she recently divorced from Gustaw Swida. Gradually the friendship developed into love and they decided to marry. He was much older, at sixty-two, but, Zofia, forty-eight, and the mother of a grown son, found the age difference of little consequence. They were married on July 6, 1939, and left immediately for the Hotel Hungaria on the Polish coast. There Ossowiecki wrote under commission to Paramount Pictures a screenplay called *The Eyes Which See Everything*—an eight-part serial based on his life. It was for Stefan and Zofia Ossowiecki, and for all Poles, a golden summer; and the last bright flicker of an age. On September 1 that age ended.

Even as his armies were invading, Hitler addressed the Reichstag, laying down the sophistries that would justify the extinction of Poland. Poland, assured of British and French support, resisted, and fought alone—Polish cavalry troops against panzers. One week later, on Friday the 8th, the first German troops laid siege to the capital city. With a kind of suicidal defiance that was to mark the entire battle for Warsaw, the Polish forces in the Wola and Ochota districts turned back these first armored encroachments. But there was never any real question of the outcome. On the 27th, with no food, water, or ammunition, the defenders, led by Warsaw's mayor, Stefan Starzynski, who symbolized the resistance, surrendered.

At first the Poles, who had a long history of being overrun by the larger countries on their borders did not realize that this time it would be different, that there would be no honor, or even humanity. After fighting with the same courage and skill that had made Polish troops desirable to commanders since the Middle Ages, the captured soldiers assumed they would be sent for "labor on German farms," as had been done in World War I. The hardworking Poles had always been valued as captives, and decency was expected, but this was not the Germany of Kaiser Wilhelm.

On August 23 von Ribbentrop and Molotov had met in Moscow and signed secret protocols. Having seen the erasure of Czechoslovakia in March, Stalin knew the borders of Europe were changing and he wanted his share of the swallowed countries. Hitler was willing to agree—for the moment. But the German troops moved faster than Stalin had anticipated, and so on September 17, as Warsaw lay under siege, Polish Ambassador M. W. Grzybowski was summoned and told by Molotov that since "Poland has become a suitable field for all manner of hazards and enterprises . . . the Soviet Government have directed the High Command of the Red Army to order troops to cross the frontier and take under their protection the life and property of the population of the Western Ukraine and Western White Russia [eastern Poland, in effect] . . . to extricate the Polish people from the unfortunate war into which they were dragged by its unwise leaders." As Germany was radically different under Hitler, the Soviet Union was no longer Russia under the Czar.

On September 28, Molotov and von Ribbentrop were able to meet again, this time to invoke the secret protocols and to announce to the world and the Poles exactly how the spoils of the past month's action would be apportioned. Poland was split in two, roughly along the lines of the Bug and San rivers—equal shares, as it were. The Germans almost immediately split their half of Poland into two segments, directly annexing the part along the German-Polish border into Germany proper, while in the other section establishing a supposedly Polish central government, later known as the Rump State, whose capital was first Kzakov and later Warsaw. This was nothing more than a charade, however, because the government officials from the first were always German; no Polish quisling could be found to front for the Nazis. Even more important, particularly in terms of establishing the context under which Poniatowski and Ossowiecki were to work in the coming years, there was almost universal citizen resistance. The city of Warsaw and the central portion of Poland were thus like a peach pit whose surrounding flesh had been devoured by Germany and Russia, but which itself still gave only the satisfaction of gall.

In spite of this resistance, and the increasingly grotesque Communist and Nazis barbarities during the fall of 1939, it was still possible to get away from Warsaw, usually by going through Italy. This took pull and diplomatic connections but it could be done, and a number of Poles did leave by this route. Ossowiecki was offered this opportunity—the necessary papers were even drawn up—and the psychic and his wife had already lost their apartment at 32 Polna and been forced to move to No. 17 Marszalkowska, but Ossowiecki refused to leave.

Today the reason given for his refusal is that he desired to help the Polish people during their captivity. This is a valid inference from his words, and even more importantly, his actions, but it is only a partial answer. He also knew running was futile because he had been told by Wrobel that should he ever take a sea voyage, it would end badly. And unless he crossed the sea, how could he escape the Axis reach? But beneath these overt reasons runs a subtler one, one that Ossowiecki himself hinted at before the invasion.

Ossowiecki remained because he was tied to an era. He knew this and accepted it; a position he could adopt because he did not fear dying and so was able to reject the undignified course of survival at any cost. And this spiritual security did something else. The help Ossowiecki stayed to give was not limited to telling families where the bodies of their missing relatives were buried, which would have been the limits of a psychic technician. This might be all people thought they were asking, but from Ossowiecki's point of view, "No deed, not even a vision, constitutes help or salvation for them. They are looking for confirmation of the fact that the spirit is a self-actualizing element, not just an inseparable function of matter. . . . It is my dearest wish that this spiritual world stop being the domain of just a few and that it open its secrets to all people . . . . I, perhaps better than anyone, know how many people are drawn to a spiritual search today." This was his destiny and his purpose, and his most important reason for staying. Ossowiecki never saw himself, as had so many researchers, simply as the psychic Answer Man. In his eyes his service to his fellows was far more complex than that.

There was a final reason why he had to stay. Even if Poniatowski did not know it yet, their work together was not finished. It seems almost beyond the limits of fiction that during a war men would concern themselves with psychic experiments in prehistory, but there is no alternative explanation to what followed. But before discussing the experiements it is necessary to establish a further sense of context.

With the partition of Poland completed, a six-year experience began that will never be fully understood except by those who survived it. The statistics are no help; they are so bizarre as to be meaningless. Perhaps the most intelligible thing that can be said is that the atrocities committed by the Soviets and the Nazis were about equal, both were oriented toward a single goal: the extermination of thirty-five million Poles and allowing for the vicissitudes of their other campaigns, the two powers did their best to carry out this plan. In Warsaw specifically, this meant a program of systematic murder by the Nazis, with the eventual aim—after the one million three hundred seventy-five thousand Polish residents had been disposed of and their buildings destroyed—of building a city for one hundred to one hundred thirty thousand German citizens.

If the full story is beyond human comprehension, individual vignettes throw some light on the horrors. Konrad Jazdzewski, for instance, recalls the occupation's effect on archaeology:

"The Nazis invasion of Poland was a catastrophe without precedent. . . . One in four Polish archaeologists perished on the battlefield or in concentration camps. Others, threatened with arrest and almost certain death, had to live in hiding; the few remaining ones were forbidden to engage in scientific work. Large portions of museum collections were carried away and destroyed. Other collections were demolished, wholly or in part, during military operations. All Polish universities were closed and some of the buildings destroyed; many specialist libraries were plundered and set alight. The printing of scientific works was prohibited, whilst some publications were burnt. The writing of scientific papers . . . could only be carried out in secret."

Nor were these prohibitory edicts idly enforced. In one year over one hundred thousand men, women, and children were methodically taken from their homes, offices, or shops, placed against a convenient wall, and summarily machine-gunned. Hitler's orders were very precise: "annihilation on the spot of the Jews, the intelligentsia, the clergy, and the nobility of Poland." One survivor remembers going to a wedding of a young scientific colleague who had been secretly teaching a few gifted students at his home. Arriving late, he was just in time to see the bride and groom come out of the church, into the waiting arms of the Gestapo. The accusation was made and the entire party, including the priest, was taken to a wall, lined up, and calmly murdered. The latecomer still has nightmares of the bride lying in the gutter, blood dripping from her face onto her veil.

In spite of these risks, after a break of twenty months Ossowiecki, Poniatowski, Radlinski, Balcer, and the others began again. On November 6 they gathered at the psychic's apartment on Marszalkowska. Only three things would be different during this second cycle of seventeen experiments. First, there was a curfew now, sometimes as early as 7 P.M. "The social pulse" that Nowina-Sapinski remembers no longer beat, and to be found on the street after hours was to be shot, so the sessions were started about five in the afternoon, often only an hour or so after Ossowiecki had consulted with the last mother or widow of the day. Secondly, in poignant counterpoint to the past, the dates of the experiments show that in spite of the summer heat, people no longer left the city for the cool breezes of country places; from now until the end experiments would be held almost every month. And thirdly, the sessions were killing Ossowiecki.

The prehistoric research had always been difficult—more often than not the psychic experienced headaches or great fatigue (the latter he had felt from the first experiments). It was obvious that this kind of contact was more debilitating to his system than even the telekinetic demonstrations he had put on in Russia. Exactly what caused the problem is unclear, but it seems to have had something to do with the very specific focus he and Poniatowski were trying to achieve. Whereas most sensitives, if they do this kind of work at all, speak of the past only in generalities, Ossowiecki by intense concentration could focus on a single activity and then fine-tune this to a single action by one individual, as he did in describing a Reutalian-culture execution:

"Now this man has the stone. I see he is holding it. This is a punishment weapon because a man he had struck on the head was brought to him by the others who held his hands. The man walked meekly, toppled and slid to the ground."

During the initial cycle of experiments, except for the first year, 1936, there had always been a cluster of contacts and then a lapse of almost twelve months, during which time Ossowiecki was able to recover with vacations, rest, and plenty of good food. Also during the period when the first sixteen episodes were held, Ossowiecki rarely had given more than one or two consultation readings the same day as a prehistoric session. Now all this changed.

He was sixty-three years old when they began the wartime experiments. Food was scarce and what there was of it lacked the proper nutriments. Even without the prehistoric research, Ossowiecki's health would have been strained, for he was giving as many as thirty readings a day for the people who came to his door. Like the only doctor in a mining town whose men have suffered an accident, he received them, mostly women. They waited quietly, clutching old photographs, a button from a uniform, a lock of hair, a last letter.

Maria Boltuc would later describe one of perhaps thousands of such readings, most of which were never recorded.

She went to him seeking information about her husband, Brigadier General Mikolaj, commander of an operational group that had fought in the battle of Kutno. She took with her a uniform cap her husband had left behind and his photograph. It was almost the end of the siege and most of Warsaw was either being bombed or subjected to heavy artillery fire. By the time her turn came, the detonations had moved so close to the apartment that everyone else had fled to the cellar. Only Mrs. Boltuc and Ossowiecki remained and she was preparing herself for his statement that a reading would be impossible. Instead, she recounts, "The Engineer closeted himself with me in a small room and, unheeding of his own peril, quietly attained a state of deep concentration. Presently he started to relate the picture before the eyes of his mind: My husband was walking with a large detachment, he saw officers beside him. He described so vividly and with great detail, that I recognized Major Kunc, Captain Kwiatkowski, and others. He saw my husband receiving a wound in the neck and falling into a ditch, trying to rise while blood spurted from the wound.

"He told me my husband needed assistance and bade me go through Zoliborz promptly, along the high road to a small bridge with white houses in the foreground and seek him there. To my father he confided that my husband's condition was very grave, nay hopeless.

"On the 25th [September], although the capitulation agreement was yet unsigned, I received from Mayor Starzynski a safe-conduct and a bottle of high-grade cognac as fortifier. Equipped with those and the instructions of Engineer Ossowiecki, I departed on my quest.

"I did not alas! find my husband alive. He had, as described by Ossowiecki, been wounded in the neck and swifty died from a hemorrhage.

"On the neighboring graves I deciphered the names of the officers whom Ossowiecki had seen near my husband in his vision."

In the face of such demands as these the very idea of conducting research seems absurd, yet Poniatowski and Ossowiecki did so, and held to the same high standards as in the prewar period. Mrs. Zofia

Podkiwinska and Ludivik Lawicki continued to select guide objects from the Majewski Museum (closed, of course), and these were smuggled out to Poniatowski, all three risking death in the process. The meetings were often held only an hour or so after Ossowiecki had consulted with the last widow or mother of the day. Meeting in and of itself was an action punishable by death, and the fact that most of those present belonged to the intellegentsia and aristocracy made the outcome, if discovered, that much more certain. Recording the material and making the drawings (prepared from Ossowiecki's rough efforts by prehistorian Dr. Janina Rosen-Przeworskie) were also interdicted activities. That the danger did not stop them is perhaps the truest measure of the importance this group of scientists attached to Ossowiecki's psychic commentaries on archaeology.

Finally on May 28, 1942, the experiments were concluded. It is typical of Poniatowski that, in the peaceful prewar days, he had "intended to carry out approximately thirty experiments with Mr. Ossowiecki" and that nothing, not even Hitler, stopped him from methodically completing his research program. In all, nineteen cultures were surveyed, beginning with pre-*Homo sapiens* and continuing by stages up to the early Bronze Age. But the third phase of his project—fieldwork—had to be scrapped.

From the beginning it had been Poniatowski's intention to conduct excavations based on Ossowiecki's directions. He had seen dowsers locate water and had understood immediately what this implied: If a dowser could find water, a clairvoyant could find and outline graves or buildings on an excavation site. Poniatowski had hoped to begin by setting up "a number of trial runs for the clairvoyant, wherein objects are hidden by third persons in natural crevices, etc.," because he correctly discerned that determining exact measurements from psychic data was a far more complex process than it appeared. He knew with certainty that Ossowiecki could give general locations for sites from a distance. Sometimes the psychic's information contradicted, and yet proved more accurate, than that possessed by Poniatowski. He also knew that the psychic could locate a modern-day artifact—say, a lost earring—in a present-time context. But when Ossowiecki was locating prehistoric material he did so from a past-time point of view, seeing the ground *as it was then*—i.e., both the geography and the artifact were always in the same time frame. To gain further insight into this process Poniatowski planned to perform a series of prefieldwork tests to discover whether an adjustment that would correct for the past-time context of the artifact and the present-day appearance of the ground could be developed, or whether Ossowiecki could learn to locate an artifact without having to go back in time. If either approach proved successful, he would then take the psychic to actual excavations.

When the war turned the Polish countryside into battlefields, military bases, and concentration camps the third phase had to be abandoned. Poniatowski regretfully decided that he would have to base his book on the research he and Ossowiecki had been able to do, even if that work was incomplete. There would be no psychic cross-check from the team of sensitives he had hoped to assemble, and no fieldwork. There was also the unresolved Magdalenian problem, and this bothered Poniatowski as much as all the rest.

It began with the sixteenth experiment, which had used as its guide object what the Majewski Museum had identified as "a Magdalenian engraving tool No. IV-472 from La Madeleine rock shelter." After being handed the object on February 24, 1939, Ossowiecki described a cremation and burial of the ashes in an urn. It was a very detailed description but it did not agree with what was known of Magdalenian funerary customs. Poniatowski knew this but said nothing. It had always been his conviction that one of the significant proofs of psychic research was internal consistency over a long period of time, so he hung on to the artifact resolving to try it again at a later date.

Almost three years later, on October 22, 1941, he put this internal control into practice. It is impossible to say what Poniatowski anticipated, but it is unlikely he expected what he got: an almost exact replay of the same cremation and burial. Ossowiecki, of course, knew nothing of the control, could hardly have remembered the earlier experiments, since he was untrained in lithics, and probably could not have recognized the tool—even if he had been able to remember exactly what he had said three years before. It was the major mystery to come out of the thirty-three experiments.

Years later a postwar Polish archaeologist suggested a solution so obvious that it may well be correct; the museum had mislabeled the now-lost stone. However, there is a more intriguing explanation: Ossowiecki was correct in describing Magdalenian funerary arrangements; archaeology was wrong at the time and has only recently begun to amend itself.

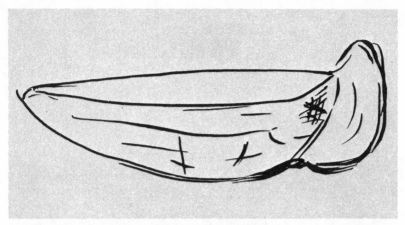

At the time that Ossowiecki gave his description of an oil lamp, and drew the lamp he *saw* in actual use, the words seemed to be nonsense. Only several years afterward, at an excavation in Dordogne, France, would his statements appear in their true light; he had been correct after all.

To begin with, artifact material of the Magdalenian culture has been found in the area where Ossowiecki placed the stone. Also, although there is still no clear evidence of cremation at that time, recent discoveries in Australia demonstrate definitely that this practice was carried out more than twenty-five thousand years ago. This, plus the almost yearly revisions backward of the invention of pottery, makes Ossowiecki's story of cremation with the ashes buried in pottery urns increasingly probable.

Nor is this the only area in which Ossowiecki's so-called mistaken impressions later proved very accurate. There are over one hundred examples, some very minor, some quite major, in which the Polish psychic's impressions in the 1930s have been supported by the orthodox research of subsequent decades. For instance, his observations about the complex hair styles of Magdalenian women now make sense in light of carvings depicting styled hair and small awllike bones that could have been hair pins. In the same vein his description of these women as having broad hips, and both sexes as having high cheekbones, now seems accurate. The discovery of Magdalenian statuettes, known as Venuses, validates the wide-hipped description.

Fortunately, Poniatowski was meticulous in his notes and all these instances are carefully catalogued. For the scholars of today, at least those open-minded enough to study this psychic

One of the most fascinating stories Ossowiecki recounted gave a detailed description of a Magdalenian woman. Particularly significant were his statements that these women had carefully dressed hair, attractive breasts, and wide hips. In light of the Magdalenian Venuses which have been discovered, his words appear to have been entirely accurate.

record, they are clues and guidance for future research. Indeed it is a measure of Poniatowski's faith in the validity of his experiments that he chose to publish the full record—including what seemed at the time inaccuracies or anomalies. Of course, the true scientist holds nothing back, and besides, most of the psychic information obtained was obviously correct, but Poniatowski presented everything because he was convinced that the next generation of archaeologists would prove even the "mistakes" to be valid observations.

Nor was Poniatowski alone in his faith. He found a publisher

Supporting the descriptions given by Ossowiecki are such artifacts as this small ivory head, and this statuette known as the Venus of Willendorf. Note the breasts, hair, and hips; all exactly as the psychic described them, with nothing more to guide him than a few unidentified pieces of stone.

willing to advance him money on the manuscript. It was an act of considerable conviction, since obviously it would be years before the work could be printed, if ever. All publishers of scientific materials—and Poniatowski intended his work "primarily for pre-historians"—had been closed down by the Nazis. By entering into such an arrangement the publisher not only risked money, he also made himself party to an act punishable by immediate execution. Yet, like Poniatowski, Ossowiecki, and the other archaeologists, ethnologists, astronomers, physicists, and mathematicians who had been involved in this strange excursion into psychic archaeology, the publisher came to view it as a contribution to man's knowledge worth any sacrifice.

Poniatowski began writing during the first weeks of the summer of 1942 and worked on into the fall. He was never to finish the book. One night (actually, it was probably about four in the morning since this was their favorite hour), the Gestapo burst into his home. He was taken to their headquarters and, according to some accounts, tortured. The charges are unknown, but once someone entered the Gestapo pipeline, he rarely returned. Poniatowski was first sent to Pawiak Prison in Warsaw, then to Majdanek, the concentration camp the Nazis had set up near the Polish town of Gross-Rosen, not far from the city of Lublin. Early in 1945 he was transferred to Litomierzyce, another German camp, thirty miles northwest of Prague in occupied Czechoslovakia. Neither Ossowiecki nor any of the others who had taken part in the experiments ever saw Poniatowski again, but somehow Ossowiecki managed to get hold of the manuscript, and so saved the report of the work to which he and Poniatowski had given six years of their lives.

The arrest of his friend and partner though marked the beginning of Ossowiecki's own end. There had been two Ossowieckis; everyone who came to know him well realized this. Now, however, it was as if the outer man was slowly being sanded away, leaving only the inner pilgrim who had always lived inside the indulgent outer form. His wife was fearful that he would be arrested at any moment and tried to convince him to be more careful, but he would not listen. Nor would he consider his health, which was continuing to deteriorate. Even though the Gestapo headquarters was but two blocks from his apartment, he did nothing to discourage the flow of people who came to him. Nor would he consider moving when Nazi officers began taking over the apartments all around him. Zofia Ossowiecki was sure the Nazis knew of his existence, and they probably did. But he did not care. It was as if he knew what he had to do and felt certain he would be allowed to finish the task.

As the German Reich began to unravel in the summer of 1944, Stalin saw his chance to take all of Poland and moved his armies into the German zone. By August 1 they were nearing the outskirts of Warsaw. Underground Command sought to stave off the Communist takeover by ordering the people to rise. The slim hope was that between the retreating Germans and the advancing Soviets, Polish patriots could seize control of Warsaw and hold it until the Allies came to their aid. Considering the military facts and Allied intentions, this was a naive scheme, but the situation was desperate and the citizenry rebelled. Most of them were unarmed, or had only small arms. The fighting was incredibly vicious and went on for two months; the SS and Wehrmacht killed over two hundred fifty thousand Poles before the city surrendered to the Germans on October 2. The Russians took over the city on January 17.

Stefan Ossowiecki, however, never saw that end. The uprising began at 5 P.M., August 1. By the next day the Gestapo had decided matters were serious enough to warrant creating a sterile buffer zone with a half-mile radius around their headquarters on Szucha Avenue; and so they began the systematic destruction of buildings within this radius and the rounding up of all non-German residents.

When the demolition explosives began, Ossowiecki and his wife realized that this time they must flee and quietly prepared to leave. Ossowiecki took with him only a small leather suitcase. In it were three things: the screenplay, *The Eyes Which See Everything;* his unfinished work on the future; and the Poniatowski manuscript on psychic archaeology. Zofia carefully packed all the money they had (U.S. gold coins she had secured on the black market) in a small flashlight and they turned and left the world they had known.

Their first stop was the building's cellar where they spent the night. In the morning German troops herded the Poles huddled in the basement into the street and set the building afire. The prisoners were quickly segregated, the young marched off to the nearby Gestapo headquarters and the older people, including the Ossowieckis, released without explanation.

A tombstone is all that remains to mark the place where Ossowiecki, and some 9,500 other Poles, were machine-gunned, and their bodies doused with gasoline and set ablaze.

The couple discussed where to go. Ossowiecki wanted to go to the home of some friends, across the street some forty steps away. His wife suggested instead that they go to the vicarage of Zbawiciel (Savior) Church. Ossowiecki gave in. Zofia Ossowiecki would later torture herself over those forty steps not taken because had they gone to their friend's house, Ossowiecki might have lived. But she needn't have blamed herself. Ossowiecki apparently knew that the next few hours could not be avoided. He had said, "I see that I shall die a terrible death. But I have had a wonderful life."

After spending the night at Zbawiciel they were preparing to leave when Nazi troops burst into the vicarage and arrested everyone. Again they were forced into the street. They were told they would be taken to Gestapo headquarters and searched, and for possession of foreign currency the penalty would be death. Hearing this, Zofia tossed the coin-filled flashlight into some bushes as inconspicuously as she could.

At Gestapo headquarters they were pushed into a milling crowd of thousands of civilians, surrounded by German guards. After a time the men were separated from the women and children. Most of the latter, including Zofia Ossowiecki, were released after two days and told to leave the area.

The men—including Stefan, still clutching his small piece of luggage—and a few women were taken to the park across the street, stripped of their clothes and all possessions, and placed in ranks. A squad of soldiers marched out; machine guns were set up. There were almost ten thousand people.* The massacre took all that day and part of the next. Gun barrels melted and had to be replaced. When it was over, the bodies were soaked in gasoline and set ablaze, the greasy black smoke rising over Warsaw to converge with the plumes of a thousand other fires. It was generally assumed Ossowiecki was amongst those killed during the uprising although, decades later, a story would emerge that he had survived only to die soon after in the camps.

About seven months after the uprising, perhaps only a few days or weeks before the Reich surrendered on May 8, 1945, Stanislaw Poniatowski was murdered at the camp near Litomierzyce. No one will ever know how many died with him.

And so it was over. Everything seemed lost. The men. The work. Until years later a shabby man in a seedy topcoat approached Zofia Ossowiecki as she walked to her apartment on Adama Pluga and asked, "You are Mrs. Ossowiecki? . . . Your husband was the clairvoyant?"

*The number of those murdered that day has been estimated by the quantity of ashes that remained.

THE SCOTTISH
GENERAL AND THE
RUSSIAN RODWALKERS

Scott Elliot and Pluzhnikov

A fifteenth-century grain-drying kiln, a Roman encampment, a Bronze Age crematorium, and a five thousand-year-old possible precursor to fabled Stonehenge—all of these were unknown until an older gentleman with abundant energy pinpointed their location, using nothing more than a little metal rod or a swinging pendulum, and then proved he was right by excavation.

"I have no interest in the psychic whatsoever. I don't even use the word. Just call me what I am. A practical dowser, that's all."

Terse, and to the point; exactly the sort of statement one would expect from a man with Scottish heritage and army training, and James Scott Elliot has both. A major general in the British Army, he found himself entering retirement in 1956 with "nothing really mentally satisfying to do." The Queen appointed him Lord Lieutenant of his native county of Dumfriesshire along the English-Scottish border, making him the area's chief executive and magisterial authority. For most men of his age and station this would have seemed more than enough. But not for General Scott Elliot.

To provide that something "mentally satisfying," he turned to archaeology and began teaching himself the exacting task of scientific excavation. Very soon he learned what all archaeologists must face—that before one can dig, one must know where the relics are buried. There are many potentially helpful scientific aids to locating sites, he discovered, but no really reliable answer to that most fundamental of all archaeological questions: Where to dig?

About this time, Scott Elliot ran across Henri de France's book, *The Elements of Dowsing,* and although "I'd never done that sort of thing before, of course . . . not in the army, you know," to his orderly mind, dowsing and archaeology taken together made sense.

J. SCOTT ELLIOT

Major General James Scott Elliot, after an important military career, retired and sought something "mentally satisfying" to do. The result was significant contributions to the art of dowsing, the emerging technique of psychic archaeology, and the history of the British Isles.

He knew nothing of the methods and successes of Bond or Poniatowski, and would have found their psychic reconstructions and metaphysical comments disquieting. He was not concerned with man's spiritual survival; to Scott Elliot, it was a simple "question of efficiency," the most logical approach to the problem. If dowsing worked, he reasoned, he would have a tool that could "lead one to the sites . . . the ones which don't show on the surface but are important . . . with no fumbling about."

Once he had determined on the course, Scott Elliot put his decision into action. He found a local man who was known as a water-

ATLANTIC
OCEAN

BANFFSHIRE

SCOTLAND

• GLASGOW

GALLOWAY
• DUMFRIES

NEWCASTLE

NORTHERN
IRELAND

NORTH
SEA

IRELAND

IRISH SEA

ENGLAND

GLASTONBURY
BUBFORD
LONDON
• SALISBURY
• BERKSHIRE
• HAMPSHIRE

ENGLISH CHANNEL

FRANCE

UNITED KINGDOM

witcher and had him explain the basic technique. Then, for the next six months, he diligently practiced it. This took a sense of humor, even a sense of the absurd. Neither family nor friends could, at first, reconcile themselves to the sight of the man they knew, a general who had led combat troops during World War II in both Africa and Italy, wandering the fields waving a little metal rod about. But he persisted, and he succeeded.

At first he simply went to sites that either he or others had already discovered by luck or such conventional approaches as selection based on proximity to known sites. He would dowse the dig and try to predict features still covered by earth. These forays into dowsing were quickly tested by excavation, and Scott Elliot found he was very accurate.

By 1958 he had taught himself the history and prehistory of the British Isles, and how to dig, as archaeologists mean that word. He had also mastered what is perhaps the most universal of all psi-related talents: He had become a proven dowser. Thus he became a unique figure in archaeology's evolution into psychic awareness—the world's first dowsing archaeologist. But Scott Elliot did not stop there; once he had determined through testing that dowsing could eliminate hundreds, even thousands, of hours of fruitless search, he resolved to make this information available to other archaeologists.

"Of course, I soon learnt they were too busy with their own things, and besides, they didn't *believe* in dowsing. So I taught myself to do the work without them and it has all worked out very well." It has indeed.

By 1961 James Scott Elliot was sufficiently confident of his abilities that when a neighboring farmer reported a large flat stone that had been turned up by the plough, the General decided to dowse the field. Before he did so, the problem was carefully discussed with archaeologists and other specialists: expert opinion holding that what the farmer had touched on was probably a cist, or neolithic stone coffin. Interesting, but not terribly exciting. However, as Scott Elliot went over the field with his rod, he became convinced that something far larger than a simple stone coffin was involved. Carefully outlining the area—which might not have been dug at all without his interest—he thus eliminated excavation based on conjecture and prepared to dig. What turned up, exactly as he had described it, was

On the strength of nothing more than a "black patch," General Scott Elliot was able to dowse a complex and fascinating site that had gone previously unnoticed, and that would probably have never been dug if only orthodox techniques had been used.

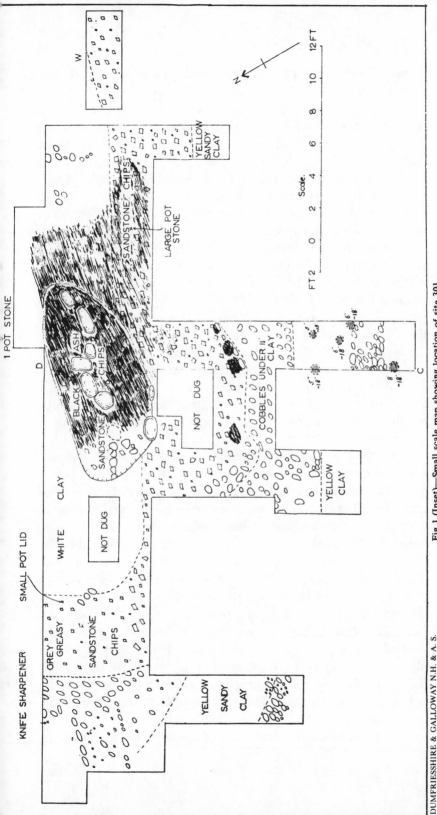

Fig 1 (Inset)—Small scale map showing location of site 301.

Fig. 2—Plan of Excavation.

DUMFRIESSHIRE & GALLOWAY N.H. & A.S.

a fifteenth- or sixteenth-century grain-drying kiln complete with tunnel flue.

The site itself was actually not that important, but the way in which it had been located and delineated was. Soon fellow members of the General's Dumfriesshire and Galloway Natural History and Antiquarian Society, friends, and even strangers were approaching him to dowse their properties to determine whether they held anything of archaeological interest. Half a century after Frederick Bligh Bond had been pilloried for his excursions into psychic archaeology another amateur was meeting with, if not warm acceptance, at least open-mindedness and friendly interest. Although he was unaware of it, Scott Elliot had already achieved a small victory.

About four years later, in October 1965, a Mr. Crosbie who worked Townfoot Farm near Dumfries came to the General and told him about a strange black patch that appeared in one of his fields whenever the earth was turned. Unfortunately, he said, the field was now stubbled over with crop stalks and nothing of the black spot was to be seen. Undeterred, the General went over to Townfoot Farm and dowsed the property. Starting in the approximate place Mr. Crosbie indicated, Scott Elliot soon determined that in an area roughly eighty feet in diameter there was a man-made structure beneath the field's surface. Under his direction a trial cut was immediately made across this space, which produced clear evidence that Scott Elliot was right again.

On the basis of this first confirmation, and with the promise of what the rod said was to come, a proper excavation was carried out later that fall, and another when the weather cleared the following spring. As the workers cut down through the yellowish sandy clay subsoil, excitement increased and Scott Elliot was further vindicated. As he described it in his official site report, there was "a large pit twelve feet long by five feet wide. The content of the pit . . . was black wood ash . . . the quantity was enormous." Around a considerable portion of the pit there was a kind of apron or terrace of crushed red sandstone chips. These were firmly packed and, in places, showed signs of fire and great heat. Further out there was a clay floor over cobbles and finally the remains of what had once been a dressed stone wall. The question was: What was the function of the pit?

A carbon-14 sample taken from the bottom of the pit yielded an

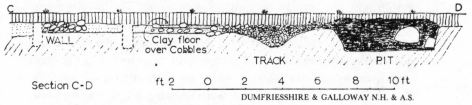

Section C-D   ft 2   0   2   4   6   8   10 ft

DUMFRIESSHIRE & GALLOWAY N.H. & A.S.

Guided by his dowsing, General Scott Elliot discovered what may well be one of the earliest sites of ritual human cremation in the British Isles.

age of approximately 4000 years± 90 years (1980 B.C.). Whatever it was, it dated from the early Bronze Age.

The riddle was complicated by the discovery, in the pit, of a heavy stone pounder, another pounder shaped like an axe head, and a very heavy stone hammer. As the General considered the evidence turned up by excavation and compared it with that from other sites in the literature, he concluded, "The site is a complex one . . . not recognizable as a known form."

Two possible explanations seemed most logical to him: that he had discovered a Bronze Age deer-roasting pit, or that here in the middle of Townfoot Farm was an ancient crematorium. If it was a deer-roasting pit, there should be post holes for the stanchions that would have been needed to hold the roasting spits. He searched but found no such holes. This explanation was further weakened by the fact that the spits would have had to have been fourteen feet long and seven feet wide—too awkward to be really practical. Finally, the track was only on one side, whereas a deer pit would have needed a "terrace" all the way around, in order to get at the spitted meat.

No, the General reasoned, it was not deer meat that had been burned but the dead members of a Bronze Age community. This also explained the relatively elaborate workmanship and the small stone chips mixed in the ash: "The sandstone chips would have heated in the fire and kept their heat so that when the fire went out, the heat continued the destruction of the partly consumed body. The difficulty in cremation by wood fire is to generate enough heat." It also explained the presence of pounders and hammers: They would have been needed to break up bone fragments for placement in urns.

Despite his meticulous work on this first major dowsing site, General Scott Elliot was denied the reward of an absolute case. No bone fragments, animal or human, were found. The acidity of the soil accounted for this; at the nearby excavation of a known grave site only half as old, the crowns of two teeth were the only remains uncovered. Still the preponderance of evidence did tend to support Scott Elliot's cremation theory, which made the dig significant from an orthodox archaeological point-of-view. Equally important was Scott Elliot's demonstration at Townfoot Farm that good archaeology and good dowsing made an effective partnership. Further he proved that he was master of them both. Not a bad conclusion to a study that had begun with nothing more than some four-thousand-year-old carbonized wood ash that had worked itself to the surface of a farmer's field. By the time he had finished with Townfoot Farm the General's reputation had spread through much of Britain and he found himself faced with more requests than he could handle.

There is a theory that holds that dowsing is only good for running water and is not a psychic ability at all but a neuromotor response to weak electromagnetic currents created by the water's movement. If this were true, then the next technique Scott Elliot began to work with, map dowsing, should have been impossible. The General had become very active in the British Society of Dowsers and in 1966 be-

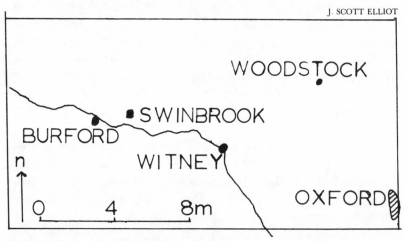

J. SCOTT ELLIOT

After he determined he could locate sites, and specific areas within sites by dowsing them as he walked across their surface, General Scott Elliot next attempted to perform these same feats with nothing more to guide him than a map laid out on a Ping-Pong table in his study. His first such "exploration" revolved around the grounds of a cottage in Swinbrook, not far from the university town of Oxford.

came its president, a position he held for nine years. In the course of his association with this group he was exposed to, and saw the results of, water location by map dowsing. Why not, he reasoned, try it to locate archaeological sites? If he could use this method, all his location work could be done in one place. He secured a simple pendulum, about three-quarters of an inch in length and pointed at the bottom, and resolved to try.

When he was just beginning at Townfoot he had been approached by a woman who owned a house known as Swinbrook Cottage in the northern edge of Windrush Valley in the Burford district. She told him that a "fey woman" had said there was something on her land which, to Scott Elliot, was "pretty thin stuff and I didn't bother with it for a long time." He had plenty of work to do, and the psychic perception of a witch did not seem the most promising avenue of research. In 1969, however, he finally capitulated to the request and visited Swinbrook Cottage.

Visual inspection showed the prospect even less promising. The cottage and its gardens were almost level with the river for which the house was named. The stream, Scott Elliot was told, tended to flood during spring thaws. In traditional archaeological judgment this would make the land an unlikely place for habitation, and even if earlier man had lived in the area, any signs of his presence would probably have long ago been washed away by the flooding waters. Nothing on the surface gave him any encouragement. All in all, an excavation seemed a waste of time. In fact, the prospects were so discouraging that Scott Elliot decided that this would be a good test for map dowsing. He told the owner he would try the site; it was the turning point in his career. With Swinbrook Cottage his work entered an entirely new phase.

He made no attempt to dowse the site while he was there. Instead, he secured a map of sufficient detail to show the garden clearly. When he got home, he enlarged this still further until there was enough space to allow for both careful checking and the outlining of even small site locations within the garden. In September 1969, after he had sketched in the surface details, Scott Elliot laid the dowsing map out flat in his study and began slowly to go over its surface with the small string-suspended pendulum he had recently purchased.

Silently in his mind he posed the question of whether there were signs of earlier human habitation. As he moved the pendulum over the map, in the western portion of what would be the cottage's lawn, the small swinging object began to rotate in a circle—positive response to the General's question. To further inquiries, the pendulum answered that there was more than one layer of artifacts, and that the site would yield artifacts from the eleventh or twelfth century, although the pendulum indicated the site was even older.

Here was a clear opportunity to validate his technique. Since no archaeological sites had been discovered in the area, and there was no evidence of such in the ethno-historic record, submerged memories of something once read, luck, and even telepathy could all be eliminated. Anything found under the west lawn of the garden would have been located by one agency—map dowsing.

Less than two months later, in November, before winter weather set in, a trial cut was made to prove or demolish the archaeological hypotheses produced in the comfort of the General's study. As his official site report describes it, this very first cut "provided ample evidence of a habitation site. Two post holes, a quantity of animal bones, and a number of pieces of pottery," which Scott Elliot immediately recognized as coming from the eleventh and twelfth centuries, were found "well below topsoil level." In one day he had demonstrated that the ancient technique of map dowsing could produce testable accurate information about location, site details, and even dating. However, Scott Elliot realized he was just beginning. Now that the trial cut had authenticated his psychic information, plans had to be laid for a careful excavation. The psychic element only provided direction; it did not obviate hard archaeological fieldwork. Scott Elliot decided to begin this as soon as possible in the coming summer.

At the appointed time the General and his crew began what turned out to be a long and complex dig. A second cut was made, and then a third. As they inched their way down, they passed level after level of man-made remains. Eventually six levels were distinguished before bedrock was reached. At the very deepest level "a beautifully laid, firm floor, constructed of small pieces of sandstone cemented together" was found, as well as two hearths, one of which may have come from a later period. Up from that level about four feet other floors were found, although these were "nothing like so well finished." There were also post molds and remnants of walls at various levels.

Post molds are a key remain for which archaeologists search. Essentially they are cylindrical cavities left when a post rots off at what is then surface level. The part of the post below ground decays, and even if it has been filled with earth and debris, it still shows up clearly during an excavation. From post molds archaeologists determine the boundaries of both single buildings and whole sites. They also use them to reconstruct past surface levels and even to obtain organic debris that may be useful in dating. The post molds at Swinbrook Cottage, and their relationship to the man-made floors of long-gone houses, provided incontrovertible evidence of Scott Elliot's dowsing hypothesis that a long sequence of buildings had stood on this ground.

The site also produced what was, for its size, "a great deal" of pottery, as well as two ivory tools and many flints and flintnaps (the flakes that chip off when a tool is made). The pottery was so impressive that the General sent it to David Hinton of the Department of Antiquities at the Ashmolean Museum in Oxford. Like all good archaeologists, professional and amateur, Scott Elliot knew his limitations and wished to have a confirming opinion.

Hinton verified the General's opinion and disclosed that Swinbrook was an unusual site. "The pottery group," his report said, "extends into the early twelfth century," although it was not "of the twelfth-century types generally found in the Oxford region."

Hinton was at a loss to say when the group began except that "it is almost certainly post eighth . . . but we have no evidence of pottery of the ninth-tenth centuries in this area yet." Elliot's date, a figure arrived at by dowsing, was thus confirmed, like all his other Swinbrook predictions.

The cottage garden later yielded a second site, which Scott Elliot located when he further dowsed the ground toward the end of the excavation of Site I. He wanted to see if there was more to be found in an area that, at first glance, had seemed a most unlikely spot to expect any archaeological results, let alone a long sequence. At this site, known as Site II, only a trial cut was made, but it also produced results. Funds were running out, however, and so the work was never completed. As the General later noted, "Could it be totally excavated, many of the present problems might be answered—though more could be added."

Swinbrook's closing thus was a beginning, for from this site General Scott Elliot emerged with a mature and complete approach to

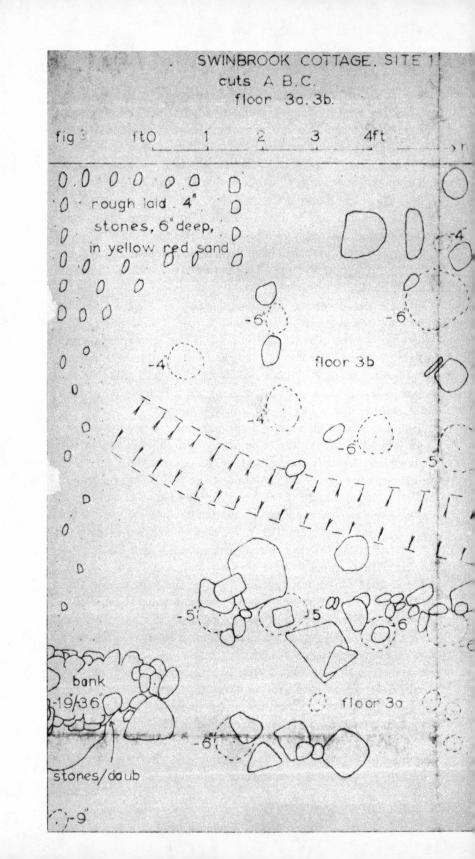

SWINBROOK COTTAGE. SITE 1
cuts A.B.C.
floor 3a. 3b.

fig 3   ft0   1   2   3   4ft

rough laid. 4"
stones, 6" deep,
in yellow red sand

floor 3b

-4

-6

-4

-6

-6

-4

-5

-6

-5

-5

-6

bank

-19/36

floor 3a

-6

stones/daub

-9"

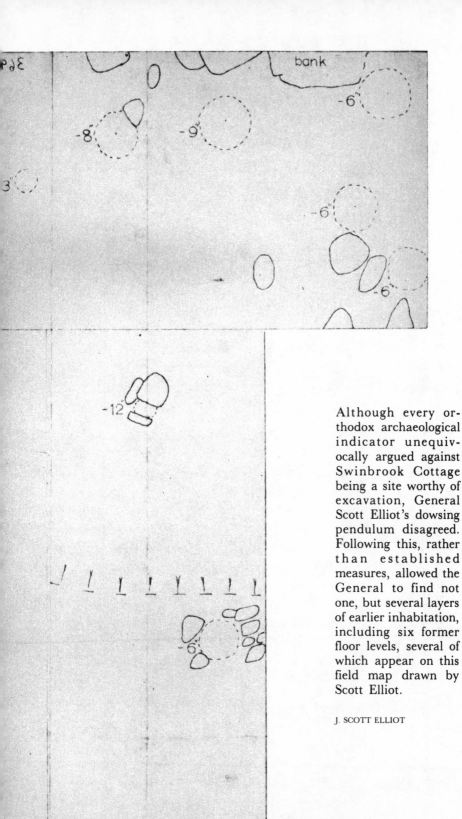

Although every orthodox archaeological indicator unequivocally argued against Swinbrook Cottage being a site worthy of excavation, General Scott Elliot's dowsing pendulum disagreed. Following this, rather than established measures, allowed the General to find not one, but several layers of earlier inhabitation, including six former floor levels, several of which appear on this field map drawn by Scott Elliot.

J. SCOTT ELLIOT

Not satisfied with his first resounding success at Swinbrook Cottage, Scott Elliot map-dowsed another section of the grounds and came up with a second, equally mysterious and intriguing site. Actual excavation again proved him to be uncannily accurate, as this field map drawn by the General himself shows.

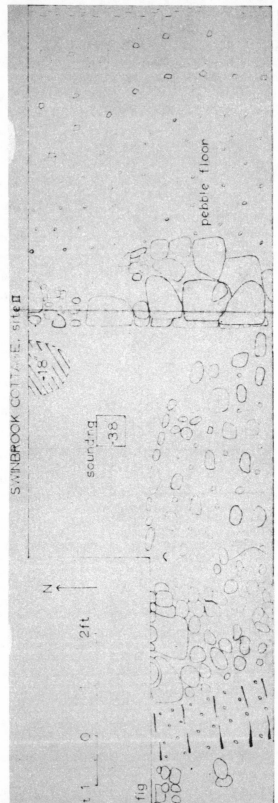

SWINBROOK COTTAGE, site II

pebble floor

sounding
-38

-18"

N

2ft

fig

ft 1   0

J. SCOTT ELLIOT

excavation. Although many other sites lay in the future, it was here that the basic technique of his psychic archaeology was forged: he had demonstrated site location, outline, artifactual details, and even dating.

His 1973 work was at Chieveley Manour, Berkshire, where he again located, predicted, and dated a long, totally unknown sequence *prior* to excavation. This included a road predating the Roman period (55 B.C. to A.D. 350), an Iron Age ditch (500 B.C. to A.D. 250), a series of Roman ditches, and Saxon and Saxon-Norman cultural remains. The Chieveley findings, and the discovery of a habitation and burial site in Banffshire, Scotland, clearly were prepared for at Swinbrook.

An entire new career has, in fact, emerged for this man who was searching for "something mentally satisfying to do." Whereas in the beginning he tended to take sites as they came, today his "main interest is in the older periods here. I started with the Roman and then got interested in the Bronze Age particularly. Now I'm even earlier. On the order of 3000 B.C. interests me now."

Of this Neolithic period perhaps no site is more interesting than the one at Nuriston Farm near Petersfield in the south England county of Hampshire. Here, using map dowsing and then testing his theories by digging, the General discovered the remains of three concentric stone circles, connected by wide shallow ditches, which he dates about 3100 B.C. Excavation is still incomplete but digging done during the first season in 1974 revealed a stone circle structure more than a millennium older than Stonehenge (1900-1600 B.C.), which is located some fifty miles away on Salisbury Plain. As a result of this work by Scott Elliot it has possibly been proven that, much earlier than anyone had suspected—before the Israelites were cast into bondage, before the mysterious cities of Yucatán were laid out, before Grecian temples rose, Stone Age Britons were constructing what may prove to be the earliest known astronomical calculator in stone.

But the General has not been content to let his work stand only on past achievements; to the present day he has continued to devote himself exclusively to the task of explicating man's early years in the British Isles. So strongly is he committed to this task, that he has moved from Dumfriesshire to a suburb of London and left his prestigious post as Lord Lieutenant.

In the fall of 1974 General Scott Elliot was visited at his East

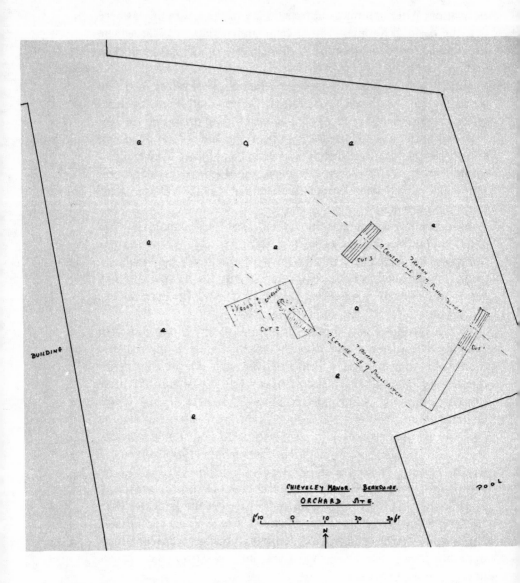

By 1973, when Scott Elliot map-dowsed Chieveley Manour orchard his skills had developed to a point where he was not only accurately dowsing material from several different epochs, but was dating the artifacts and ruins more surely than was possible by the famous Carbon-14 test —which he used to check his own work. Whereas Carbon-14 always

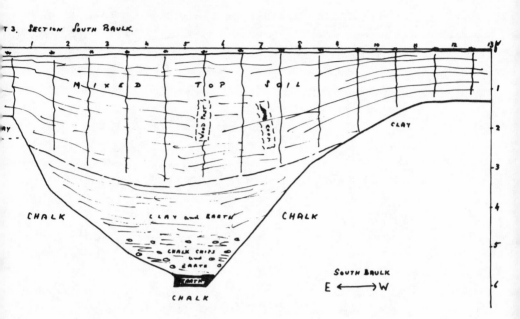

CHIEVELEY. ORCHARD SITE. CUT 3.

T 3. SECTION SOUTH BAULK.

MIXED TOP SOIL

CLAY

CHALK   CLAY and EARTH   CHALK

CHALK CHIPS and EARTH

EARTH

CHALK

SOUTH BAULK
E ←——→ W

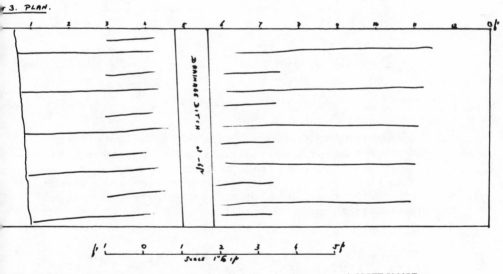

T 3. PLAN.

DRAINAGE DITCH

Scale 1"=5'

J. SCOTT ELLIOT

has a plus or minus factor of decades and sometimes centuries, the General's dowsing brings his datings down to a single year. Depth means nothing, nor, as the Chieveley Manour excavation makes clear, does a complex history of inhabitation. This preliminary site map served Scott Elliot as a first-stage evaluation of the area.

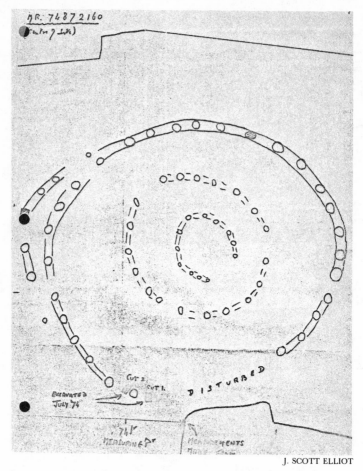

J. SCOTT ELLIOT

Guided by nothing more than a dowsing map, Scott Elliot was able to outline what may well be a precursor to fabled Stonehenge. Preliminary excavations appear to support conclusions developed in his study months prior to the first spade of upturned earth. Using dowsing, the General has demonstrated the ability to solve one of archaeology's continuing problems: Where to dig?

Finchley home, north of London, by Christopher Bird, co-author of *The Secret Life of Plants,* trustee of the American Society of Dowsers, and a leading historian of dowsing. Bird was amazed to "discover the extent of General Scott Elliot's work, the precision with which he carries it out, and the depth of understanding he brings to the task. I was taken into a large study where he works and there, laid out on a Ping-Pong table, were meticulously drawn archaeolog-

ical maps revealing details of most of the known—and some unknown—cultures of the British Isles. Just a cursory examination of
this material would have taken a week, and a study in depth much
longer."

By the time they moved to the dining room Bird was convinced
that "here was a man whose contribution to both archaeology and
dowsing was immense. He showed me only those finds that were
proved by digging. The General does not make unvalidated presen-
tations. He is an outstanding example of a peculiarly British
type—a gentleman who denies all claim to professionalism and yet
is no amateur."

Bird's observation is strikingly apt. Like Lieutenant-General Augustus Pitt-Rivers, who was the originator of so many archaeological techniques, General Scott Elliot seems destined to hold a special
place more for his avocation than for his professional career.

If there is any blight on the picture, it is the all too familiar one of
professional uninterest. Despite the quality of his work, however,
and despite his clearly recorded successes, Scott Elliot still digs
alone. If he has not been subjected to the ridicule Bond was, this
may reflect more his social standing and military achievements than
increased academic tolerance. From the beginning the General has
funded all his own digs and paid for the carbon-14 tests to check his
dowsing dating. With the exception of one academically affiliated
scholar at Newcastle University, no professional archaeologist has
taken part in any of his excavations, although he does think that recently he has made a convert or two. He hasn't really tried to make
converts though; that is not his form. Nor does he feel any bitterness
about his academic neglect; he is certain that eventually his work
will make his case. He is absolutely certain that "young archaeologists ought to be trained, if they have any possible talent for it, in
using the rod or the pendulum. It would save enormously in their
work. In time. In the limited money available, in . . . well, in everything. This goes for all sorts of other things as well. The oil industry
is obvious, metals, minerals, too. That sort of thing hardly needs
mentioning. But a professional archaeologist who could use a rod
would be something really worthwhile . . . I can tell you that."

If Scott Elliot's words have had little effect on academic archaeology and geology in England, the Russians view the subject of
dowsing very differently.

At about the same time as the United States Geological Survey

(USGS) was issuing a restatement of its more than fifty-year-old position on dowsing—"as far as scientists are concerned, the subject is wholly discredited"—the Soviet scientific establishment was holding its first interdisciplinary seminar on the subject.

In the face of increasing evidence that there is merit in dowsing, Soviet researchers in the early 1960s organized the Interdepartmental Commission on the Biophysical Effect. There was nothing spiritual, or even parapsychological, behind this decision. In fact, the old Cold War shibboleth about "Godless Communism" may, for once, be a valid consideration. Since they were undeterred by—or more accurately, uninterested in—the metaphysical implications of psychic research that have played such a large part in American scientists' aversion to psi exploration, Soviet scholars could make a simple pragmatic judgment. If dowsing could find molybdenum, copper-bismuth, bauxite, and tin-tungsten—to name only a few of the raw materials that make our high-technology civilization possible—the Russians would dowse.

A vast body of literature began to grow up, produced by some of the best minds in Eastern Europe. By 1971 this interest had escalated to a point where, in March, researchers from forty separate institutions, and fourteen cities, gathered in Moscow for a conference. There, some fifty abstracts on the use of dowsing and its applications in scientific research were presented. Little or none of this was known in the West, except perhaps to intelligence specialists, until Christopher Bird in his capacity as trustee of the American Society of Dowsers became interested in the subject.

Bird was perhaps the only nonintelligence community individual to gain access to the 1971 Moscow conference papers. Impressed by what he read, he attended the First International Conference on Psychotronics held in Prague during the summer of 1973. Because he was fluent in Russian and other Balkan languages, and had a solid reputation in the dowsing community, the doors opened for Bird. When he left Prague he took with him a large stack of papers on dowsing and a long list of scientists to contact in the future. The conference was evidence that a policy decision to investigate the scientific uses of dowsing had been made by the Soviets, and it was being carried out with the awesome force that only a centrally controlled nation can muster.

"It's perfectly obvious what they have done," Bird said, after returning to his home in Washington, D.C. "They simply decided that this was too accurate and helpful a phenomenon to disregard. So they changed the name to Biophysical Effect, based on the work begun in the 1920s by Leonid Vasiliev at the Institute for the Study of the Brain and Nervous Activity in Leningrad, called the actual technique the Biophysical Method, and began a serious academic study of the subject. The name change, of course, came because as professional scientists they couldn't work with something known by the unscientific word *lozakhodstvo*, literally rodwalking. Much the same thing was done in France. There they coined the word *radiesthesia*, which means perception of radiation. Since there may not be any radiation, however, that term could well be totally unscientific. It would be a marvelous irony, wouldn't it?"

Most of the papers Bird brought back dealt either with geologic dowsing location research or work on the nature of the effect itself. (Unlike a single individual such as General Scott Elliot, the Russians had the means to research the "how" of dowsing.) Buried among these papers was a presentation with profound implications for archaeology, architecture, and history.

Written by Aleksandr Ivanovich Pluzhnikov, docent, candidate of technical sciences (a docent is a lecturer and the candidate degree is the Russian version of a Ph.D.), the paper described several years of research Pluzhnikov had conducted "at the request and with the cooperation of the Scientific-Methodological Council for the Preservation of Cultural Monuments of the USSR Ministry of Culture, the All-Russian Society for the Preservation of Historical and Cultural Monuments, and the Shusev Central Scientific Research Museum of Architecture," to give only a partial list. (This is roughly equivalent to an American scientist being funded and supported by the Smithsonian Institution, the American Museum of Natural History, and some branch of the Rockefeller philanthropies.)

Pluzhnikov's research was designed to "search for and to describe the contours of subterranean architectural and historical objects, no traces of which show above the surface of the ground," using the "Biophysical Method." It is a succinct and simple statement of psychic archaeology in action.

Because of the interests of his sponsors, Pulzhnikov's general area for research was not his to decide; he would study the Moscow oblast, an administrative territory corresponding in size roughly to an American county. By 1970, with his initial planning completed, Pluzhnikov, supported by a team of archaeologists and historic architects, took his "operators" (dowsers) and began marching over this land where Russian history had witnessed so many turning points. Here Russians had stood against the Tartars, against Napoleon, and against the Nazis. Here the Russian Orthodox Church, born in Kiev, grew to power, and here the Revolution of 1917 turned Russian against Russian until the Czar fell. Here Pluzhnikov sought history's archaeological remnants with what are affectionately known in the Soviet Union as "steel wire whiskers"—the thin steel rods dowsers use. Pluzhnikov even learned to dowse.

For the first time in history work on psychic archaeology was being carried out not by eccentrics like Bond or well-respected amateurs like General Scott Elliot, but accredited academics. In Russia, psychic archaeology was receiving support from a scientific establishment. All the controls were present, all the assets needed available, and all the tests required performed. What were the results? To begin with, one of Russia's greatest historical and archaeological enigmas was successfully dowsed.

The story has its beginnings at two o'clock in the morning, on September 12, 1812, outside the town of Borodino, 124 kilometers west of Moscow, as Napoleon just delivered Correspondance 19182. He himself had written the words:

"Soldiers!—This is the battle that you have longed for. Victory now depends on you; it may be ours. . . . May it be said of you: 'He was at the great battle beneath the walls of Moscow.'"

Spread out before him in the dark night were 135,000 men, thousands of horses, and 587 guns. A sort distance away, partly shielded by thin woods, were 132,000 Russians under the command of General M. I. Kutuzov, about the same number of horses the French had, and 624 guns. The battle had been heated on the 5th, but diminished to skirmishes on the 6th. Both sides knew that the coming engagement would be the final one. For the Russians, the task was to hold out as long as possible and to inflict as many casualties as they could. For the French, the need was for prisoners and a sufficiently overwhelming victory to break the defense at Moscow. The key to the battle was the Rayevskii Redoubt, a battery of eighteen guns dug in atop Kurgannaya Hill.

At dawn, some three and one-half hours after Napoleon's message, the battle began with an attack by Prince Eugène de Beauharnais, Viceroy of Italy. At 9:30 the first charge was aimed at what the French called the Great Redoubt; another was launched at

11:00. Both were repulsed by the Russians with enormous loss of life on both sides. Napoleon paused, regrouped, and shifted an additional thirty-five thousand men into position against the Redoubt. During this time the crucial fortification on Kurgannaya Hill was altered to withstand what would unquestionably be a last massive assault.

At two o'clock in the afternoon the attack came. By four o'clock the battle was over, the Rayevskii Redoubt was controlled by the French. But the losses were incredible. As many as 58,000 Frenchmen were dead or dying, including 47 general officers. The Russians had lost 44,000 men, including 23 generals—38,500 on the final day of battle alone.

Napoleon would write, "Of all my battles, the most terrible was the battle I fought at Moscow. The French showed themselves in this battle to be worthy of victory but the Russians won the right to be called invincible."

It was a hollow victory, for the French Army had received a blow from which it would never recover. This battle, the frigid Russian winter, and impossible logistics would soon be Napoleon's undoing. And for the next 158 years students of the Napoleonic campaign in Russia would argue the role of the Rayevskii Redoubt. What changes had been made between 11 and 2 o'clock on September 7? What did the Redoubt look like exactly? How, specifically, was it placed? There were no answers. Or at least none until Pluzhnikov and his dowsers went to work.

Within hours of arriving at the site Pluzhnikov was in possession of the outline of "former trenches, palisades, and parts of a fortification belt of the foxhole type." All had been located and contoured. Additional dowsing revealed "detailed studies of the majority of old fortifications . . . [and] a number of lost mass graves (in particular, French ones) were located." Even the damaged part of the nearby Kolotsk Monastery was outlined. Painlessly, and with great speed, Pluzhnikov's "operators" had provided him with information not just on the Rayevskii Redoubt, but on the whole battlefield. Questions a century and a half old were answered—every point was confirmed by excavation. In a battlefield where virtually nothing had been visible on the surface, now everything was known, and understood.

On the strength of this success Pluzhnikov next attempted to obtain the information necessary to undertake the rebuilding of a long list of Russian Orthodox monasteries. Situated throughout the Moscow Oblast, in the cities that ring the Russian capital, these religious centers have played a crucial role in the nation's history. Most date back to at least the thirteenth century. Many, such as the ones at Volkamsky and Serpukhov, were fortified—an important consideration since they had been key defensive positions in almost every major invasion of Russia for those of the Tartar princes Tokhtamysh and Edigei (in 1382 and 1408) to the campaign of Adolf Hitler. Some are still standing and are now museums; some are mostly ruins. All had either "former fortifications, towers, outer moats, subterranean moats and passageways, buildings, porches or burial grounds," whose location and shape were lost. As at Borodino, the unknown soon was known, and confirmed by fieldwork.

Pluzhnikov next took a kind of side trip into the city of Moscow itself. Here he was able to clear up a number of ambiguous details about earlier forms of the famous Krutitsky Palace, home first of the Russian Orthodox Metropolitan of Moscow, and then the Primate of the church.

With this task out of the way, Pluzhnikov felt that his research team was ready to bring its skills to bear on one of the major Czarist unknowns facing archaeologists, historians, and architects. They would try to unravel the location and "contours" of a summer palace built by perhaps the greatest of the Moscovite Czars, Boris Fedorovich Godunov, whose reign was a major turning point in Russian history, and whose life would later serve as the inspiration for a play by Pushkin and an opera by Modest Moussorgsky.

The Godunovs were Tartars who had come with Batu Khan's Golden Horde—the Mongol army that had swept across the Eurasian plains during the thirteenth century and turned Russia into a Tartar suzerainty. The family stayed, settled in, and by Boris's time (1551) had made themselves a powerful force in the country. Boris began his career in the court of Ivan the Terrible, went on to become the most powerful boyar (a kind of Russian baron), and on February 21, 1598, became the elected Czar. During his seven-year reign Godunov secured the interests of future historians because he opened up Russia to the West, and raised the Orthodox Church to equality with the other Eastern Christian churches.

But from the archaeologist's point-of-view Godunov is of interest because he built a Summer Palace known as Vyazemy Bol'shie (literally Sticky Earth) which eventually grew into an entire complex of buildings, fortifications, churches, and an unusual series of bell towers. Over the centuries, however, this fascinating "ensemble of buildings" (as the Russians style such things) decayed, and eventually whole sections vanished from both view and memory. Indeed, when Pluzhnikov began his work, the best source of information was not Russian but Polish because during the time of the Polish Czars a considerable study of the Vyazemy Bol'shie had been made.

Pluzhnikov had the scholars of his team begin, as they always did if such information was available, by searching through the ethno-historic record. His approach was not unlike that of Frederick Bligh Bond and, in fact, the Summer Palace excavation was quite similar in scope and size to Bond's Glastonbury work. After carefully doing this homework, Pluzhnikov then had the dowsers brought in to go over the site so that he and they could attempt the crucial psychic location work. To their eyes, the Palace complex appeared to be the scattered remnants of what had once been two-story-high walls, a single separately standing bell tower, and one white stone church that still bore Polish inscriptions dated 1611, 1618, and 1620. What would happen that day, after Pluzhnikov's team added their psychic contribution to these fragmented remains was perhaps the most astounding single achievement in the history of psychic archaeology.

As the dowsers walked over the site, Pluzhnikov and the other researchers had expectations supported by past successes. They were prepared for good results but not for the fact that *"within eight hours"* they would have "the foundations of [the] wooden and stone palace, the wooden walls of the estate, the location and course of a moat, the outer parapet, the foundations of pillars running from the palace to the existing church, and the foundations of the totally lost porches of that church." In addition, they also located six bell towers—an especially significant find since as late as the 1951 second edition of the *Bol'shaia Sovetskaia Entsiklopedia,* in the last entry about the palace in the capricious *Soviet Encyclopedia,* only five towers were suspected. As in every other case, preliminary excavations demonstrated the dowsers' accuracy. In eight hours the work of months—perhaps years—of conjectural excavation had been accomplished. The digging is still going on, verifying day by day the work of the Pluzhnikov research group. Reconstruction will take years, if not decades, but all of it will be based on the work of that eight-hour survey.

Pluzhnikov attempted one last project before issuing his report in

1973; a site somewhat different in that much of it had already been restored by conventional archaeological and historical architectural techniques. Still, there were enough questions about Starocherkassk, the former capital of the Don army, to make the project interesting. Again, the team was able immediately and unhesitatingly to locate "the wooden walls of the former capital . . . the exact location of the separate low towers of the fortifications . . . as well as some locations of the outer wall." What is most important, however, is that at Starocherkassk, as at the Summer Palace of Czar Boris, dowsing information "cast doubt upon how [the] historical complex has been restored." Excavation resolved in favor of the dowsers.

The team had become so proficient, in fact, that, when searching for basic locations, they could drive at speeds of almost fifty miles per hour and get the same results as when they walked over a site. The dowser would sit in the speeding car while researchers posed questions he then repeated in his mind. Indeed, the only limitation Pluzhnikov encountered was, as he wryly noted, that the car had to have "a commodious interior" to keep the dowsing rods from banging into things or poking someone's eye out when there was a positive response.

By June 1973, after having dowsed and tested by excavation ten separate sites, Pluzhnikov felt completely confident in presenting a paper at the Prague Psychotronic Conference.

In the beginning he had said his purpose was to "search for and to describe the contours of subterranean architectural and historical objects, no traces of which show above the surface of the ground." His conclusion was equally straightforward: "The use of the Biophysical Method in searching for underground anomalies, and the study and restoration of historical monuments as related to architecture, has great scientific and practical value. *The method is simple, quick, and gives reliable results.*"

But perhaps the most important point for science was that for the first time an interdisciplinary team had approached psychic archaeology with unprejudiced critical judgment. Here was what Bond had been hoping for, what Poniatowski had wished to achieve but was precluded from doing because of war, and what Scott Elliot did not have the funds to accomplish. The only limitation to Pluzhnikov's work seems to be that like all the Russians, until he was introduced to an American dowser named Frances Farelly, at the

1973 Prague conference, he did not understand or believe in map dowsing, and he has made no attempt at anthropological reconstruction of his sites.

These criticisms are minor, however, and Pluzhnikov's achievements major. Indeed, nothing like them was to be attempted by a university-affiliated scholar in the West until an elder archaeologist in Canada decided to embark on a new phase of his already eminent career.

CHAPTER
IV
THE TRANSITIONAL MAN

J. Norman Emerson

The white-haired archaeologist with the paternal expression looked out over the podium at his colleagues. "It is my conviction that I have received knowledge about archaeological artifacts and archaeological sites from a psychic informant who relates this information to me without any evidence of the conscious use of reasoning." Violently readjusting emotions were portrayed on the faces of his listeners.

When a respected member of any profession turns to the psychic, and begins to talk about it publicly, more than casual attention is paid by his peers. Anything involving the psychic creates controversy, of course, but when the experimenter is the national father of his discipline, it is more than a controversy, it is an event. Thus the calm statement delivered by Professor J. Norman Emerson in March 1973 at the annual gathering of Canada's archaeological establishment did more than dispel the prevailing late afternoon audience ennui.

Certain people in the room knew it was coming, had known about Emerson's growing interest in and increasing commitment to what he variously called psychic or "intuitive" archaeology, but almost no one expected a statement quite so unqualified by modifiers ("weasel words"): "It is my conviction . . . ." It was such a flat assertion that some shaken listeners may not have heard him state that his respondent's accuracy ran to "about eighty percent." However, hardly anyone missed the conclusion; it was even less compromising than the opening: "By means of the intuitive and parapsychological a whole new vista of man and his past stands ready to be grasped. As an anthropologist and as an archaeologist trained in these fields,

J. Norman Emerson, senior professor of anthropology at the University of Toronto, founding vice-president and former president of the Canadian Archaeological Association, is considered by many "the Father of Canadian Archaeology." His strong support of what he calls Intuitive Archaeology has provided an umbrella for several graduate students. As a result, Canada stands on the forefront of this new technique.

it makes sense to me to seize the opportunity to pursue and study the data thus provided. *This should take first priority.*"

With this statement by Emerson, Canadian archaeology moved into a new epoch and placed itself on the blade edge of archaeological research. Not since the tragic termination of the Ossowiecki-Poniatowski experiments in Warsaw had the psychic approach had such an influential patron.

It seems fitting that it was Emerson who performed his country's first psychic experiments studying the past, for he is a unique figure in Canadian archaeology: senior professor of anthropology at the University of Toronto, the oldest and most respected department in the country; founding vice-president of the Canadian Archaeological Association; and, since 1946, fresh from the University of Chicago and newly returned to his native Toronto, teacher of Canada's anthropologists and archaeologists. Close to ninety percent of that nation's professionals in those fields have trained with Emerson at some point in their careers. His former students include other University of Toronto anthropology faculty members, directors of the National Museum of Canada, professors at other universities, and most government archaeologists. Because this is so his impact on Canada's research in these disciplines has been immense and as he spoke that afternoon these credentials stood like a praetorian guard behind him—unreferred to, but visible to every mind's eye. Yet, as Emerson read his paper, there was an unvoiced question: Why would a scientist, in the last years of an influential career, suddenly desert thirty years of orthodoxy for something like the psychic? The answer to this lay, hidden by its utter simplicity, in Emerson's life long motivation as a researcher. The psychic was attractive because it allowed him to better serve the imperative he had always served: to know the past as fully and accurately as possible. He had always cared more for explaining than finding.

"Traditional research into prehistory," he would later say, in explaining his motives, "is pretty solid. We are able to trace, with a great deal of confidence, the story of what happened and at what time in the history of man such events took place, as well as how extensive such happenings and developments were over a geographic range and territory. But traditional research into human prehistory still has a major weakness. There is a real lack of humanity. Real men and women hardly ever tread across the pages of our site reports, conference papers, and written books.

"We move from questions of when, how widely, and what happened in the past—where we have some confidence in our findings—to questions of what did it all mean and of what value was it, with less and less assurance and greater speculation. As the realms of art, symbolism, social meanings, and individual and societal values are encountered, our ability and confidence vanishes. Yet these are all the questions which make such a difference when one tries to understand a living person and his culture.

"I have to confess that after thirty years of work and, let's face it, as one of the few real experts in my special field of Ontario Iroquois Indians . . . well, prior to 1971, if you had asked me, I'd have had to say of those questions there was no way I could even have attempted to answer them. Today, however, I would reply that, yes, it may well be possible to do so—with the help of psychic persons."

Emerson's trek into the realm of the paranormal began in the 1960s, when his wife Ann became interested in the subject. A friend had given her a biography of Edgar Cayce, and when she said this had fascinated her, followed it with others. Ann Emerson then joined an Edgar Cayce Study Group and there met Lottie McMullen. Their mutual interests caused them to become friends and Mrs. McMullen confided that her husband seemed to have "psychic abilities." Gradually the two families became friendly. The Emersons visited the McMullens' summer cottage and the two men began to go fishing together.

In spite of the growing friendship, however, in the beginning Emerson had listened to his wife talk about the psychic, about which he was "almost totally ignorant," with "less than my full attention." He was, after all, "a hard-nosed researcher who had spent most of my adult life trying to apply the best scientific methods to my chosen field." The psychic was "O.K. for a Wednesday evening meeting, but had nothing to do with me."

His wife felt differently, however, and, worried about her husband's deteriorating health, she asked McMullen if he could "see" anything that might be helpful. The taciturn, "totally average" McMullen at once began to talk about Emerson's health "with great specificity and considerable authority," she recalled. "He said things he couldn't have known about Norm and made recommendations which a man with his grade-school education couldn't possibly have guessed." Mrs. Emerson wrote these down and showed them to her husband.

Emerson found himself suddenly interested in McMullen's "unorthodox abilities" because if any subject besides archaeology concerned him at that time, it was his health.

"George's suggestions made sense and nothing else up until that time had worked to restore my strength. Better yet, when I tried them they worked. When you are confronted with the prospect of an extended serious illness and then you try something and get well—it makes an impression. You respect the source no matter how crazy, imponderable, or un-understandable, in terms of normal human experience, it may be."

Fascinated by the potentialities of such an information source, Emerson asked if George could do psychometry. He said he could, and so, with no idea of what might happen, on New Year's Day 1971 the professor asked his new friend to try his psychic ability on

A self-styled "average guy," George McMullen has nonetheless demonstrated an extraordinary ability to go back in time. Through his psychic guidance, archaeologists in Canada, the United States, Israel, Egypt, and Iran have gained insights which would have otherwise been denied them.

some Indian artifacts. McMullen's information proved to be extremely accurate. "I was fascinated not only by the validity of what George told me, but also the dimension it added to the limited data to be derived from artifacts by the usual archaeological means."

Still Emerson's commitment remained quite tentative until one day, after talking to a group of graduate students about his new views, one of them came up and presented him with the best-selling book *Croiset—The Clairvoyant* by Jack Harrison Pollack. Emerson took it home, read it, and crossed a threshold.

Gerard Croiset is a Dutch psychic in his middle years who has been studied for more than two decades by one of Holland's and Europe's preeminent parapsychologists, the director of the Parapsychological Institute of the State University of Utrecht, Dr. W. H. C. Tenhaeff. His research with Croiset is famous throughout the Continent, particularly the popular police work locating lost children and solving murder cases, but until Pollack's book Tenhaeff's research was inaccessible to most English-speaking readers because very little had been translated from the Dutch. It was not murder or lost children, however, that mesmerized Emerson but a selection of experiments Pollack presented involving archaeology and its sister disciplines, physical and cultural anthropology, and history.

This work first began in 1953 when Tenhaeff's old friend Dr. Marius Valkhoff, dean of the Faculty of Arts at the University of Witwatersrand in Johannesburg, South Africa, passed through Utrecht. A Dutchman by birth, Valkhoff had been rector at the University of Amsterdam, and later a visiting professor at Utrecht in his specialty, medieval Romance languages. A contemporary of Tenhaeff's, he nevertheless viewed the Utrecht psychologist as "a former master" and had studied with him at the Institute in the 1940s. He never passed through his native country without stopping at the typically Dutch glazed brown brick row house at No. 5 on the quiet side street of Springweg, where the Institute had its headquarters, to call on his friend and parapsychological mentor.

This time, however, as he mounted the steep stairs at the rear of the former residence, he had more than their long-established camaraderie on his mind. Dr. Valkhoff had an interest in the past, and a challenge for both Tenhaeff and Croiset. He had resolved to see if a paranormal source could be used in practical experiments, particularly to gain the intimate reconstructive and explanatory information that artifacts and even documents so rarely provide. He had brought with him four objects: a small bone, four fragments of limestone, a fossil, and a small marble-shaped piece of lead, each carefully selected to test his theory.

The experiments began on December 17, 1953, when, as Pollack recounts it, Dr. Valkhoff was at the Parapsychology Institute with

Gerard Croiset (standing), perhaps Europe's best-known sensitive, talks with the man who has done the most study of his talents, Professor Willem H. C. Tenhaeff, the first man to hold a formal chair in parapsychology. With them is Dr. Tenhaeff's long time assistant, Miss N. G. Louwerens. In the 1960s, Tenhaeff, Croiset, and Professor Marius Valkhoff of the University of Witwatersrand performed experiments in psychic archaeology that convinced Professor Emerson to begin his work.

Number 5 Springweg, the home of the Parapsychological Institute of the State University of Utrecht in Holland. It was from these quarters that Dutch sensitive Gerard Croiset reached out with his psychic perception to a spot thousands of miles away and hundreds of years back in time. With nothing more than a fragment of animal bone to guide him, he produced detailed descriptions of the initial period of contact between whites and the primitive Bushmen of South Africa.

PARAPSYCHOLOGICAL INSTITUTE OF UTRECHT

Professor Tenhaeff. On the table lay a small open box containing many little pieces of stone. Buried among them was a tiny piece of bone of approximately the same color, which made it very difficult to differentiate. Yet when Croiset entered the room, he instantly walked to the table, "drawn to the bone like a magnet." Without even greeting the two professors, Croiset lapsed into what Tenhaeff described as "a kind of half-trance" and, without hesitation, "took the small bone out of the box."

Although he had been told nothing about what he would be given, nor where it might have come from, Croiset had selected the one object Valkhoff wished to have examined. After rolling it about in his hand for a few moments, he began to talk about it in the hesitant, almost jerky manner that marks his psychic discourses; verbalizing what he, like Ossowiecki, "saw" as an inner vision not unlike a movie scene. The scene was a cave in Africa framed by the frenzied music and firelit faces of a human sacrifice.

As the psychic talked on, both Valkhoff and Tenhaeff sat mute (the South African especially was almost entranced). The scene Croiset was describing was specifically accurate on many points; the small bone had been found by Valkhoff personally at the infamous Cannibals' Cave near Mamathes in the South African protectorate of Basutoland. A human sacrifice was the least grisly of the events associated with this cave.

When Croiset finished that day he indicated he was not really through and asked to take the small bone home with him; Valkhoff was only too glad to agree. Exactly a week later the psychic elaborated on the sacrificial scene, saying, "A whole troup of people are passing. . . . I see them with their hands tied behind their backs. They are lashed on their heads or their necks . . . and are thrown into the fire . . . half burned, they are flung from the place over the precipice," which Valkhoff recognized as the one lying at the cave mouth, "thirty feet in height."

Both readings amazed Valkhoff not only for their consistency, correctness, and detail, but also because the bone that Croiset was using to focus on Cannibals' Cave was not human but animal, a fact that Croiset had correctly discerned at the very beginning of the first session. In view of this overall accuracy the South African felt the experiment merited more than just a passing grade. Only two points nagged at him: first, that Croiset might have read his mind (although much of the information was consciously unknown to Valkhoff at the time), and second, a strange reference Croiset had made to non-Negroid observers at the sacrifice.

". . . Negroes. *But also people who are much whiter—people of lighter color* [emphasis Pollack's]. These men of a lighter color carry lances in the hands and shields. They are remarkably tall. The shields are oval and pear-shaped. They have plumes around their heads. Not as the redskins wear, but a kind of plumage on their heads."

This and other details about the observers did not fit. Valkhoff would have dismissed the description as one of those fanciful illusions that clog almost all psychic readings and make the use of such data unacceptable to those unwilling to sort the wheat from the tares except for one thing—Croiset seemed quite certain of the description and it was one of the most detailed sequences to come out of either session. Though Valkhoff could conceive of no acceptable explanation for this information, he was not willing to dismiss it so he just filed it away as the major question to come out of his first contact with Croiset. It would be a year before the issue was settled. In September 1954, after pursuing the matter through correspondence, Valkhoff received a letter from one of South Africa's best-known archaeologists, Professor C. van Riet Lowe.

Dr. van Riet Lowe said the answer might lie in the rock art of the Bushmen, a small-statured and distinct race of dark people, but neither white nor black, who inhabited lower Africa before either of the other two races were present. Particularly, he felt that paintings discovered in a cave known as "Ha Khotsa near Theko about sixteen miles east or south of Cannibals' Cave," might be important because this prehistoric rock art portrayed just such a group of visitors. They were even dressed and armed just as they had been seen by the twentieth-century Dutchman as he sat in the third-story Utrecht office. "In it," van Riet Lowe wrote Valkhoff, "we see seven local naked tribesmen armed with bows and arrows repelling the advance of six men armed with spears and wearing jackets or breastplates and helmets or turbans."

The archaeologist went on to note that for many years he had interpreted this sequence of drawings as evidence of Egyptian-Bushmen contact. But, "We now know firm trade links existed between Shiraz and Sofola about a thousand and possibly more years ago." He concluded that it was quite possible that light-skinned traders coming from what is now Iran could "at some remote time in the past have penetrated as far south as the Union or even into Basutoland. The turbaned figures (seen as a kind of plumage by Croiset?) might represent Persians, who in the eyes of your psychoscopist were 'white.' His interpretation may therefore not be quite so fantastic as it might at first sight appear. . . ."

Valkhoff now knew it was impossible for Croiset to have read either his or Tenhaeff's mind because neither of them knew anything of such information. Therefore the psychic's insights had to have come from another source.

Before the letter from van Riet Lowe was received—indeed, on that same Thursday in December 1953—Valkhoff ran off experiments with the other three objects he had brought along: the limestone fragments, the fossil from Makapan Cave, and the small marble-shaped piece of lead. He was amazed at Croiset's stamina and ability to switch not only from object to object but also from geographic place to place and epoch to epoch. It seemed almost as great a feat as his accuracy.

There was no doubt in Valkhoff's mind, either then or four years later in 1957 when he returned with a sixteenth-century manuscript whose translator he was trying to identify, that "Croiset perceived information by other methods than normal perception and reflection—by means of paranormal cognition. All these impressions passed through Croiset's subconscious mind and may have been slightly altered or even confused with memories and other impressions. Nevertheless," he concluded, and this was the, main point, "it cannot be denied that here we have a new source of knowledge which may prove fertile in the long run. Object reading done by a gifted psychoscopist can certainly provide us with useful tracks to follow up. . . . Gerard Croiset gave us here a lead which appears to conduct us somewhere else where we could not have arrived ourselves."

As Norman Emerson sat reading in his favorite chair in the living room of his Toronto lakeside house, almost twenty years after Valkhoff's experiments and close to a decade after Pollack's book was first published, this final statement was illuminated on the page. He found himself filled with a strange mixture of emotions, both relief and exultation: "I'm not crazy for thinking of using this stuff in a practical way. Here not one but a series of scholars, including archaeologists, had used it or commented on its accuracy." Smoking one of his ever-present cigarettes, Emerson decided that "if men like Tenhaeff and Valkhoff were willing to do such experiments and later talk about them openly . . . keep at it year after year . . . the least I could do was spend a couple of months and see where it led. If Valkhoff was right, I could find out answers to questions I never thought would be solved."

Emerson's previously rather casual experiments subsequently took on a new direction. The experimental sessions, often conducted in his living room, now had a purpose—nothing less than discovering the true prehistory of Ontario. The "Father of Canadian Archaeology" very quietly initiated a new phase in archaeology. His senior status assured that he would command at least *a* hearing.

In George McMullen, his first and still primary psychic respondent, Emerson discovered a willing ally. Over his years as a bush guide McMullen had come to believe that the Canadian Indian was neither fully understood nor correctly represented in the formal history of the province. This had given him a painful sense of inequity, for which, until the Emerson experiments began, he had found no means of release.

As further impetus towards cooperation with Emerson, McMullen like Croiset, whom he resembled in many ways, had a strong desire to see an academic imprimatur on his work. Until Emerson began to ask questions that were important to his research, and to acknowledge forthrightly the psychic's ability to help him, McMullen had never received such recognition or approval. Like many sensitives, even his family had rejected his "differentness." The warmly extended support of a man of stature in a professional community salved many wounds.

From his youth, McMullen had had a physical handicap—he was deaf in one year. He also had strange flashes of insight about things that were going to happen, deeply disturbing to a young boy of strict fundamentalist background. One case still carried the sting of a bad memory. A neighborhood boy went off for a holiday weekend on his motorcycle and George innocently told all their friends that none of them would ever see him again. Quite predictably they laughed and teased George until they learned that the boy had been killed in an accident. To McMullen's bewilderment and distress, he was treated as if he were in some way responsible for the death.

His mother was even more disturbed than his friends about these prophecies and beat him as if physical abuse would drive out some motivating demon. As a final touch to this sad drama, she took McMullen to their minister, who stood the boy before him and railed at him about consorting with the devil. Misunderstood even by God's deputy, he turned inward and stopped talking about what he saw, but never gave up his conviction that both his mother and the minister were wrong. This feeling got reassurance from an unexpected source.

In his neighborhood there was a man everyone thought of as

slightly crazy because he sat by the hour on his porch in a rocking chair staring silently into space. McMullen somehow knew he could talk to this man about his differentness so he was not surprised to find the older man's eyes search his and then nod. He told the boy something he too had kept to himself: The reason he sat silently rocking on his porch was that he could communicate with what he described as beings who did not have bodies. To McMullen's unspeakably intense satisfaction, this older man then set about teaching him how to channel his psychic abilities so that he too would be able to see and talk with these beings. With practice, the youth soon found he could sense and communicate with entities he saw as lights.

To prove to himself it was not a delusion, McMullen tested these entities with questions and became convinced that they existed when the information they gave him proved correct time and time again. But not until years later, when his friend "the Doctor," as he has always called Emerson, began handing him bits and pieces of broken pottery, worked stone, and twisted metal did his power achieve a special purpose.

Influenced by the format and purpose of the earlier Dutch experiments, and aware of "enough sites to keep me busy through three careers," Emerson was only secondarily interested in using psychic information in problems of location. He was primarily concerned with the questions of reconstruction and explanation, which neither he nor any other expert archaeologist had been able to answer. He began by trying to explore the parameters of McMullen's ability, because in "no way did I know what could be done with this sort of information source."

In one session the psychic was handed what looked like a little tapered clay cylinder. McMullen just looked at it and felt it, then seemed to get a kind of "abstracted or disoriented look on his face and suddenly began to talk." He told Emerson it was the stem of a smoking pipe, described how it was manufactured, correctly located it as coming from a site along Black Creek in what is now metropolitan Toronto, and gave its age. He went on to tell about the artisan who had made it and what life was like in his village. Finally he took a pencil and paper and drew first the shape in which the clay was formed and then how it was bent to form a pipe bowl—as well as how it was decorated. Emerson immediately recognized it as "a typical Iroquois Conical Ring Bowl pipe, one of the popular types recovered from the Black Creek site; and one of the predominant types to be found in Middle Iroquois times."

During this same exploratory session Emerson handed McMullen a small fragmented bowl-like clay object excavated at a site along the shores of Bass Lake in Simcoe County near Orillia, Ontario. Following the same procedure, McMullen again correctly gave its age, location, function, and a wealth of reconstructive detail. Once again, he drew a picture showing the artifact plus the piece that was missing. Emerson was again impressed because the psychic had accurately drawn "a typical Pinch-faced Human Effigy." It was exactly the sort of artifact an archaeologist might expect to find at "late prehistoric and historic sites in the Simcoe County area."

At first Emerson wondered if George was simply reading his mind—a fascinating ability in its own right but not very helpful as a tool for advancing archaeological research. It seemed unlikely since in the case of the pipe bowl, Emerson was not sure himself of its type, but the possibility could not be denied. It was clear to the professor that some kind of test in which the answer was totally unknown to him would be required. As it turned out, the challenge was soon presented in the form of a visit to a prehistoric Iroquois

This drawing is a reproduction of the original psychic drawing produced by George from the Black Creek pipe stem.

a. Indicates the actual stem held in his hand for psychometry.

b. Indicates the shape of the moulded clay in the process of manufacture.

c. Indicates the final shape and decoration of the pipe bowl.

J. NORMAN EMERSON

site near the prehistoric village of Quackenbush, about eighty miles from Toronto. Emerson wanted to see the site and asked George to accompany him. After they arrived, the two of them walked out over the land and Emerson asked McMullen, who knew no more about this site than he did about any other—which meant almost nothing—what he psychically saw.

Thus began what was to become a familiar pattern for the two of them when in the field. "First," said Emerson, "George just sort of takes in the lay of the land, rapidly walking around, noting what is there today. Then he seems to become abstracted and begins talking about what he is seeing, which seems to be the site as it was as a functioning Indian village."

Emerson, following with his tape recorder, was at first fascinated by George's commentary but then disappointed to hear him state that these people did not cultivate corn, beans, and squash. "I found this very hard to accept, you know, because well, these are the crops Iroquois tribes used as staples."

This failure to perceive something as well established as the village's basic foodstuffs did demonstrate George's ignorance of prehistoric Iroquois, but that validation was small recompense for the professor's vanishing hope of using his psychic friend in fieldwork. In an effort to check George's claim, Emerson questioned the site's archaeological field director, who assured him that the excavation had turned up "abundant evidence of corn, squash, and sunflower seeds."

As Emerson sought some way to reconcile the psychic information with both the excavation record and the traditional assumptions about the three crops, he went back over George's taped description and was struck by the emphasis the psychic had placed on the village's extensive trading with other Indians. He also remembered the investigating archaeologist saying he "felt that he could make a good case for trade in stone." Perhaps, Emerson thought, trade was the factor that would enable him to reconcile the psychic information and the findings of traditional archaeology. To test his theory, he had a pollen analysis of the site carried out.

Samples were sent off to the laboratory; Emerson impatiently awaited the results. "When the study came back it reported the relative abundance of a whole variety of trees, plants and grasses. But only one corn *pollen* grain had been found, and that was listed as *problematical*. The results were not at all what you would expect had corn been cultivated at that site. Nothing at all showed of the other two crops." So, the presence of the grains indicated that the Indians had eaten the traditional Iroquois diet, but the absence of flower pollen was a strong "although not absolutely conclusive, since the tests were not quite extensive enough for that," argument in favor of McMullen's insistence that they did not cultivate the crops.

"To me, it seemed as if George had not only steered us in the right direction but had led us into thinking about the Quackenbush village site in an entirely new way. I think George was right and we were wrong and I believe more conclusive tests will prove that to be the case." Professor Valkhoff's assertion that Croiset had provided "a lead that appears to conduct us somewhere we could not have arrived by ourselves" seemed to Canada's senior archaeologist to be true of the psychic approach in general.

By the beginning of 1973, after almost two years of experiments fitted into an already overcrowded schedule, Emerson felt he was ready "to go public." He chose the annual meeting of the Canadian Archaeological Association scheduled for March, asked George to go along with him, and requested time to deliver the paper that began, "It is my conviction that I have received knowledge about archaeological artifacts and archaeological sites from a psychic informant . . . ." It took a special kind of courage to risk an orthodox career of thirty years by embracing so unequivocally a technique widely regarded as questionable if not lunatic. The poststatement track record of men and women in science who had done this was not encouraging. Emerson realized and accepted this and was, thus, doubly pleased to note that among the many faces marked with frank skepticism were some that looked thoughtful and a few that appeared cautiously enthusiastic. One whose face seemed to reflect all three states at once was Jack Miller, from Port Clements, British Columbia.

Miller came over that afternoon and asked whether Emerson thought McMullen would be willing to try his hand at solving a little mystery. Emerson said he thought George probably would, so Miller brought to the banquet that evening something made of black stone, about two-thirds the size of an average man's palm, flat on both sides but carved along the edges. Emerson thought it looked like the head of an ape or primitive man. A friend of the owner was

On the basis of this simple carved piece of argillite, George McMullen and several other psychics were able to produce information that eventually rewrote a chapter in the history of British Columbia.

sure it was an unfinished pipe blank. It was generally agreed by those present (and Miller confirmed the assumption) that the stone was argillite, a shalelike rock about the hardness of soapstone that is found on the Queen Charlotte Islands off the coast of British Columbia and that, in both prehistoric and historic times, has been used as a carving material.

Miller told Emerson privately, before McMullen was invited over, that he knew where the stone had been found—at the bottom of a post hole he had excavated at a site near the town of Skidegate on the Queen Charlotte Islands—and the time period in which it had been worked. At the beginning of the experiment Miller asked McMullen to ignore location and time. What he needed, Miller said, was archaeological context; what did the carving mean, what had it been used for, and who made it?

Miller then gave the piece to McMullen, who handled it briefly, while Emerson looked around the group trying to appear both encouraging and unconcerned. The psychic began to speak. If the Quackenbush village controversy over corn had been a shock, it was the merest tremor compared to what George said this time.

The stone, McMullen stated, with the certitude possible only to one totally ignorant of intellectual information on a subject, was carved by a black man from Port-au-Prince in the Caribbean from whence he had been brought to Canada as a slave.

Emerson was appalled, as he admitted later. "Here I had just presented a paper on how good George was and how you could use psychic data, and just a short while later he was saying something patently ridiculous. As an archaeologist, I realized there was latitude for several possible explanations, but the stone was clearly British Columbian black argillite, and that a black from the Caribbean could have carved it was not an idea within the parameters of the possible."

Emerson covered his shock with a lot of qualifications along the lines of "you never can tell . . . unfavorable conditions . . . needs further work . . ." and asked Miller to lend him the carving so McMullen could study it again in calmer circumstances. Miller agreed and, putting on his bravest face, Emerson said he would report the results. Polite conversation was made all around and the group quickly broke up.

After returning to the Toronto area Emerson did indeed ask McMullen to "read" the carving again. "I thought, you know, the banquet, all the excitement, who knows what energy was around that night. Now back at home in his own surroundings we would get things straightened out. The real story would emerge." But on the second reading George stuck to the story he had told at the annual dinner and actually expanded on it.

The black man, he said, had been born and raised in West Africa where he had been captured by slavers and taken to the Caribbean islands. There he worked for quite a while before being sold again, this time to the English. He was transported on an English ship. (George knew nothing about the location of Miller's excavation) to British Columbia "around Bella Coola . . . south of Prince Rupert." At this juncture the psychic insisted the man had escaped from the ship and found an Indian tribe that accepted him, in whose village he married and lived out his life. There he had carved the black argillite. Emerson was in no hurry to report back to Miller and relate what this second experiment had produced.

As he sat in his study patiently transcribing the tape recording of George's comments, Emerson replayed the problem in his mind. "I couldn't prove, or for that matter disprove, a word of it, but I sure could say it defied my imagination. And yet George was so sure . . . not once but twice. He wouldn't give an inch on his explanation. Suddenly I found myself thinking, 'Forget about any preconceptions about the story being right or wrong. Start on the assumption George is right.'" But how, he wondered, could you ever test such a hypothesis? You couldn't, he thought, unless bones were found or something in the ethnographic record equally evidential turned up.

For several weeks the puzzle seemed to afford no resolution, but then a chain of events occurred that culminated in the development of the two-pronged psychic method in archaeology Emerson has followed ever since.

To begin with, Emerson's daughter Lynn came home from an extended stay in Europe to be married, and her university roommate from Ottawa arrived to be a bridesmaid. In the course of her visit Emerson learned that Lynn's friend could "read" Tarot cards, and began to wonder whether the cards were only an occult prop for a psychic ability essentially similar to McMullen's. He told the girl that if she could read cards, she could probably psychometrize objects, and then asked her if she would try. When she agreed, he began testing her with Indian artifacts from his personal collection. At first she felt the need to lay out the cards, as if it were a Tarot reading, but soon she found she could get highly accurate impressions simply by concentrating on the objects Emerson passed to her. The professor, then, handed her the argillite carving without comment. To his astonishment and excitement, she began immediately telling him that it had been carved by a black man from Africa who "had been the victim of a vast, sweeping slave trade and had been brought to the New World." As she went on, she gave many details about incidents that McMullen had not mentioned, but she also corroborated many of the features he had described.

To Emerson, the fact that her reading supported McMullen's earlier psychic commentary was almost as important as the reading itself. It occurred to him that if he asked a number of sensitives to comment on the same object, without telling them anything about the artifact or the fact that others had already read it, he might be able to develop a psychically produced matrix of data that would cross-check.

Ann Emerson, meanwhile, had been making contacts among those in the Toronto area who were interested in parapsychology. Through them she was able to put her husband in touch with the people he sought; individuals who appeared to have psychic abilities. Emerson began giving artifacts to various people who visited their home, asking what impressions they got from them.

He would begin with an Indian pipe stem or bowl, something he knew about in considerable detail, and if the person did well, would

Professor Emerson would try out likely candidates with artifactual fragments such as these examples of t·ade goods offered to the Iroquois by
French explorers: if they did well, he would then go on to show them
the argillite carving. In this way, he was able to build up a psychic team
whose comments, taken together, provided major new insights in
Canada's past.

follow this up by giving him or her the argillite carving. Through
this winnowing process, Sheila Conway emerged as an excellent
psychometrist, and a near-relative, Jim, also showed promise.

Emerson soon discovered one of the major problems encountered
by those who wish to do practical research with psychic sources: if
more than one psychic is used, the readings by the different sensitives may not completely jibe—although all are plausible. The researcher then has the problem of deciding which version to believe,
even tentatively. But, Emerson thought, if this apparent difficulty
was turned back on itself, it could be a potential point of validation.
"It seemed to me that, even allowing for the possibility of mental telepathy or the more subtle problem of having each succeeding psychic read the impressions left by an earlier one . . . well, even with
that if they *agreed* on a number of points, you could still assume you
had something. You would have a check for information you
couldn't get any other way. If, on top of that, some of the points
brought out psychically *could* later be checked by a trained researcher using orthodox methods . . . to my mind, you had the making of a
sort of psychic archaeological team. You had a case [that was]
worth listening to."

Over the next few months Emerson put his idea into practice and tape-recorded the results. When he transcribed and compared the results he was "elated to find many similarities and reasonably few contradictions. If the stories were not the same, it was usually a case of one person seeing something the others did not, rather than fundamental disagreements." So many were the correlations, in fact, that he listed them on the topic of the argillite carving:

Jim: "He came from Africa, as I said before, about halfway down the west coast of Africa and about thirty miles inland."

George: "There was a certain amount of water where he came from—there's a waterfall, quite high waterfalls—the central, west central it seems to me. It was very heavy, very thick jungle, it was very heavy, very dense, very wet, very damp."

Sandy: "He was from the interior of North Africa. Now there was a lot of French influence . . . but he wasn't in that area . . . he was more in a jungle area going from the desert into the savannah area—savannah land."

Sheila: "The jungle is behind me here. It's not really jungle country, though. It's hot. I feel as though I'm up on a high plateau—high up, and the ocean is *miles* down below."

On the question of the race and station of the carver whom George had described as "a black man":

George: "I don't know whether he was picked up in a group or whether he was sold by other people. Anyway, he ended up in slavery; he came over in a slave ship . . ."

Sandy: " . . . and he was the victim of a massive, sweeping slave trade insofar as people went into the interior because they needed men to come to the New World to work for them."

Jim: "They were raided by a renegade African and his cohorts who captured quite a few of his village, including his own family, to sell them as slaves to the English or the Americans."

In response to the question of where the man had finally ended up and where the piece was found, which George had described as "up in B.C. . . . around Bella Coola and Bella Bella . . . up in that area . . . south of Prince Rupert":

Jim: "He made the carving out of the black rock from the mountains there nearby . . . in the western part of the North American continent . . ."

Sheila: "Kind of looks like the kind of stuff that comes from the Queen Charlotte Islands—what is a black man doing in the Queen Charlotte Islands?"

Sheila also reported being conscious of Russian sailing ships in what was probably the Hecate Strait running between the Queen Charlotte Islands and the Canadian mainland and described them as "broad-beamed and tubby." She even provided the name of the slave ship on which the carver had been brought to the New World—*Majoree.*

As Emerson reflected on this mass of data, often through the long hours of a cold Toronto night, he realized that "no matter how preposterous George's original story sounded, I had heard not one but a whole list of other psychics read that same rock cold—no preparation or background of any kind, or even any knowledge that others had tried it—and to a person they had psychically reported the same facts. With the entire globe to pick from, they had all settled on the same area for his early years—northwest Africa, perhaps the Gold Coast; the same type of geography and vegetation—a plateau near a jungle; and the same part of Canada for his escape and association with the Indians."

It was either accept the psychic composite as a working hypothesis or "you had to take refuge in the even more improbable explanation of a mass hallucination commonly shared by everyone who picked up the artifact even though it was days or even weeks between the individual incidents. I never heard of, nor can I imagine, such a hallucination, and would find it harder to accept than what any psychic team said."

Only one thing remained—to tie the psychic commentary into orthodox research. Emerson had close to two hundred pages of tape transcriptions, and there were many points, such as the ship name, that could be checked. Still he felt the first and most critical step was a traditional evaluation of the artifact itself. Here again, an inexplicable coincidence solved his dilemma. At the end of April, just over a month after that banquet reading by George McMullen, Emerson found the person to make this evaluation: a former student, Allen Tyyska. After an absence of months he stepped back into Emerson's life by paying an unannounced call to his home. Now an archaeological planner in the Historical Planning and Research Branch of the Ontario Ministry of Culture and Recreation, Tyyska, in 1973, had just left the University of Toronto where he had been under Emerson's tutelage.

"I had taken a short-term job with the Royal Ontario Museum and was busily cataloguing a number of collections, including an extensive one of West African art, pieces from the upper Volta, the Niger, and generally that area of West Africa between Sierra Leone and the Cameroons, along that coast ... Nigeria, Gold Coast, Ghana." Africana was "certainly not my specialty as an archaeologist, but after reviewing the literature and working with it intensively for six months, I was certainly not ignorant of it either. So when Professor Emerson, who knew nothing of my recent work, handed me that carving with no comment other than to ask what I thought of it, I felt that I had something legitimate to say."

Tyyska's statement could hardly have been more appreciated:

"This little piece of argillite looks like what is known as a 'passport mask' such as is usually owned by young West African males. It fits into the art styles I was cataloguing ... there are a number of things about it that would belong somewhere in that general area ... I would say with pretty fair confidence that this object was carved by someone trained in, and familiar with, the art techniques found specifically in the Gold Coast area of West Africa ... more precisely the upper reaches of the Volta. The only difference is that it is in argillite as opposed to wood. But then, of course, argillite is native to B.C. and was often used as a carving medium.

"The mystery to me was, and remains, how such an object with such a clear West African motif turned up in a Queen Charlotte Indian village. No orthodox archaeological technique could provide an answer to that puzzle."

Tyyska's caveat did not matter to Emerson. He now had a technique, and with the young government archaeologist's judgment and psychically cross-checked composite, he also believed he had the answer. Tyyska agreed: "I would entertain it very seriously."

Even with all this, the story of the argillite carving was not quite over. Almost two years later a team of physical anthropologists totally unconnected with Emerson went to British Columbia to do blood analysis of the Indians in the area. Their report, when filed, contained what was, for them, the disturbing observation that in an area where no blacks were supposed to have been until modern times, one tribe showed unmistakable evidence of a black forebearer. The tribe in question was the one into which Emerson's psychic team had said the escaped African had married!

The bewilderment of that banquet night had turned to triumph. It was more than a single object explained, more than a fascinating insight into Canada's past. And it was more than the fact that with the argillite carving, on top of the Quackenbush village success, Emerson now felt confident of his psychic respondents even when their reconstructions "seemed a little flaky at first." The main point, Emerson suddenly realized, was that he had devised and validated a technique for utilizing psychic data in archaeology, a model that could be applied in any situation. Ever since then he has used an approach employing multiple psychic responses to the same questions interpreted by trained ("I hate to say it but that is absolutely necessary") researchers, buttressed "by the best orthodox procedures applicable to the situation. You don't throw out the old to ring in the new; you blend the best of both to get something greater than either."

Emerson was now vulnerable on only one front: His psychic team had yet to locate previously undiscovered features of a site. Since he himself had no current plans for surveying or digging, it could not be his dig; some other archaeologist, open-minded enough to allocate precious excavation time to such a location experiment during Ontario's short archaeological field season, would have to be found. Yet again a former student came to his aid.

C. S. Reid had studied under Emerson at Toronto, and then gone to McMaster University in Hamilton, Ontario, to do graduate work. He had two digs that were an integral part of his degree program going, and they were only a short distance from his former professor's house. Learning of Emerson's recent work, Reid invited his former professor and McMullen to visit the sites to see if George could indicate anything that might help him with his excavations. The results of these visits put Emerson's final insecurity to rest.*

Convinced now that psychics, working with trained researchers, could successfully solve problems over the entire spectrum of archaeological and anthropological challenges, Emerson now had no qualms about devoting increasingly large blocks of time to his psychic research even if it meant spending long hours on extracurricular transcription and study.

In the summer of 1973 he began working, as opportunity presented itself, on what he feels is his life's work: a complete reevaluation through psychic/archaeologist teamwork of Ontario's past. "Much of what I learned merely provided another unusual collaboration in

*See Chapter 6 for the complete stories.

support of what was already known. And much more just filled in chinks. But there was a substantial portion where nothing at all had been known before, new knowledge was gained, or our traditional reconstructions demonstrated to be wrong."

Sometimes he found that "the information I was getting was full of those little human details archaeologists say they want to get but rarely talk about because there is no real way to prove or disprove the data. That's one reason archaeology is so often unhuman: We can only talk about the stuff we can prove, and that has nothing to do with motivations and feelings. Well, in most cases, I couldn't prove the data any better than anyone using traditional methods, but I came to have confidence in what the psychics were saying because they tended to agree, and I had learned that this meant something."

Using agreement as a probity factor, Emerson felt he was "making progress although there is still a vast amount to be checked . . . no way I'll ever be able to do it all." But sometimes he got help from unexpected quarters. One such incident involved some of his "oldest" work with McMullen.

For many years the professor had been interested in the very earliest inhabitants of the region. The archaeological record indicates some kind of man moved into Ontario to stay about 11,000 B.C. By about 5000 B.C., or perhaps a little earlier, another group came to the area, thus beginning what is known to the discipline as the Archaic Period. Although the presence of these groups was well documented, very little was known about them.

"We think they had no pottery—that's a later stage of development—and yet the later ones, the Archaic People from about 5000 B.C., appear to have been quite facile in working with copper from native deposits. So good were they, in fact, that they exceeded the capabilities of many later inhabitants. But as for their religious belief or even what kind of houses they lived in . . . we really had to say we didn't know much of anything . . . they came after the last ice age, were hunter-gatherers, and later [they] had copper . . . that's about it."

Quite by accident a cache of stone tools came into Emerson's hands and he, wondering whether something could be learned using psychic techniques, asked McMullen for psychometric readings on them. George immediately launched into a description based on these artifacts from approximately 9000 B.C., saying the first group should properly be called "the Wind People." Unfurling their story like a flag he portrayed a life ruled "by howling winds," the people's skins almost blackened by exposure and wind chafing. Suddenly, Emerson realized McMullen was talking about people whose pit formations on the north shore of Lake Superior he had spent two bleak summers studying.

"I started to get a whole new point of view," Emerson said. "Here was a group of human beings, surrounded by these winds, rocky landscapes, very few trees, the ice age is still present but receding, subarctic conditions—and yet they're still surviving. They walk hundreds of miles with the inner knowledge, like animals, of where the fish that are the key part of their diet will be. They remove shore rocks to create pits that trap the fish (the very type of pits I studied almost twelve thousand years later). When they catch the fish, they eat them raw. Then it's back on the move, and if you misjudge where to stop next, perhaps starvation."

It was all very interesting, but how could one prove something that took place perhaps twelve thousand years ago?

Talking about this peculiar reading on the Wind People of Lake Superior with a geologist friend, Emerson was surprised to "hear him tell me, 'I have no quarrel with that, Norm . . . it makes a lot of sense. I don't know whether you know it or not but there are a lot of sand dunes up in that northern area. The only way they could have gotten there is through the action of constant terrible winds.'"

As the months progressed, Emerson began to accumulate more and more such material. He also began to study how to get the best results from his psychic collaborators, data that could "be plugged into either excavations or the ethno-historic record." With increasingly successful results to back him up, Emerson also began to deliver further papers and to talk quite freely to any colleague who "would listen with an open mind." He had no "desire to make converts as such but I did think people ought to know what was going on, what could be done."

Despite his desire to get the best information though, Emerson had thought very little about what the psychic experienced when he made his contact. For this, help came from his wife Ann.

"A lot of people, I know, think Norman is kind of far out. What they don't realize is that he is really quite orthodox about his work. He keeps his scientific detachment and is interested in getting further material to compare with what he already knows, or to locate a site or something like an Indian longhouse within a site. Things that might take months to find otherwise, maybe even years, considering how short our digging season is sometimes. I'm more curious; I want to know how the psychics get their information. And so I asked questions about how they perceived these things and how it affected them. Often I pick up different information than was given to Norman, because he is asking other questions."

It was Ann Emerson who discovered what was going on inside McMullen's head when he went to a site, what happened to him during his period of disorientation and running around—which was all a nonpsychic observer could see. The only comment the Professor makes about McMullen's psychic dynamic is, "At no time would he go into what could even remotely be called a trance."

McMullen has explained what different processes obtain, depending on what is being asked of him. When he goes to a site, he says, "I project myself up in the air . . . I'm looking down on what's going on below as if I were up in a helicopter or perhaps a tall tree." Once he has attained his out-of-body perspective, he sees things not as they are but "as they used to be . . . the time period the professor is asking about." When this is happening, he cannot see the site as it is now. This is why he first walks around to orient himself.

To make contact with this past time period and to answer certain types of questions, McMullen maintains he has nonphysical informants. He sees them as beings of light and hears their voices directing him. "It is as if there were a kind of overlay," Emerson says, "Water level, maybe major rock structures or hills and valleys are still the same, although the trees are bigger and the ground surface may have changed due to erosion and such. You know, he'll ask me, 'Is there a mound or point over there?' when it would seem all he has to do is look over and see it. That was how I came to realize that he does not see the present reality when he is perceiving the past—it wasn't there in the time he is now seeing."

Having accomplished his aerial survey, McMullen says he "returns" to the point of view of his body and then often begins rapidly walking around the site landscape. "I used to try to follow with my recorder," Emerson recounts, "but couldn't keep up or get a decent tape. Now I wait until he returns and then we either talk or we walk along, at a reasonable pace, back over the ground, this time recording his impressions. It's as if he had two pictures, one present and one past, and has 'synched' them by triangulating certain extant geographical configurations and big trees that are still alive."

Once he begins, McMullen states that he can "see their campfires, their houses, sense the feelings people have ... whether they're happy, angry, or expectant, what a ceremony they may be performing means, and how they feel about it." He can even smell them. He has told Ann Emerson he can see colors and actually hear people talking. In fact, she remembers "he has told me, and I think Norman also, more than once that sometimes so many are talking that it takes him a while to sort things out, to know [which one] to listen to."

To both Ann and Norman Emerson, the role of McMullen's informants on problems involving on-site survey is still unclear. McMullen says he usually sees two people, whom he identifies by their voices as a man and a woman. In Ontario at least they generally are the same ones and, as Mrs. Emerson says, "If you ask George if they are 'dead' people he tells you they say, 'Who is dead? You are the ones who are dead because you are so unaware.'"

These informants are not the only or even the primary source of McMullen's information. "I don't know that we are totally clear on this," Mrs. Emerson says, "George is very frank and open when you ask him but it's hard to know what to ask or how to interpret what he is trying to tell you. Sometimes it seems there are just no words to say what he is trying to explain."

To Emerson, it seems that "George is getting information from several sources. First, he is overhearing and empathetically picking up the mood of the people he is viewing, with a sort of passive reception. The kind of thing you would hear and sense if you went to a dinner and just looked around and sat quietly listening. Second, he seems to be able to ask questions sometimes, and those answering the questions appear to be people who lived a long time ago. On one occasion, for instance, George and I went out to a site only to be met by what George said were six Indians who said, 'How come it took you so long to get here? Tell the professor between the six of us, we represent two centuries of history at this place. We believe we can give you the answers to any questions the professor is going to raise. . . .' Now that seems to me to be a case of a mediumistic phenomenon involving Indians who inhabited that site over a long period of time. I know it sounds weird but, through George, these people have given me pretty good answers.

"The third source are his informants, although I personally am not clear how much information they provide or whether they just sort of guide the process.

"Since I don't see all this, I really can't say what is actually going on and, to be perfectly truthful, beyond a certain point I don't care. It is for scientists with training different from mine to measure it; you need to know your limitations. My interest and expertise [lie in] archaeology."

When no on-site work is required and McMullen is simply asked to psychometrize an artifact, the process, although similar, becomes even stranger. Then, if asked for the artifact's site of discovery, the psychic says he goes out until he is "way up in the sky and can see the earth," from which vantage point he "knows" or is told the location, as he was in the case of the argillite carving which he described as having come from "southwest of Prince Rupert."

In other cases, he has told the Emersons, his informants have to "go" and get more information before answers can be given, and he himself sometimes has to go, out of body, back in time, to see the artifact's location. Once he has held an object, he can "always tune back in on it or go to the place where it comes from to get [additional] information."

With another of Emerson's psychic informants, Sheila Conway, the process is somewhat different. As Mrs. Emerson explains it, Mrs. Conway "actually goes into a light trance, although unless you were paying close attention, you wouldn't notice it." Once in this state, she says she concentrates and then sees before her eyes something like "a teeny-tiny film strip of the period in question." As she focuses on these pictures, much as Ossowiecki and Croiset focused, they begin to move and grow bigger, although, unlike with Ossowiecki, they do not seem to run backward—that is, from the more recent to the more ancient past until the desired epoch is reached—and then run forward. Instead, Mrs. Conway suddenly finds, as in a futuristic film, that she is "surrounded by what is going on, hearing it, seeing it in three-dimensional color, participating in it."

Unlike McMullen, she does not seem to converse with those in the retrocognitive sequence. Also unlike him, she has an acute awareness of their emotional state and the general ambiance of the place and the event going on. Perhaps because she has had considerable musical training, her ear is particularly attuned to nuances of speech and music and on one occasion she has reproduced, for Emerson to tape, a chant being sung at a sunrise ceremony.

While in this state, her voice, which Mrs. Emerson describes as a "very pleasant soprano," becomes "rather flattened in tone and is more halting—although, as with George, it is always her voice." She also can answer questions and seems to be able to control her point of view so that she can, for instance, see "either the inside or the outside of a building, although not both at once."

The other respondents Emerson used on his argillite carving team, Sandy and Jim, seem to have a more generalized talent, limited to picking up impressions or fleeting visualizations, although, the professor hastens to add "these were mystifyingly accurate."

By 1974 Emerson had begun what has become his most tedious and trying task—the transcription, cross-correlation, and comparison of his psychic information with traditional ethno-historic, geographic, climatological, cultural-anthropological, and archaeological sources. It is a mammoth task involving a collection of over two hundred tape recordings of psychic information. He also continues to collect new data. "If I had two assistants and no teaching schedule for a year, we'd make a good start," he says. But before he could even begin to plan that task, one of his papers, which were by now reaching even nonarchaeologists, caused a response that was to open up an entirely new and unanticipated chapter in his career.

V
THE CANADIAN AND
THE SEER'S SON

Emerson and Cayce

Hugh Lynn Cayce, eldest son of Edgar Cayce, the most famous psychic since Nostradamus, was in his sixties and had achieved most of his goals. He had taken the Association for Research and Enlightenment, which for most purposes had ceased to exist when his father died in 1945 (people could no longer obtain readings), and built it by the 1970s into a viable organization with thousands of members throughout the world. A multipress in-house publishing arm had been established and there was a million-dollar library and conference center that any small university would be pleased to own.

If researchers of academic parapsychology had never paid much attention to the Cayce material, others had; sales of almost ten million books about Cayce attested to that. So did the growing list of

Hugh Lynn Cayce, chairman of the board of the Association for Research and En-lightenment.

This rare early photo of Edgar Cayce and his wife Gertrude dates from the period when he began to provide his unusual and controversial chronologies and histories of Persia and Egypt.

ARE PHOTO

physicists, physicians of all kinds, geologists, and chemists who had become interested in using the readings in their research. Almost the full spectrum of American scientists was represented and through their efforts, particularly in the field of healing, almost irrefutable evidence of Cayce's accuracy had been established.

Only one achievement still eluded the seer's son. Although his father had spoken what became literally volumes of archaeological and cultural-anthropological data, no responsible effort by trained archaeologists had ever been directed at validating or refuting it. One of the major continuing commentaries in the almost fifteen thousand readings the ARE had on file remained a great question mark.

Although many periods were touched on, perhaps eighty percent of the archaeological material fell into fifteen epochs, and within that number, the majority were found in five time periods. In

Cayce's world view these seemed to be of particular significance, and for them he provided dazzling detail: legendary Atlantis; a century of Egyptian history; a contemporaneous period in the Mexican state of Yucatán and neighboring Guatemala; Persia at the time of Zoroaster; and the Near East at the beginning of the Christian era.

In all, these five periods seemed to constitute what New York University Professor Henry Bamford Parkes, in reference to a related but chronologically different sequence, once described: ". . .at rare intervals in history, factors in human affairs make for the emergence of novelties that cannot be satisfactorily explained by any acceptable theory of causation . . . considered together [this] appear[s] to constitute a group of mutations in man's spiritual development comparable to what happens when a new species emerges in biological evolution."

Hugh Lynn and a few others recognized this quality even as Edgar Cayce was delivering the information, but since they had no background in the prehistoric and historic topics discussed, they had little or no idea of how to develop the themes or how, while Cayce was still alive, to prepare for the major problems his information would later face. The situation was exacerbated by the manner in which the psychic's perceptions were presented. Except for a handful of readings devoted exclusively to such matters, most of his observations emerged piecemeal, almost as asides, over twenty-three years, so the composite picture was lost in a bewildering morass of individual facts. No attempt was made to tie them together, let alone correlate Cayce's words with more traditional sources; history and archaeology were far down the list of research priorities. It was not until several years after Edgar Cayce's death, when Gladys Davis Turner and Jean McCullough had indexed the fourteen million words of transcribed commentary, that even the outlines of the mosaic emerged. And it was not until years later, when the organization was fiscally viable, that genuine attention was paid to the real implications of Cayce's words in these areas.

However, by the mid-1950s Hugh Lynn was able to begin at least a preliminary effort. Two possible approaches presented themselves to him: The ARE could mount a dig itself or it could find someone or some other group willing to do so. The first option, Hugh Lynn realized, was impractical. He was not an archaeologist, and the ARE did not have the money to underwrite such an effort, or the ac-

ademic status to get the necessary excavation licenses involved even
if the funds had been available. This left only the second option, and
to achieve it he had nowhere to begin but with his natural con-
stituency, the ARE membership, particularly those with a professional background in science. For the next twenty-five years Hugh
Lynn tried to encourage any member who showed an interest in
testing his father's archaeology to do so.

Unfortunately very few of them were specialists in this particular
field, and even those who came from a related discipline such as geology usually felt that public involvement would cost them their careers. The only real response Hugh Lynn received centered on the
most controversial material of all, Atlantis, and he soon learned that
developing meaningful research on Atlantis was a highly problematical undertaking. It is difficult enough to verify archaeological data
dating back a few thousand years, and Edgar Cayce had said that
the most recent sinking of Atlantis was thirteen thousand years ago,

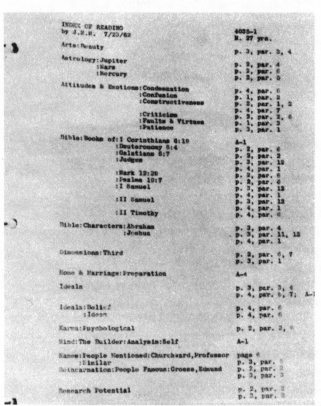

A typical Edgar Cayce
Reading Index. Con-
tained within hun-
dreds of such readings
are the sentences and
paragraphs which,
when taken together,
provide the conception
of world history lying
at the core of much of
the psychic's commen-
tary on man's history.

and that much of the evidence was now underwater. It was hardly the most practical point at which to commence archaeological research in the readings, particularly since no orthodox findings existed against which to measure and correlate.

In three other areas, Yucatán, Egypt, and Iran, the problems were somewhat different. Basically, they revolved around the fact that while Cayce's descriptions were internally logical and quite detailed, they made almost no immediate sense when compared with accepted chronologies and reconstructions for these locales. Also, with certain notable exceptions in the Middle East, no specific locations had been asked for or volunteered, and when there were some indications, these directions alone were inadequate to assure discovery. Consequently, should an archaeologist become interested in these tantalizing hints about the Middle East, there was very little for him to bring his skills to bear on.

It made a difficult situation for Hugh Lynn, or anyone else willing to study the Cayce readings seriously. The origin of the problem was not hard to understand when viewed in its proper context. To begin with, the Cayce readings were not primarily oriented toward producing such information.

Second, Cayce himself was unschooled and, in contrast with such clairvoyants as Rudolf Steiner, who was an academic and attracted academics, Cayce during his lifetime rarely drew scholars. The people who came to him were mostly small businessmen, schoolteachers, and householders, and generally they were in the throes of a physical or emotional trauma; Cayce's psychic advice was usually a last resort.

Third, those asking questions simply did not have the background to inquire intelligently about archaeological matters, even had they been so inclined; nor did they have the knowledge to follow up effectively and seek meaningful clarification of what was voluntarily given. As an example, Cayce described a site in Peten, Guatemala, on the Yucatán Peninsula, that could conceivably be the famous Piedras Negras complex excavated over almost a half-century by Professor Linton Satterthwaite. Unfortunately, being nonarchaeologists, none of the individuals posing questions had ever heard of Piedras Negras and so nothing that might have tied the two together was ever asked.

Fourth, while questioners had a layman's familiarity with Meso-American, Egyptian, and other archaeologically established chronologies, they did not have the detailed comprehension to see that Cayce's dates were almost completely different from what was often a very firm dating sequence. In fact, this question of chronology has been the major stumbling block to scholarly acceptance of any of the Cayce reconstructions almost from the beginning.

No criticism should be directed at the questioners, however, who made no pretense of being experts. It was simply one of parapsychology's tragedies: a great, almost unequaled psychic source died before any properly trained researcher could be found who was willing to ask him the right questions. Even when testable information was supplied, the probability of getting a hearing, let alone cooperation, from a foreign government or some accepted institution would have seemed the merest chimera to Hugh Lynn except for one thing. In the fifth major area his father had discussed, the Holy Land, under what amounted to triple-blind conditions, Edgar Cayce had pinpointed the location and described the life and interests of the Essene community at Qumran!

It is very unlikely that any of the scholars who ultimately became involved with the Dead Sea Scrolls or the excavations at the sites where they were found had any idea when they began their work, in 1948, that their research would validate the words of an unschooled but clairvoyant Kentuckian. Yet eleven years earlier Cayce had anticipated what they would discover as he lay in a sleeplike state on his couch in Virginia Beach.

He had, in fact, begun talking about a group he called the Essenes as early as 1934, but those around him paid little attention to his comments, mostly because they themselves knew almost nothing specific about the Middle East, let alone an obscure religious community located there. Nobody seems to have cared enough to even ask Cayce where this group was located. Fortunately, on Tuesday afternoon June 22, 1937, Cayce told them anyway. He placed the community "on the way above Emmaus to the way that 'goeth down towards Jericho' and towards the northernmost coast from Jerusalem"—that is, along the road running east from the town of Emmaus (known today as Imwas) to the Dead Sea, where the road turned north to Jericho; the entire area within land taken by Israel from Jordan in the 1967 war.

Nor was this the only evidential statement made that day. The sixty-two-year-old woman for whom Cayce gave the reading was told that she had been an Essene woman during the life of Jesus, and had run a school at this location.

The ARE members listening to this reading had small appreciation of the implications of Cayce's words, but had it had been a matter of burning interest to them, they still would have been hard pressed to learn more because, in 1937, there was almost no definitive information on the Essenes. A considerable number of scholars held that the Essenes were either a myth or, at best, a small inconsequential schismatic branch of Judaism, a band of monastic celibates. Even those researchers who were willing to admit that the Essene community had existed as a group of any size would not have accorded it the status of major influence or teaching center. And the most radical academic would have rejected the idea that the Essenes numbered women among their members in any position, let alone as significant contributors to the community's intellectual and spiritual life.

As to the location Cayce gave: Had a Near East scholar been at Virginia Beach, he would have said that the only site corresponding to the directions given was Khirbet Qumran, where a small ruin had been known about for years. Since it was believed to be the remains of a Roman fort, he would hardly have credited it as the coed community of mystics described by Edgar Cayce.

These opinions were destined to change when sometime in the early spring of 1947 (the exact date has never been determined) a fifteen-year-old bedouin of the Taamirah tribe, named Muhammad adh-Dhih (literally "the Wolf") and several friends went up into the caves above Qumran and there discovered jars, many of them broken, filled with leather-wrapped linen cloth. Thus the Dead Sea Scrolls reentered history.

As excavation proceeded and the scrolls were studied it was soon apparent that Qumran had indeed been an Essene community. Its use as a fort came later, after the Emperor Vespasian's legionnaires overran the site about A.D. 68. Women were members, it is now generally agreed, and as the Scrolls themselves demonstrate and Cayce had stated, the Essenes had a strong interest in astrology.

Armed with this unequivocal proof of one of his father's psychic hypotheses, Hugh Lynn asked: If Qumran proved out, why not also Egypt and Iran? Surely researchers must now give the material a hearing. But even with this clear proof he found doors not only closed but locked. "When I showed it to American archaeologists they still wouldn't believe it. They either just wouldn't talk, or they politely, and sometimes not so politely, implied I had made it up." Gradually, though, after overseas trips that began as early as 1952, Hugh Lynn found a few individuals in the academic world of the Middle Eastern countries who were at least willing to listen. It wasn't easy, however, and in Egypt he found his task to be especially formidable.

Beginning in the 1920s, almost with Edgar Cayce's first words on any topic other than physical/medical matters (which almost exclusively dominated the readings fot the first twenty-one years he gave them), the psychic had talked about Egypt. It was perhaps the richest reconstruction Cayce ever supplied.

Stripped of its deep spiritual overtones, which for all but archaeological purposes are the real heart of this data, it told the story of a man named Ra-Ta, a priest who, along with a tribal people, migrated to the Nile Valley from the Caucasus Mountains. A range lies between the Black and Caspian seas, largely in what is now Russia and traditionally has been considered the geographic border between Asia and Europe. Ra-Ta, who is the critical personality in Cayce's entire Egyptian saga, was apparently not a man of the Caucasus people but had joined with them and their ruler Ararat to fulfill what Cayce indicated was a mission of personal testing. Ra-Ta appears as a powerful and yet secularly involved religious leader on the order of Islam's founder, Muhammad.

In what is now Egypt the immigrants met and conquered an even more primitive people, although according to Cayce these native Egyptians had the most sophisticated culture in that area, which is the reason they were selected by Ra-Ta as the most potentially fruitful group with whom the Caucasus group could involve themselves. In physical description the native group seems similiar to the Kalahari bushmen, whose origins science still does not know beyond the fact that they migrated south from northeastern (Egyptian?) Africa at some prehistoric time.

The ruler of the native Egyptian people was an elderly man named Raai, and he, like Ra-Ta, seems to have had a higher vision; consequently, although there was some resistance, the conquest was largely bloodless. Raai retired, leaving Ra-Ta to merge the two cultures in an attempt to create a new culture which recognized the worth of the individual.

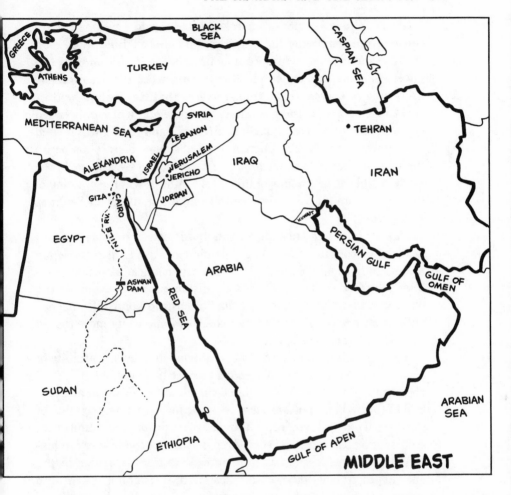

At about this same time, according to Cayce, a third and more advanced group migrated into the area, remnants of an already ancient Atlantean civilization. After a period of competition with the Caucasus immigrants most of the Atlanteans also were willing to join forces and a racial and cultural triumvirate developed. Despite continuing opposition from a group of Atlanteans who wanted to establish a sort of feudal system over the native Egyptians, all three peoples began to build a new society with Ra-Ta as the spiritual leader and the Caucasian ruler Ararat as the secular head. When jealous factionalism continued to threaten this budding effort, Ararat magnanimously passed on his power to his sixteen-year-old son Araaraart.

There followed a period of relative peace and development resulting from combining the best from all cultures. But all too soon intrigue and dissension broke out again, causing Ra-Ta and 231 followers to retire into exile. This further disturbed the situation for, according to Cayce, the various factions that Ra-Ta had held in check rose up and a nine-year civil war broke out, which only ended when the populace and all factions tired of it and Ra-Ta was invited by Araaraat to return as unchallenged absolute ruler. Thus began the culminating chapter of the saga. According to Cayce, one of the major visible achievements of this final synthesis was the building of the Great Pyramid with Hermes as its architect and Hept-Sept as supervisor.

Cayce's tale is an amazing and fascinating story running to thousands of words. Just the telling of it piecemeal, over decades, while maintaining complete internal consistency, is a feat almost as troubling to orthodoxy as any of the long-distance medical diagnoses the psychic made. Unfortunately, it had almost no points of tangency with what orthodox archaeology considers one of its most tightly pinned down reconstructions.

To begin with, there is an almost insurmountable discontinuity of chronology. Edgar Cayce placed his reconstruction at about 10,500 B.C., from which time he also dated the building of the Great Pyramid of Giza. This is inconsistent with the majority interpretation of the archaeological record, although, significantly, a number of scholars do postulate an earlier, predynastic, epoch of Egyptian history. Cayce also does not seem to agree with records written by the ancient Egyptians themselves. Scattered throughout Egypt are stelae, plaques, and wall inscriptions—literally thousands of them—with specific references to the length of a ruler's reign, the duration of a campaign, a description of the number of years it took to build a tomb or construct a temple. Many were carved or painted contemporaneously with the event described. This reportage of who did what, and when, is not particularly unusual; there is, for instance, hardly a federal building in the United States without a bronze or granite tablet carefully listing the date built and the then-in-office U.S. President.

By adding up all these Egyptian dates using what Egyptologists call the "minimum date method" (that is, the least length of time a reign could have endured, or campaign have lasted), a date for the first dynasty (as opposed to the predynastic "Time of the Gods") of approximately 3400 B.C. was arrived at. This figure was further refined by astronomers. These specialists, working from the careful Egyptian star observations also included in many inscriptions, determined, by computing the ascension of the star Sirius on Thout I (Egyptian calendar) with July 19 (Gregorian or modern calendar), that the 18th Dynasty began in 1580 B.C. ± 3 or 4 years, thus lending additional support to the emerging chronology.

This sequence received further validation as the discipline of archaeology developed. Eventually the techniques of stratigraphy (the examination of accumulated material in relation to geologic strata), seriation (the chronological arrangement of artifact material using some logical principle other than that things are simply on top of one another), carbon-14 dating of organic material, dendrochronology (a chronology based on tree rings), and pollen analysis were brought to bear on the problem.

Archaeology thus produced from many sources what many researchers believed to be a clear picture of civilization springing from a beginning in approximately 4500 B.C., with the first dynasty taking power over a united Upper and Lower Egypt about 3100 B.C. Toward the end of this dynasty, about 2890 B.C., the primitive use of masonry began. Before that date the people were presumed to be, first, hunter-gatherers, and then villagers living in buildings in some respects similar to Southwestern Indian adobe houses. According to this chronology, the Great Pyramid at Giza was built in approximately 2600 B.C., or just under forty-six hundred years ago, by the second or possibly third king of the Fourth Dynasty—Khufu (called by the Greeks, Cheops).

There is no mention anywhere in the official chronology of Ararat or any of the other individuals in the Cayce reconstruction, but there is clear material evidence of the existence of Khufu and seemingly impregnable testimony that the Great Pyramid was built during his reign.

This revelation came in 1837, when Colonel Richard W. H. Vyse rigged demolition charges in the ceiling of a low space immediately above the King's Chamber of the Great Pyramid and successively blasted his way upward through four other spaces. Known collectively to Egyptologists as the Relieving Chambers, all but the lowest one—from which Vyse began—had been built without an entrance.

Their sole function was to carry the immense superstructure weight of three hundred feet of masonry out laterally, thus saving the pyramid's most important void from being crushed.

Vyse's methods were crude but effective, and when he was through, he had revealed four rough limestone rooms with granite roofs, roughly thirty-nine inches high. Not terribly interesting. But of enormous interest and value were the quarry marks clearly daubed on some of the blocks. They were interpreted as indicating that the stones in these hermetically sealed rooms had been placed there in the seventeenth year of the then thirty-three-year-old Khufu's sovereignty. Accordingly, the pyramid would have been built in the last third of Khufu's twenty-three-year reign—or almost eight thousand years later than the date given by Edgar Cayce. It is the most devastating single piece of evidence against the psychic's Egyptian material.

All this was gradually, and sometimes pointedly, brought to Hugh Lynn's attention. Logically he should have given up, but he believed in the material and for more than thirty years he had seen it prove out in other fields against great odds. Besides, he had a trump card. His father had given the specific location of a spectacular site—a hall or chamber of records that the psychic insisted would provide evidence of both Atlantis and the Ra-Ta story. It lay, he said in his sixteenth reading for one man, "as the sun rises from the waters—as the line of the shadow [or light] falls between the paws of the Sphinx . . . between the Sphinx and the river." If his statement were true, a single excavation could rewrite the history of the world! That promise was enough to salve the wounds left by an unlimited number of rebuffs.

If anything was going to be done about testing the archaeological material, however, Hugh Lynn needed more than the psychological support of past successes in other fields. He needed allies. In 1972, after lending a hand to a man in his twenties, he got them.

Mark Lehner had been an antiwar activist counterculture commando. By the seventies, though, like so many of his contemporaries, he had not so much dropped as burnt out. In an attempt to sort through things he made contact with the ARE, and from it learned of a tour leaving that summer for, among other countries, Egypt. During the trip he found the purpose and challenge he sought.

"For me it was not a case of becoming a 'true believer.' I was interested in the philosophy, that's true, but I also thought it was time to open this material to the outside world. . . . To put it to the same tests as any other body of scientific data, however it was derived. I also admired the fact that Edgar Cayce in effect had given us a substantially validatable clue and that Hugh Lynn was willing to, in effect, stake the Edgar Cayce Egyptian story on the turn of that single card. Imagine what it would mean if he was correct!"

With the help of the ARE, Lehner went to Egypt, learned Arabic, enrolled in the University of Cairo, and began the tedious process of making himself an academician. All this done so that he could work from the inside instead of just tossing in comments from the perimeter. Soon he had made contacts with archaeologists Hugh Lynn had already met, and arranged for others to meet Hugh Lynn. If these men did not embrace the Cayce readings, neither did they dismiss them out of hand.

In 1973 Lehner published a monograph, *The Egyptian Heritage,* based on his studies to that date—the first responsible research ever done on Cayce's Egyptian material. This complex manuscript wove back and forth between the readings, and more traditional sources, which made it difficult for many ARE members to follow. But according to one reader, "It was the first thing I had read about Cayce [that] wasn't the usual pseudoscience you find when the psychic is involved."

Richard Roche (a pseudonym he uses in his writings about Cayce to protect himself from retribution) is a doctoral candidate in the anthropology department of a major California university and had never even heard of Edgar Cayce until a few months prior to the publication of Lehner's work. He began his exposure to psychic material as a social scientist/observer of the dynamics of a Cayce study group in preparation for a paper on the sociology of religion. After one meeting someone handed him a copy of Lehner's book.

Roche became as intrigued as Lehner with the internal consistency of the Ra-Ta saga. It was difficult to dismiss a story that had been doled out by sentences over some two decades and yet contained few contradictions. Lehner had put these hundreds of reading extracts into a logical and chronological order, for the first time creating a reconstructive sequence. Roche wondered if it would be possible to take that sequence and, using established research techniques, tie it into the traditional sources archaeologists accepted—in other words, complete the process Mark Lehner had begun.

Motivated by simple curiosity, he started correlating the readings with classical literature and Egyptian myths. Soon he realized that this project he had begun casually enough was "a genuine challenge; there was something there." Roche was particularly struck by the resemblance of the Ra-Ta material to Plutarch's account of the Osiris legend. He also examined the traditional chronology "without any of the preconceptions that dominate the orthodox, largely British, view of Egyptian history." Since he was training in the multi-discipline anthropological archaeology that predominates in the United States rather than the single-focus art history of classical archaeology, he began studying the Cayce story from a linguistics frame of reference, making a survey of possible alternative interpretations of certain Egyptian hieroglyphic symbols.

For over a year Roche was immersed in his studies, totally unknown to the ARE. Finally he surfaced with "the general outline of an alternative hypothesis, or a reinterpretation of accepted existing sources, in which the Cayce chronology and the Ra-Ta story fit." He then decided it was time to talk to someone in Virginia Beach.

Violet Shelley, in the summer of 1974, was editor of ARE Publications, with an office on the third floor of the old Cayce hospital situated on a well-cultivated sand dune overlooking the Atlantic Ocean. "I looked up one day and saw a man walking across the parking lot carrying one of those knapsacks students seem to consider part of their uniform. Five minutes later he walked into my office. He was extraordinarily handsome, just like Lehner, and even more intense. He sat on the edge of a chair and began to pour out this story of what he was doing . . . all accompanied by many photographs and drawings of hieroglyphics.

"He told me he had hitchhiked all the way across the country to see me, and I felt bad about that because frankly I didn't understand most of what he was saying. But I was sure of one thing: here at last was someone to finish the job Mark Lehner had begun . . . another chance to tie the Cayce material into some real research . . . to open up the door to accepted science. Except for the medical material, we'd never had that before. I told myself, 'You can't let this man get away without getting him to agree to write it as a booklet.' "

After an hour of conversation Roche agreed, provided he could use a pseudonym; Mrs. Shelley was only too happy to agree. Having settled this point, she asked him if he wanted to meet Hugh Lynn, or perhaps Gladys, Cayce's lifelong secretary (they are both considered tourist attractions). Or maybe he would like a tour of the library or the vault where the original readings were stored?

"He said, 'No, thank you anyway,' that he'd better get back to California . . . and a few minutes later I saw him walk back across the parking lot to Atlantic Avenue and stick his thumb out. All the way from California, here an hour, and on his way back. It was the most extraordinary thing that ever happened to me the whole time I was editor up there."

After another year of work, during which only a few letters passed between them ("his mostly limited to one or two sentences"), Roche fulfilled his previous summer's commitment. *Egyptian Myths and the Ra-Ta Story,* as it would be entitled, arrived one morning on Mrs. Shelley's desk through the mail. It was an even more complex and densely reasoned manuscript than Lehner's effort. Its fifty-three pages included many illustrations, over two hundred citations, and an eighty-two-reference bibliography. And these were not pseudoscientific references, but exactly the same sources a doctoral candidate would submit to his committee along with his dissertation.

When Roche apologized for "the very preliminary nature of this brief manuscript" and explained that he intended to continue his research, Mrs. Shelley found his attitude amusing. She was concerned that no one but another professional in the field could read it. Roche's *Ra-Ta Story* was the most scholarly writing the ARE had ever published.

Essentially Roche had attacked the problem of the Cayce material at its three most apparent discrepancies from the standard Egyptological reconstruction: chronology, lack of any other record of such names as Ra-Ta, and the building dates of the Great Pyramid and the Sphinx. In response to these challenges he had first reinterpreted the standard classical sources every other Egyptologist used and then advanced the theory that what had previously been considered myths—the predynastic "Time of the Gods"—might, in fact, be accounts of living people simply garbled over time, since the tales were ancient when the first known dynasties began. He suggested that the first dynastic period was really a *re*unification of Egyptian civilization and not its beginning. The approach was not as novel as it sounds; indeed, there was a successful precedent—just the sort of success to make Roche's arguments carry weight.

In the nineteenth century, against almost unanimous opposition, the German-born Heinrich Schliemann had taken a very similar approach. Convinced that the Homeric poems were not myths but the recounting of actual events, places, and people, he had used them to find the city of Troy, as well as several other controversial locations. Although Roche never mentioned the parallel, certain other archaeologists, who read his work somewhat uncomfortably, were all too aware of it. Perhaps the mythical age of Osiris, Horus, and Set was real.

Roche's research also turned up the fact that he was not the first to walk back through the maze of Egyptian myths. The well-respected German scholar K. Sethe had also done this and, to Roche's elation, had concluded, "There was a predynastic foundation to the Egyptian culture." Wondering why Sethe's commentary on this subject was not included with his other work when it was presented at the university, Roche finally concluded that "since it didn't fit the accepted scheme, it was ignored. Not even subjected to criticism—just avoided altogether."

It was with the names themselves that Roche began his next step, and again using standard sources, he offered a reinterpretation. In the case of Osiris, for example, Roche was inclined to believe that because of the way in which the hieroglyphic for this name was formed by a scepter and an eye, "Osiris does not have to be translated as a proper name but rather as a title that denotes 'foresight' or the abilities of a 'seer.' The scepter denotes royal office and the eye a psychic ability, so the literal translation would be 'the Royal Seer.' In all probability Osiris is the Hellenized form of that holy office's title." Looked at from this point of view, Roche felt a very good case could be made, by virtue of the parallelisms in the two stories, that Osiris and Ra-Ta were one and the same, and an actual person.

Only in the case of the construction dates did Roche seem to have difficulties, and although he was convinced that "the case for chronology in Egypt is nowhere near as strong as most Egyptologists would have you believe," he admitted that the inscriptions naming Khufu, which were found by a man who could not read them in the Relieving Chambers, did "cause some real problems." He also acknowledged that "the proof of my or any other theory lies in the evidence produced by actual excavation."

To this end, he hoped that the "hall of records" information would "be acted upon," although, since "that would be a major undertaking," a first effort might be carried out at either of two other sites: the ancient first capital city of Egypt, Heliopolis, or, even better, the temple site known as Jebel Barkal. The first location, Heliopolis, was established as the original capital of Egypt by Sethe in 1908, but despite almost universal agreement about its importance, it has never undergone a major excavation. This, however, would be an enormous project, so Roche felt Jebel Barkal would be the better choice.

"As the cultural center of Kush (located immediately south of Egypt) and the site of what may be a very large temple—for these reasons alone, forget about the Cayce material—Jebel Barkal is worth excavating. It has been known about since 1949, but there never seems to be either funds or the interest to do the job," Roche said, and such archaeologists as A. J. Arkell seemed to bear out his judgment: a temple "of the style of Abu Simbel* and of even greater size," was the way Dr. Arkell put it.

Roche was of the opinion that opening the site might provide clear "evidence that Cayce's 10,000 plus chronology was in fact correct, and that the 'gods' and 'goddesses' of the predynastic period were real men and women. What I have come to believe is that both orthodoxy *and* Cayce are essentially correct—excepting the traditional dismissal of the "Time of the Gods" as myth. What we really have is two *separate* epochs, both with actual events. Cayce, by the way, implied this was the case but the point was never clarified by questions." These convictions only increased his frustration at that lack of interest in excavating Jebel Barkal. "The dig could hardly be simpler. A child could do it. All that is called for is the opening of the entryway to the temple. An excavation into the hill between the central colossi would, in the opinion of myself and other researchers—who know nothing at all about the Cayce angle—lead to the temple entrance, and possibly major discoveries."

With the work of Roche and Lehner, the ARE position was considerably strengthened, for these two men had provided for the first time a link between the orthodox and the psychic. But, as always, it all came down to digging, and in the case of the "hall of records,"

---

*Abu Simbel is the fabulous temple complex, famous for its enormous seated figures of Ramses II, which, thanks to an international fund-raising campaign, were cut out from their rock base and raised to preserve them from inundation by the Nile backwaters after the completion of the Aswan Dam.

the Cayce information was not adequate to meet the task of location. Another psychic would have to lead the way. Hugh Lynn began to search for such an individual, but before he found him, another chain of events, which would affect his Egyptian efforts, began to play itself out.

At about the same time in 1973 that Lehner and Roche had begun to develop the Egyptian material, Hugh Lynn had opened an Iranian front, "not with the idea of digging" but of putting "this information into the hands of a reputable archaeologist in Iran, prior to the time any future excavations are carried out there." To this end he was able to arrange a correspondence introduction to Professor Ezat O. Negahban, senior archaeologist at the University of Tehran and, for his excavation of an Elamite city mentioned in the Bible and dating to about 3000 B.C., the most famous Iranian researcher in his discipline.

What led Hugh Lynn to this effect was again that as for Egypt, Cayce had supplied a complex and fascinating story for Persia.

The time is an undetermined number of years after 10,500 B.C. but before 7000 B.C. Egypt is the nexus of civilization; her culture dominates the world. The Great Pyramid has been built and the library at Alexandria established. Something resembling peace prevails, with one exception.

In what is now Iran (although parts of the Arabian Peninsula as well as parts of the Aegean and eastern Mediterranean seem to be included in this term) primitive, largely nomadic tribal groups war both with each other and, to the irritation of the civilized world, on travelers and traders crossing between Asia and the Mediterranean. Only one leader seems to have recognized authority over the land as a whole—Croesus—and he, after a number of the nomadic tribes band together, is overthrown.

Finally, a man known as Uhjltd arises, takes power, and with his wife Llya, a niece of Croesus', establishes the "city" of Toaz or Is-Shlan-doen, or as it is most often referred to in the readings, "The City of the Hills and Plains." It is not really a city but a collection of semipermanent tents clustered near a spring that Uhjltd develops into a place of healing. Nearby is the more permanent village of Shushtar.

Uhjltd and Llya have two sons, Ujndt and Zend, and the family, using Uhjltd's Egyptian training in administration, turns the city into a major religious, commercial, and healing center. So great are its attractions that the Greeks become covetous and overrun the city. At about this point Uhjltd dies and after a short period of coauthority his sons Ujndt and Zend have a falling out that leads to the withdrawal of Zend and his people for a time. Eventually, they both return and play prominent roles in the city's affairs: Ujndt handles civil matters and healing, and Zend oversees religious concerns. Some local autonomy is maintained, although the Greeks are the ultimate rulers. Later Zend fathers a son whom Cayce calls the first Zoroaster. Whether this individual or a later one begins the religion we know as Zoroastrianism is unclear.

Again, a fascinating story, but one at variance with traditional history as it is supported by the majority interpretation of ethnohistoric and archaeological discoveries. But there are some beguiling clues in Cayce's Persian commentary that are not present in the Egyptian material. To begin with, there are historical personages. A king by the name of Croesus, for instance, is an established figure. Solon speaks of him, as does Herodotus and Plutarch. So too is Zoroaster. There unquestionably was also a Greek invasion. The problem is that Cayce's chronology is again out of synch.

The fact that Uhjltd is unknown really does not matter; the identity of many lesser Persian rulers is lost to us. What is important is that Cayce states that he came to power by allying himself with local tribes and overthrowing a king who lived in a fabulous palace made opulent by his great wealth. The psychic calls him Croesus II. Traditional historical sources record only one Croesus, and his story is almost point for point the same as the one outlined by Cayce. The historical Croesus did have great wealth, he did live in a fabled palace city, Persepolis, and he was overthrown when Cyrus the Great aligned himself with local bedouins, one of whom could easily have been named Uhjltd. The difficulty is that Cayce places this overthrow between 10,000 and 7000 B.C., whereas classical sources provide a firm date of 546 B.C.

Strangely enough, a case *can* be made for Cayce's dating in relation to Zoroaster. He states that there was more than one religious leader by that name but that the one who founded Zoroastrianism should be dated about 6600 B.C. This disagrees with the dates usually presented by Western scholars today—either 630–53, 628–551, or 618–541 B.C., and there is evidence to support a sixth-century date. An ancient Iranian oral tradition, as interpreted in the West, states that Zoroaster lived 258 years before Alexander; the assumption is this means 258 years before the Macedonian conquerer overran the Persian capital of Persepolis in 330 B.C.

But under it all is the compulsion for an elegant solution. If Zoroaster can be dated to this period, then most of the major world religions (excluding, of course, Christianity and Islam) were founded almost within a single century, during what historian Karl Jaspers calls the Axial Period, roughly the eighth to the sixth century B.C. Few people besides theological historians seem to realize that Confucius (555-478 B.C.) and Buddha (567-487 B.C.) were almost exact contemporaries of Zoroaster, as was, according to the best approximations, Lao Tzu, founder of Taoism, and Mahavira, who is the most probable founder of Jainism.

As Professor Parkes notes, "In Palestine the line of monotheistic prophets which had begun with Amos of Tekoa halfway through the eighth century, reached its culmination near the end of the sixth century with Deutero-Isaiah. Among the Greeks the sixth century saw both the inauguration of philosophical speculation with the work of Thales and his successors and the establishment of democracy in Athens."

So powerful is the conjunction of coincidences that Professor Parkes felt obliged to make the statement cited earlier, in reference to Edgar Cayce's world view, about certain periods being a time of spiritual mutation as real as a biological one.

But a strong case can also be made for a date agreeing with Cayce's; that is, about 6600 B.C. Frank O. Adams, a retired Army colonel who has spent some two decades in a very intense and detailed study of certain portions of the Cayce material and their relationship to traditional sources, began outlining this position as early as 1964.

"When Zoroaster lived is a moot question. His era is lost in the shifting sands of the east. Hermodorus and Hermippus of Smyrna both place him five thousand years before the Trojan War, and these Greeks were certainly closer to the event than we are today. The same can be said for Xanthus, who puts him six thousand years before Xerxes, or Eudoxus and Aristotle, who say he lived six thousand years before the death of Plato. To my mind, Aristotle's position is particularly important, for if we place Zoroaster in the mid-fifth century, as many do today, this puts his death just about two

ARE PHOTO

This little-known photograph shows Edgar Cayce's study at his Arctic Crescent House in Virginia Beach. From this room, a man with little formal schooling slowly elaborated, sentence by sentence over a course of years, a history of the world which, while internally consistent, was sharply at variance with that proposed by formal academic studies. As strange as his words sound, in an increasing number of instances it is his version of the past which has proven ultimately correct.

hundred years before Aristotle's own passing, or less than that from when Aristotle made his estimate.

"I find it hardly creditable that one of the outstanding intellects in all of man's history could have placed one of the world's great prophets at 6000 plus B.C. if, in fact, this man who had influenced hundreds of thousands of people, even created a state religion, had actually died less than two centuries earlier. I just don't see how such an error would be possible."

The followers of Zoroastrianism agree. They believe, as Dr. Framroze Bode, scholar and high priest, states, "Zoroaster was *not* a sixth century figure. He was of 6000 B.C., and arguments to the contrary are based on incorrect interpretations of ancient records and teachings by those who will not or cannot go back to the original sources."

The differences between the various reconstructions, obviously, are far more involved than just an argument over dates, but an open-minded conservative position—always a good starting point where the psychic is concerned—seems to be that we are ignorant of the background of either the faith or its founder.

Because there is such confusion, it is possible to offer a third hypothesis, one that reconciles the traditional material with its seeming contradictions, and still considers what Cayce had to say on the subject. Suppose there really were two figures named Zoroaster (or that some close name variation existed that later got confused), one who lived around 6600 B.C. and another who lived in the sixth century B.C. It is even possible, as Roche postulated about Cayce's Egyptian account, that the name and the title—Zoroaster—became one in the sixth century when the religion was at its zenith. It would have been a practice not so very different from our calling the Pope today the heir of Peter, or referring to Peter's throne. Given this reasonable postulate, orthodoxy is mollified and the Cayce story begins to make good sense. Confusion now could be explained as caused not by inaccuracy on the part of the psychic but by the lack of sufficient background in his questioners for them to see the problem and to ask for clarifications.

A determined skeptic could defend another proposition: that both the Egyptian and the Persian accounts came from Edgar Cayce's subconscious mind. In the case of Egypt, while this is a highly tenuous position because of the internal consistency, details, names, and pinpointing of the location of a "hall of records," it is not an impossible one—at least until excavation proves Cayce erroneous. Certainly a rural Kentucky schoolteacher would have known about pharoah; Egyptology has always been archaeology's most flamboyant daughter, and stories of its discoveries, even in Cayce's childhood, had penetrated the educational hinterland of the United States. There is also the Bible, with its exotic tales of bulrushes and parting seas. Without doubt Edgar Cayce's unconscious *had* been exposed to the spore of Egypt. This plus the fact that his story exists largely in a vacuum makes charges of fantasy all too plausible. Even Richard Roche's careful analysis does not entirely dispel the miasma of confusion, as he is the first to admit.

But the Persia tale is different. A Greek invasion, fortified palaces, Zoroaster, warring nomadic tribes—except for the name Croesus, limited to an anecdote—it is all very unlikely knowledge for a man of Cayce's training. And how can one explain Cayce's description of a *caravanserai* city near or perhaps commingled with Shustar, described as an ancient city beyond whose borders in the hillside there were caves and an ancient *well* or *spring associated with healing?* Where would Cayce have learned such things even if he had set out to deceive? From the tiny summer-resort-oriented public library of Virginia Beach in the 1930s? And having learned them, how could he have kept all the details straight through more than a decade? More than that, although there are cases where Cayce was wrong, in the three-quarters of a century since he began his readings in 1901 to the present day, thirty-three years after his death, there has never been a shred of evidence to lend credence to the idea that he was a trickster. Even those who felt there must be a nonpsychic explanation of the readings after examining the man never questioned his personal integrity.

Ultimately it all came back to digging. There were many testable points of geography, and many descriptions and locations of archaeologically retrievable artifacts and records. Hugh Lynn felt he had a good case to make, but realized its limitations all too clearly when, in 1973, he first contacted Professor Negahban.

Having initiated contacts in both Egypt and Iran, he was faced with taking the next step: validating the material, at least tentatively, so that someone like Negahban would feel there was some supporting evidence. As Hugh Lynn said, "How could we ask him even to consider it—the money, time, and energy—without something else to prove it?" The answer, in stages, began to develop in his mind.

First, since he was not an archaeologist himself, he needed an archaelogist willing to consider what the readings said. Preferably this person would be a professional who had stature in his peer community; ideally, he would be an archaeologist who had already worked with psychics. Second, he needed to locate another psychic, an individual with proven archaeological capabilities. It would be most desirable if these two people, the archaeologist and the psychic, had already worked together. But where would one find such an exotic combination? Hugh Lynn had never met either person singly, let alone working as a team, and he had been in the psychic research field and looking for many many years. In 1974 his luck changed. By then Norman Emerson had published several articles and delivered a number of papers. He was on record as engaging in and supporting psychic research, and reverberations from these statements had sent pulses along the parapsychological community's network all the way to Virginia Beach. And Norman Emerson got a chance to pay back his debt "to the material that got me started in psychic archaeology in the first place."

On Sunday, October 20, 1974, Hugh Lynn and Emerson met. During their conversations the statements Emerson made about his friend McMullen seemed to promise everything Hugh Lynn had hoped for. But after years of checking psychics he knew that individuals capable of providing the level of detail he needed were very few. He had no choice but to ask Emerson if McMullen would be willing to undergo a test, and was pleased to hear the professor reply that he thought George would be more than willing. Further discussions then and later established the experimental protocol.

The material would be of Cayce's choosing, the place McMullen's home at Nanaimo, British Columbia, and the date June 16, 1975. Accompanying Hugh Lynn would be Arch Ogden, president of the Edgar Cayce Foundation, another of the nonprofit organizations set up, along with the ARE and Atlantic University, to work with the readings.

"I took a set of eight groups of sherds," said Hugh Lynn, "pieces of pottery that Arch, Charles Thomas [Dr. Charles Thomas Cayce, his eldest son], and I, independently of one another, had picked up in various parts of the world. We grouped them A through H—eight sets of these, wrapped them all up in plastic bags, put A B C on them, turned on two tape recorders and said, 'George, what do you get?' "

To a group of sherds that Hugh Lynn later found out were from Tiberias, on the shores of the Sea of Galilee, a city that has changed hands and that was the only city associated with any of the eight groups on the table, George got, ". . . and the city has changed hands many times. It is a city on a large inland water."

He also correctly located a site on Mount Carmel and gave specific details on terrain, animal and plant life, buildings, dress customs, and lifestyles. In all, George "hit" four groups with uncanny accuracy, giving details that were unknown to those present and that later proved to be correct. Even without these postexperiment corroborations, Hugh Lynn heard enough that day to realize his search was over. He not only had an archaeologist and a psychic, but a

ARE PHOTO

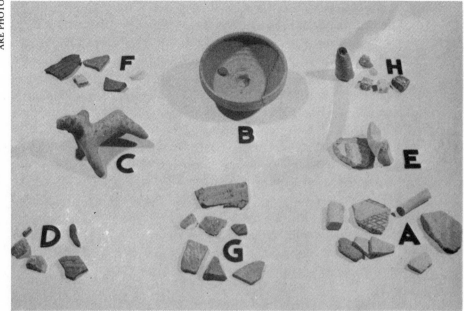

It was George McMullen's correct identification of many of these sherds which convinced Hugh Lynn Cayce that he had, at last, found another sensitive who might provide additional psychic guidance to augment his father's words.

proven psychic archaeological research team, just the combination he needed, the one for which he had so painstakingly looked. He asked both Emerson and McMullen if they would be willing to go with him and Ogden to Egypt and Iran. Equally intrigued, they agreed.

Back in Canada, McMullen had second thoughts and told the Emersons that he felt the Middle East was not "his element," that he had "an intense feeling of loneliness and many misgivings." Whatever George did to overcome these feelings he has never been able or fully willing to explain, even to the Emersons, but finally he announced that he was ready. Meanwhile Emerson had arranged for a leave of absence and Hugh Lynn put a traveling party together. On Tuesday, October 7, the Emersons and the McMullens flew to New York to join a group that at various times over the next twenty days was to number from ten to fourteen. It included Lehner and his Egyptian wife, who lived in Egypt.

For Emerson, the Egyptian arrangements were something of a letdown; they had been made by a nonscientist and there was, in the professor's eyes, "no way I could call it a scientific expedition as I had always meant those words." Still, both he and George entered into the proceedings with good spirits, which were considerably tested as the days went by. The unstructured approach was difficult to accept for a man of Emerson's self-discipline. He began to feel "like a spare wheel. I was an archaeologist but there was no archaeology going on, at least not as I know it." Worst of all, once they were in Egypt, a tension seemed to develop between Emerson and McMullen; conversations between them almost ceased. Finally, able to tolerate it no longer, Mrs. Emerson extracted from McMullen that he "was being subjected to his own frustrations."

The psychic told her again of his feelings that Egypt and Iran were not his natural "territory" and that his paranormal perceptions were being controlled. He explained that he felt "as if I

ARE PHOTO

Like Edgar Cayce, the Canadian psychic George McMullen feels Egypt has a past extending farther back in time than the chronology usually advanced by orthodox researchers. Only excavation of key spots he and Cayce have pointed out may finally settle the question. Here George (left) provides psychic insights for ARE Chairman of the Board, Hugh Lynn Cayce.

were walking down a corridor and, only for brief periods, am I allowed to see what lies beyond." The Great Pyramid particularly disturbed McMullen. After going into the King's Chamber for only a few moments he said, "It was as if the top of my head was being pulled upward and I knew that if I didn't get out immediately I would lose all my psychic abilities."

Hugh Lynn, for his part, suddenly realized there had been a major misunderstanding: The Egyptian portion of the trip had never been planned as more than a survey and get-acquainted trip; no experimental controls, or even technical discussions with the Egyptian archaeologists, had ever been intended. Although Emerson would later feel some opportunities had been missed, better communication gradually smoothed things over and he and McMullen became more at ease.

Despite all the confusion, a picture supportive of the Cayce readings emerged. George began by doing some uncontrolled work (that is, no effort was made to keep either Egyptological or Cayce background material from him) and what he said, while hardly "clean" psychic research, was of interest. Of greatest significance were his perceptions regarding the Sphinx and the Giza Plain. There he outlined underground and as yet undiscovered water systems, pointing out the location of channels, pools, and fountains in an area that had been largely overlooked by Egyptologists (although drilling in front of the Sphinx earlier had hit water).

Most important of all, McMullen confirmed that there was indeed a body of records and, just as Cayce had predicted years before, felt that the way to pinpoint the spot was to observe the shadow cast by the Sphinx. The fact that McMullen supported Cayce in this matter was interesting, but to Emerson's considerable satisfaction he did more than that. In this, the most important information he gave while in Egypt, the psychic was operating at least in part along the lines by which Emerson felt a psychic archaeological experiment should be conducted.

He could not be sure his friend had not been told about the Sphinx's crown. Since the Englishman Thomas Shaw had clambered to the top of the famous statue sometime between 1720 and 1733 and found a hole in its head (which he thought might be the entrance to a tunnel), Europeans had known something might have been placed atop the monument. It was not long afterward that scholars determined that it was a "crown" of some sort. But what did it look like, and how big was it? Very little was, or is, certain. For this reason, when McMullen began to describe its shape and give specific dimensions to the ARE group, Emerson was very pleased to record his data, not only for what they meant but also because it was previously unknown material which could be proven. Here was a testable psychic hypothesis.

McMullen then went on to provide yet another piece of testable information, this time far more controversial. He stated that the Great Pyramid once had a capstone and then gave *its* dimensions. Whether there was a capstone—and if there was, whether it was a single block or was built up of many blocks—is disputed by scholars. Thus George's comments were not only of interest but verifiable, and the way in which they could be tested tied the Pyramid, the Sphinx, and the "hall of records" together into a tidy interrelated package.

Although there is little probability of ever finding either the crown or the capstone or stones, this did not matter. Because of the way in which the psychic had phrased his words, they could be checked by orthodox research. For, similar to a schoolmaster putting a problem to his class, McMullen had posed a question in plane geometry.

He said the "hall of records" could be located "on a day late in October as the sun sets, the shadows cast by the Cheops Pyramid and by the head of the Sphinx—allowing for the crown and capstone— will overlap, merge, and coincide at a single point on the flat pavementlike plaza area in the direction of the Nile."

It was a fascinating, seemingly straightforward presentation, but like so much psychic information, it was far less amenable to correct interpretation than at first appeared. From what "period" was McMullen seeing this event? What shifts in the earth's axis, earth movements, or a thousand other variables had to be considered? No one, then or subsequently, with the training to make an evaluation has ever considered the challenge in McMullen's statements. Should they ever be proved correct, almost as an aside to the main point—the discovery of a fabulous cache of records—the problems of capstone and crown dimensions would be solved. Further, McMullen's comments on their appearance, and the fact that there was a single capstone, would, by virtue of his tested success, be the most plausible explanation advanced.

More was in store for the group concerning George's descriptions of the Sphinx. After completing the survey of the Giza Plain, the party drove down to Luxor and from there made a side trip to Abydos, one of the most venerated necropolises of ancient Egypt. While they were there, Lehner introduced Emerson to an elderly lady, Mrs. Dorothy Louise Eady, an archaeologist who, although officially an amateur, was recognized by professional colleagues. She told Emerson that years before she had been involved in an excavation around the base of the Sphinx that had uncovered stelae that gave details on the size, shape, and position of the missing crown. Her information, unknown to either Emerson or George at the time the Giza tape recording was made, supported the Canadian psychic in almost every particular.

To Emerson, "It is all very interesting. I cannot stress too firmly, though, that the proof of all this lies in the digging. But after all that George has done so successfully for me, and after Mrs. Eady's corroboration . . . well, I take my friend's words very seriously and would advise others to do the same. The excavation and measurement to determine where to excavate are not uncomplicated. . . . We'll just have to see."

If Emerson and George were pleased, Hugh Lynn could hardly have been less so. Here was clear-cut support for his father's words on the most testable portion of his Egyptian world view. Although the trip had begun with some confusion and misunderstanding, by October 14, when they prepared to leave Egypt, everyone involved considered the first leg of their journey at least a qualified success. As it turned out, it was just a prologue for the saga of Iran.

After warm-up visits to Professor Negahban's Museum at Haft Tepe and the Zoroastrian site at Susa (best known as the site of Daniel's tomb), the group went for several days to the ancient city Shushtar. Here, although George still felt a sense of pressure, Emerson began to get a "sense of control over what George and I were being asked to do." For one thing, the reconstruction of Iranian archaeology was far less complete; George would be facing fewer well-known ancient temples whose museums and displays might "pollute" him before he was asked questions—a situation that had caused Emerson extreme frustration in Egypt. Instead, there were only arid washes (or *wadis*, as such things are known in that part of the world) and rocky outcrops where nothing but his psychic intuition could serve him. The entire situation was closer to a double-blind—in some cases, triple-blind—experiment, and Emerson was happier.

The next morning, with glaring heat already pounding down, the party loaded into four-wheel-drive vehicles and made a jolting dusty drive out into the country. It was agreed between Hugh Lynn and Emerson that the first objective would be to see if George could "tune in on and possibly locate the site described by Hugh Lynn's father as 'The City of the Hills and Plains.'" Secondarily, they were interested to see if McMullen could confirm both Cayce's earlier reconstruction and his statements about a cave with records.

Cayce had said it was seven and a quarter miles from the city. With eyes on the odometer, the party jounced along. Spirits were particularly high, despite the heat and dirt, because one of McMullen's best "hits" the year before, when Ogden and Hugh Lynn had visited his home in British Columbia, had been from this site.

Just after passing a herd of goats McMullen, in the front jeep, pointed off to the side and asked the driver to stop. There, he said, back up a *wadi* was the cave and in it a body. He added that when the site was dug, artifacts of a "survival-level technology" would be found. "George went on," Emerson said, "to describe the area, much as Cayce had depicted it earlier as a coordinating nexus; a center for the activities of passing caravans. Additionally, he stated firmly, and this also agreed with Cayce, there was a healing center here."

To Emerson's trained eye, "Although I have no illusion about being a Near East expert, archaeology is archaeology no matter where it is practiced and George's words made pretty good sense. It was obvious that the area was littered with an abundance of artifactual material, particularly potsherds, but no sign of the permanent structures you would expect. People lived here, left no buildings, but plenty of other evidence of habitation. What set of circumstances would account for that? A large semipermanent tent city would explain the artifacts and the lack of signs of buildings. The reconstructions of Cayce and George agreed, and we had the beginning of evidential archaeological data from traditional accepted sources to support them."

Returning to their hotel, each deep in his own thoughts, Cayce and Emerson felt there was good reason to believe that there was a growing case for the existence of "The City of the Hills and Plains," and that very probably they knew its location. Final proof would come through digging.

Up early the next day, their third in Iran, the party again drove out beyond the city, this time to the east to see a healing spring, known locally since ancient times, and mentioned by Edgar Cayce—who could not possibly have been consciously aware of its existence. No sooner did the jeeps stop than George was out and Emerson immediately recognized the jerky movements, the temporary disorientation, as the first stage of "his psychically moving back in time and locating himself." After surveying the spring, George was off at his usual run, scrambling over the rocks up the wash to locate what he said would turn out to be, if excavated, yet another cave, also with a burial. At this site, McMullen added, tablets would be discovered.

The others saw that the cave would require much more work to excavate than the first one. Sometime in the past there had been major sloughing of the cliff face and the cave entrance now lay buried beneath large boulders. When Emerson mentioned this, McMullen assured both him and Hugh Lynn that the effort would be worth it, as a substantial part of the cave would be found intact.

Indeed, before that day was half over it seemed that most objectives of this first tour had been amply met. McMullen had supported Cayce on most key points and, equally important, had rounded out, through careful questioning, the earlier man's descriptions, thus resolving nagging questions. For the first time there was Egyptian and Iranian psychic data detailed enough so that actual digs in three locations (in Egypt, near the Sphinx, and in Iran, the two caves) could be carried out with the reasonable expectation of success.

Their survey of Shushtar completed, the group moved by stages to Teheran, Iran's capital city. There Emerson experienced the high point of his trip. To begin with, he was able to sit down and talk at length with a kindred spirit, a fellow professional. Both McMullen and Emerson liked Negahban immediately, and although he was younger, like Emerson, he had done his doctoral work at the University of Chicago. Since the archaeological fraternity even today is a small one, major contributors know quite a lot about one another. As they conversed, the two men felt like Southern cousins who have never met but know, in their bones, the same family history and what to expect of each other.

They discussed comments McMullen had made to Emerson about Negahban's current work, and the Iranian, who had not been present at Shushtar, was intrigued. More than that, if the president of the Canadian Archaeological Association was willing to stand behind this technique, and if the information about his own work was as specific as the little he had heard, then he would like to see it in action. He asked if Emerson could suggest an experiment. He could indeed; let Negahban gather a group of sherds and they would give them to McMullen to read. Negahban agreed and the experiment was set for that evening after dinner.

At the appointed time they met; a group made up of several of the ARE party, in addition to Negahban, Emerson, McMullen, and R. Mouzami, who had acted as the party's translator at Shushtar. Without preamble the Iranian archaeologist presented George with a plastic bag full of potsherds saying, "Put them in order." It was a test of seriation—the chronological arrangement of artifacts. After spilling them out, George began to do just that, starting with the oldest and moving forward in time with each sherd; all the while providing a running commentary on what each cultural period was like and the purpose for which each object was used. No one offered encouragement; Negahban said not a word until—in just a matter of minutes—it was over. With great enthusiasm the Iranian exclaimed, "Your classification is perfect, George . . . aside from just a few points. This is *magnificent* because you picked them exactly as they came together in the same period."

To Emerson, while George's success "was not really surprising, it was, as always, a great relief." More than ego was involved in his feelings.

"I had always felt on that trip that we ran before we walked. That is, we were locating things before we had run an on-site test case under controlled conditions with an expert in those areas of archaeology present. Now we had done that, it had worked, and we had a baseline from which to begin evaluation. Details like that are important in any kind of archaeology, particularly the psychic variety . . . half the battle in archaeology, at least good archaeology, is keeping control over the details."

Although they realized that only the first steps had been taken, everyone present experienced a feeling of fulfillment. For Hugh Lynn Cayce, it had come a day or two earlier when McMullen had made his comments on the city first described by his father. For Emerson and for George, it came in the hotel room when the psychic had obtained the immediate confirmation that he could work in another country accurately.

Emerson thought, "We are just at the beginning. There will be digs, I'm sure of that. And I may have the privilege of playing some role. I have no delusions, however, that I am that mythical figure of popular imagination ... you know, Omniarchaeologist. The guy with the white pith helmet who leaps from site to site, plane to plane ... a little Meso-American here ... Mesopotamia there ... a little Ontario prehistory for seasoning. I've always seen myself first and foremost as a teacher, a starter, a beginner of things. I think I was able to do a little of that on this trip. When you can share with your fellows a technique that makes archaeology a better discipline, and let us look a little deeper and truer into man's past, for me that is just as important as any find."

VI

CHILDREN OF THE CHANGE

Garrad and Reid

Charles Garrad had been digging by himself for days, searching for a Canadian village he was sure was there. "By four o'clock on Thursday I had reached my limit of exhaustion. I was totally alone. Very cold and very tired. It came at the end of the day, when my physical exhaustion, mental exhaustion, and hunger all hit me at once. I got utterly fed up with the situation and became enraged, my fury boiling over. Whether that caused it or not I cannot say."

Garrad's area of specialty is very specific—a single segment of southern Ontario known as the Collingwood Blue Mountains, a single band of Indians known as the Wyandots (Samuel de Champlain called them the Tobacco People, *Nation du Pétun*), and a single period of time—1600–1650 A.D. For years he had been trying to sort out the various Wyandot villages to determine which ones the Jesuit missionaries had recognized by name, which ones Champlain had visited, which ones had been the sites of mission stations, and which ones had been attacked by the Wyandot's sometime enemies, the Iroquois.

By that fall of 1972, although he was a nonacademic archaeologist, Garrad had developed a fairly conclusive interpretation of his area of interest, one that his fellows in the Ontario Archaeological Society recognized as authoritative. There were still loose ends, however, and he had taken a week off to try to tie some of them up.

By late afternoon it seemed the week's work had been for nothing, and so, with rage driving him, Garrad walked back a short distance from where he was digging to what he thought of as the middle of the village, and looking carefully around to make sure that nobody was likely to overhear him, shouted out, "Look here, you goddamned Indians. I'm tired, and I'm cold, and I'm bloody hun-

198

gry. What do you think I'm here for anyway? It's certainly not for profit. Who wants to spend his bloody holiday this way? You'd better give me something in five minutes or I'll quit!"

Stumbling back to where he had left his shovel, Garrad picked it up and tried to thrust it into the ground. He missed. The shovel glanced off, waving over the soil top and blowing up some dust. "But while the dust was lifted, I plainly saw in the ground several [smoking] pipes, and then the dust settled and the pipes were obscured from view."

To Garrad, it was astonishing. In all the years of his archaeological career he had never found, complete, a ritual pipe used by the Wyandots and other Indians in their ceremonies. Then, suddenly, at the spot where he had been digging for days with no luck whatever, there was not one but several pipes.

With a strength recharged by adrenalin, he began to dig again.

Typical Indian site excavation.

Only something was wrong. Pipes that had been easily visible with only the raising of dust were now almost a foot deep in the ground; it took hard labor to get them out. As he bent to pick up the three pipes, a strange chill passed through Garrad and he felt contrition for his harangue of an hour before and a strange compulsion to make amends.

Going back to the place where he had stood and shouted, he carried one of the pipes. Once again looking around to make sure he was alone, he satisfied a strong urge.

"I didn't know what to do, but I lifted one of the pipes to the sky, and to the ground, and to the four winds, saying aloud, 'Thank you.'

"That remark was greeted by laughter. I looked over to where the sound was coming from, and there, several hundred feet away, on the other side of the field, was a cluster of Indians. There were three boys in the front, and four adults in the back. It was just a blinding flash, they weren't there for more than a maximum space of a second. I could just see the right hand of the nearest male rested

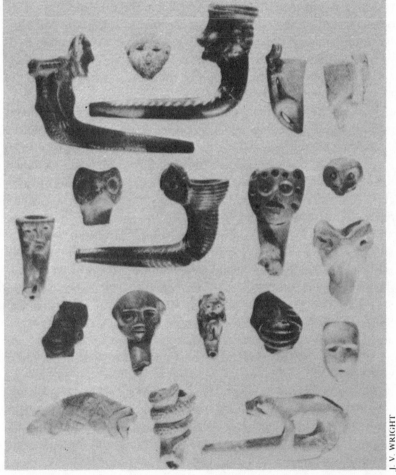

J. V. WRIGHT

Effigy pipes of the type that Charles Garrad discovered. Such pipes represent one of the major artistic achievements of the Ontario Indian Tribal groups.

on the left shoulder of the boy in front of him. They were arranged as though posed for a photograph."

Garrad says he considered it more important that their laughter was not derisive than that they were not "real" as he had always thought of reality before. How he knew their feelings, or why he was not bothered by their phantasmagoric presence at the time, he cannot say. He was sure of one thing, though: the Indians were relieved that he had made this find, and "in some way I did not understand they had been helping me . . . they approved of what I had been doing."

As quickly as they had come, the feelings and the experience were over. Garrad stood in the middle of an empty field, in the day's last light, an exhausted man, confused yet exhilarated. What had happened to him was a violation of his beliefs and his training as a scientist. Yet there were the three pipes and, burned into his mind, the image of those seven Indians.

Pushing the amazement out of his consciousness, Garrad tottered over to where his equipment lay, picked it up, and walked like a drunken man through the night across the fields to where his car was parked. Driving home, not knowing that he was joining a long list of other archaeologists who have had experiences with the psychic, Garrad decided that the experience was something that would never find its way into his reports.

Then, a year later, Professor Emerson began to talk and write about psychic archaeology. Although Garrad did not know Emerson well personally, like virtually every other archaeologist and anthropologist in Canada, he had once studied under him. Emerson was the founder and first president of the Ontario Archaeological Society, the organization that had been sponsoring Garrad's work. To hear Professor Emerson openly discussing the kind of experience he, Charles Garrad, had personally undergone was a profound experience.

After one lecture Garrad rushed up to Emerson in great excitement, saying, "I too have had something happen to me on an Indian site and, until this evening, I was afraid to tell anyone what had happened, or even to think too much about it, for fear that they would think I had been hallucinating."

Emerson assured him he probably had not been hallucinating, paid serious attention to his description of what had happened, and offered to introduce him to some of the sensitives with whom he was working.

With Emerson providing an emotional and intellectual bulwark, Garrad began for the first time to consider seriously using the psychic as an information source in his fieldwork. He soon decided that the correct approach lay not in relying on spontaneous unpredictable psychic incidents in his own life but on the disciplined use of trained sensitives. After some preliminary psychometry sessions, Garrad began considering the list of sites in which he had an interest, searching for one where he might best try a first field experiment. He settled finally on one known as Ehwae:

"Ehwae was significant for several reasons. First, it was the principal village of the Wyandots for some time; second, the Jesuits had been very active there, naming the village St. Peter and St. Paul sometime in 1639; and third, the village was abandoned very rapidly for reasons no one seemed very clear about."

The decision to begin his work with Ehwae placed Garrad in the forefront of what was to become the third generation of archaeologists working with the psychic. Thanks largely to the umbrella of respectability Emerson's seniority and status provided, in Canada at least psychic archaeology took a step toward entering the mainstream of research.

The sensitive whom Garrad chose was Sheila Conway, the same woman who had worked with Emerson on his Cahiague site and on the argillite carving. On Saturday, August 4, 1973, Garrad picked up Mrs. Conway and together they drove out to Natawausakik Township in Simcoe County, Ontario.

"It was windy. We started off into the woods; I didn't really know what to expect. Mrs. Conway began in a kind of conscious clairvoyant state but, as we walked, she slipped lower and lower, deeper and deeper into trance, finally ending up in a very deep trance."

As she had for Professor Emerson at his Cahiague site (the Iroquois capital) Mrs. Conway seemed able not only to read surface impressions but also to smell, hear, and see Indian life as it was lived sometime in the 1630s:

"There is a lot of activity, a lot of noise. I can hear dogs barking ... children are running and laughing. The children haven't any clothes on. A lot of birds singing; even around the encampment, the

birds sing. Small wild animals are scurrying around . . . as if they
have no fear . . . as if they have a perfect right to do this thing. Very
busy people. Constantly making things, making designs . . . the
same design was woven into bands, woven on garments . . . a diago-
nal three straight lines signified the house they lived in . . . signified
shelter.

"They know when people are coming. It's as if thoughts are sent
through the air on the wind . . . they have trained practitioners who
can do these things . . . they also know the kinds of people who are
coming. The greatest thing in a war party is the element of surprise.
It isn't always successful because of the mind carrying . . . I suppose
it is their chieftains who do this. This is why they became chieftains.
It isn't only medicine men.

"The medicine men understand healings. Many of them can effect quick healings . . . healings are not carried out with a lot of
dancing and chanting and shaking of rattles. The real healing is
done quietly . . . it is done with all this preparation but the real
healing is done in a sudden hush. The sick are often carried from the
place where they live. They are carried into the forest . . . the healing is done in the forest. Women give birth to children near running
water.

"A sacred fire . . . a ritual fire is in the middle of the village . . .
for every member . . . men, women, children, young and old alike.
When the fire is doused, it signifies the beginning of a new travel. It
is part of their travel ceremony. When the camp is ready to move on,
many things are broken . . . these things symbolize the life of that
camp and they are now going on to a new one. The old things must
be made new."

All this and more poured out of Mrs. Conway as she and Garrad
walked through the summer woods. Suddenly, however, she
stopped, and as Garrad looked on helplessly, she stared through
him as if he were not there and burst into great racking sobs.

"The babies died. It struck the children. And when the children
died, they said the birds sang no more."

Over three hundred years after it had happened, a middle-aged
white woman cried for the deaths of Indian babies, and an archaeologist got a chance to pull aside the veil and step vicariously into
the culture that was his obsession.

Garrad realized that here was not only priceless information
about the Wyandots, but also the probable reason they had aban-

doned Ehwae. The French priests had brought more than religion: worked into the very fabric of their clothes were pathogens against which the Indians had no resistance. As would eventually happen to Indian villages all over Canada, European diseases destroyed much of the native population—the children suffering the worst. Mrs. Conway's description was totally consistent with what Garrad already knew about Indian-European contact. He felt it very probable that she was right; the village was not so much abandoned as fled from.

He knew, though, that this isolated conclusion could never be proved absolutely. And so he began to search through Mrs. Conway's words for other verifiable data, working on the academically acceptable theory that if other data could be proved accurate, the points that could not be tested would, in context, have considerable validity.

Her statements about healing were of immediate interest. Anthropology has traditionally viewed the medicine man, or shaman, as an aberrant or psychotic individual who either consciously or unconsciously manipulates his fellow tribe members through meaningless but emotionally powerful rituals. The assumption is that any healing that takes place occurs in purely psychosomatic illnesses.

Although this is still the majority opinion, recent research at the Menninger Clinic in Kansas, using a Shoshone medicine man, Chief Rolling Thunder, has shown how superficial and erroneous it is. Rolling Thunder has performed numerous healings on gross physical traumas such as gashed legs, first in front of an array of doctors in an operating theater at Menninger and then elsewhere, including the Edgar Cayce Foundation at Virginia Beach.

More importantly in terms of Mrs. Conway's statements, he effects these healings exactly as she described the process to Garrad. And her statements in August 1973 could hardly have been colored by Rolling Thunder's demonstrations, for he had just begun working with researchers and the experiments were being carried out without any publicity or publication of reports.

Rolling Thunder is also more than a healer. He is an articulate spokesman for the Indian perspective on the world; and his statements on the silence necessary for healings, and the Indian concept of "psychic hotspots," coincided with Mrs. Conway's perceptions. He has even supported her observation about Indian women trying to give birth near flowing water.

As Garrad surveyed the parapsychological literature, he discovered that researchers in that field took a somewhat different view from anthropologists of nonmedical healings, and that, regardless of culture, psycho-spiritual healers all seem to seek quiet and natural surroundings.

The account of the fire and the breaking of old items also intrigued him, and although he did not find any specific references to either in the literature about the Wyandots, the concept of breaking the old when departing and making the new upon arriving, and the "hearth" fire, were within acceptable parameters of what was known of Indian behavior.

One other topic Mrs. Conway touched on did produce results that made up for not finding specific references to the fire and breaking account. About halfway through her psychic discourse the sensitive began this description:

"Many people come to visit . . . many visitors . . . there are important people who come wearing woven grass capes . . . with "colored quills . . . the size of the cape, the length of the cape, denote the importance of the persons wearing it."

Garrad had never heard of such an Indian practice, although it was an ancient Japanese custom to wear grass capes as foul-weather gear. Consequently, he was particularly excited to discover, "looking at the ethno-historic sources, that one group with whom the Pétun people were friendly was an Ottawa or Algonquin group by the name of Cheveralev, and that no less an authority than Champlain himself described how they wove mats using grass and corn stalks, corn leaves and wild rice stems—that sort of thing."

Garrad felt he was on fairly safe ground in concluding that a people who made mats might well use them not only for sleeping on but as wraps. "So what we're seeing here are Ottawa people." Just the visitors one would expect to find in the major village of a group of friendly "cousins."

After carefully evaluating everything he was told by Mrs. Conway, Garrad felt, as had Emerson, "everything this psychic has said is entirely within acceptable realms of knowledge. Much of it cannot be absolutely substantiated from archaeological sources, but it is certainly in keeping with what we believe probable from archaeological and ethno-historic sources." Charles Garrad was convinced.

Actually, his case is stronger than he feels it to be. Although the description Mrs. Conway provides of medicine men and chieftains "reading thoughts on the wind" sounds fanciful, Mrs. Conway may have erred on the side of conservatism when she told Garrad about Indian telepathy.

In the late 1950s a researcher named Stiles, of the Museum of the American Indian, made one of his periodic visits to a tribe of the Algonquin people, known as the Montagnais, who lived in eastern Canada. He brought back from that trip an observation that, being an anthropologist with a healthy sense of professional survival, he did not choose to publish but did present to Professor C. W. Weiant who, a few years later, entered it into the literature.

What Stiles described was a situation in which "certain members of the Montagnais habitually repair to the woods, set up a log shelter about the size of a telephone booth, get inside and, when the power is sufficiently strong, make contact with a friend or relative who may be hundreds of miles away. A two-way conversation is carried on seemingly by clairaudience. If no contact is made, it is assumed that the person with whom contact is sought has died." Stiles went on to say that when "the process is going on, the shelter shakes."

It might be possible to dismiss Stiles's account as overly credulous were it not for the fact that the exact same observation has been made by two other, independent sources: a trapper who spent sixty years in eastern Canada among the Montagnais, and a Roman Catholic missionary to the same tribe, who would not so easily credit the effectiveness of what to him must have been a pagan ritual. And these two witnesses were not recounting events occurring hundreds of years ago but in our own century.

Garrad did not know about Stiles's commentary then, and did not need to. What he was able to verify was more than enough proof for him. Shortly thereafter he took Mrs. Conway to the other major site in which he had an interest: the village of Etherita.

Sometime about 1640 Ehwae was abandoned and the Pétun moved to a different section of the Collingwood Blue Mountains, where they established a village they called Etherita. This village is regarded as having a particular significance because it is known that the Jesuits built a church there, which was presided over by a Father Ganier. Unfortunately, after the massacre of 1649, the exact location of Etherita was lost.

Garrad decided to tell Mrs. Conway nothing about the site, but to take her there and see if she could psychically produce "any evidence of its being Etherita, perhaps by perceiving a church there, or a Jesuit priest."

What he got was even better than he hoped for; Mrs. Conway described not only a church, but also a log house for the priest. She then went on to give a picture of Father Ganier himself:

"A kindly man. He seems to have very round pink cheeks. It's a black robe he's got on . . . leather belting, leather belting which ties, no buckle . . . there is a cross . . . a cross tucked under the belt. A leather thong around his neck . . . and there's a pouch . . . filled with something aromatic . . . not tobacco. It's an herb of some kind . . . He is in some physical difficulty. There is much pain in the knees, in the hips. I feel, too, that he has had many dental problems. I feel pain from my left jaw shooting behind the left ear. . . . He . . . he is in much physical distress.

"This man with the ruddy cheeks seems to spend a lot of time with the women. I have the feeling that he really isn't well liked for this. Even his brother priests think he is wasting his time . . . showing them things. They say it is a waste. 'What better things for the Lord could be done,' so they say . . . ."

Here was the kind of fascinating information no artifact could provide—except psychically: the story of a priest who, in an age in which both European and Indian cultures were male dominated, spent time showing things to women. But was it evidential? Certainly the description of a ruddy-cheeked priest could not be so considered; anybody who lived out of doors most of the time in the bitter Ontario winters would have ruddy cheeks. The belt without a buckle was suggestive, but there was no way to prove such a personal habit of dress.

The comment about the pouch of aromatic herbs, which Mrs. Conway had said seemed to her "wet," however, had a certain evidential touch, especially in light of her related comments about his state of pain. What could it be? Garrad made only a cursory attempt to answer this question for he had an even more tantalizing clue to work with. But it seems reasonable to conclude that Mrs. Conway was describing an asafetida bag.

As early as 1398 these pouches of a concentrated resinous ("wet"?) gum were mentioned in literature by John de Trevisa. The substance has an almost nauseatingly pungent aroma and, as late as the first decades of the twentieth century, was worn around the neck to ward off illness. Exactly what a seventeenth-century European, in poor health and isolated from doctors, might use. Again, Mrs. Conway's remarks, while not provable, seem not only acceptable but peculiarly detailed in a way that pure fantasy would hardly provide.

But it was not asafetida bags or buckleless belts that caught Garrad's attention. He was riveted by his sensitive's statements about "the leather thong around his neck . . . there are also little . . . they look like buttons . . . no, medals . . . they're made of metal . . . dangling from the leather thong around his neck."

Years before a senior colleague and friend of Garrad's had discovered just such a medal at the same site, and it was unquestionably Jesuit since on one side there was a representation of St. Ignatius Loyola, the founder of the Society of Jesus (Jesuits), and on the other a representation of St. Francis Xavier, one of the order's most venerated saints.

It was too great a coincidence to ignore and, consequently, the small religious device was generally thought to be the personal property of Father Ganier by those who accepted Garrad's contention that this site was indeed Etherita. Garrad thus had artifactual, ethnographic, and psychic support for his location of a site that was critically important to the understanding of Indian-Jesuit contact.

That there had been more than one medal on the thong was not unreasonable. In fact, the very first psychometry session Garrad had participated in—an experiment involving three psychics—had produced a consensus that the medal was one of three worn on a leather thong. Mrs. Conway knew nothing of this earlier experimentation.

After evaluating what he had been told by Mrs. Conway, Garrad felt even more secure in his belief that what she next said to him, about the church itself, was also true. As they walked over the ground together and Garrad asked her about the chapel, her answers gave him the eerie feeling that time had reversed itself and that through his sensitive he had become a dweller in two worlds. His eyes saw only trees and fields; to her eyes, the view was very different:

"I see tall wooden palisades . . . in the enclosure is a small wooden building. Surmounted by a cross. A building set on logs which raise it from the ground. A high step into the front door of this narrow little wooden building with the brown wooden cross. A peaked roof at the front. A dark musty interior . . . windows in the logs covered by skin flaps.

"At the far end of this narrow building, an altar built of half log rounds. A polished surface . . . as though oiled. A rough wooden cross like the one on the roof, and in the center of the altar earthen bowls, full of oils and rough candles."

The psychic settled deeper in her trance, by now totally oblivious of the time Garrad thought of as "real." In the middle of a bright twentieth-century afternoon she relived a Christian communion that had taken place more than three hundred years before. Here was life as it had been lived by a lonely French priest in an isolated Indian hamlet, as strange to him as Tibet would be to most of us.

"The owner of the medal is lighting the candles. He wears around his neck a very large cross . . . and a thong, a very narrow leather threadlike thong upon which hang three medals. He genuflects before the altar, before turning to leave it. . . .

"There are no benches upon which one may sit in the little chapel. Those who come to service must sit cross-legged on the floor. I see the priest moving to the back of the altar, and he brings out the bread . . . the host . . . presently he walks to the door of the little chapel and some Indians come . . . of the men . . . not the women. The boys may come . . . if they are of some age, some look ten years old, some eleven or twelve years old. The priest stands aside, and the men and boys file into the chapel and sit on the floor. The priest moves among them . . . he lays his hands upon the heads of the boys and upon the shoulders of the men. And, as he bends forward to do these things, the three medals glint and swing in the light of the candles. It is murky and dark, for it is the end of day. The sun is setting, and sunlight comes through the hollowed-out window on the left. . . ."

What was it like to be a priest in a foreign land? This is the kind of question traditionally derived archaeological information can never answer. Psychic perception did provide an answer and supplied some understanding of why the church missionaries maintained with such fervor the purity of their ritual. For its religious soldiers, posted to desolate duty stations, the ritual, no matter how alien the congregation, was a link with not only the church but also their entire past. Is it any wonder then that some came to depend too heavily on that link and intolerance and fanaticism were the result?

For Charles Garrad, "a straight archaeologist before I had my little experience," there is now no turning back. To do so would be like suddenly perceiving color in a color-blind world and then spitefully giving up such vision because it had not been there before.

At about the same time that Garrad began working with Sheila Conway, another of Professor Emerson's former students also began to consider the results his teacher was getting in the latest phase of his long career.

C. S. Reid is a good-tempered Irishman known as Paddy, whose professional life was just beginning when Emerson issued his first paper on psychic archaeology in March 1973. He had done his undergraduate work at Toronto and then gone north to McMaster University for graduate work.

In 1972, under the auspices of the Ontario Archaeological Society, he had begun excavation on two Iroquois sites. One, the oldest, was a small 1.1-acre village of the Pickering branch of what is known to archaeologists as the Early Iroquois Tradition. Radiocarbon dating by Reid placed it at A.D. 975 ± 120 years. Known as the Boys site, this prehistoric village became the subject of intense investigation by Reid, who carried out a second year of fieldwork in 1973, this time under the sponsorship of McMaster University.

The other site was much larger, about ten acres, and more modern; a fairly substantial village that preliminary analysis of ceramics had assigned to the Middle Ontario Iroquois period, about A.D. 1300. This location was known as the Sewell site. (The names refer to modern farms on which the sites were found.)

By the spring of 1973 Reid had settled on these two sites as the subjects of his master's thesis, and he was busily at work excavating to that end. The results were encouraging but perplexing. At the Boys site two middens (essentially Indian garbage heaps) had been located and excavated, as well as the east wall of what Reid had designated House I. The fire pits, spaced down the middle of Iroquois longhouses, had also been located. On the debit side, despite considerable fruitless test trenching, Reid had been unable to locate the village's palisade. This wooden fence, which runs around almost all Ontario Indian villages, is an archaeologist's first priority, since it

212

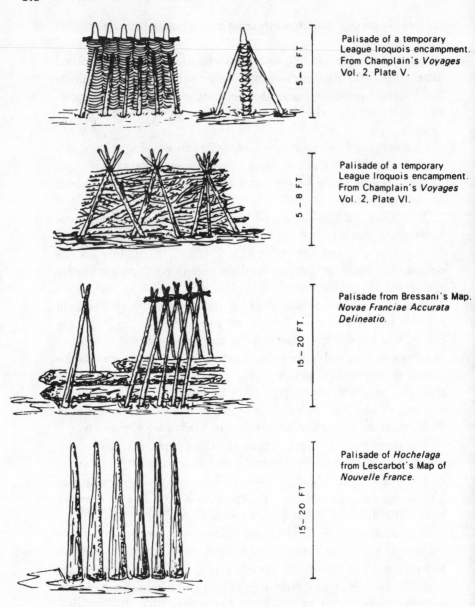

Palisade of a temporary League Iroquois encampment. From Champlain's *Voyages* Vol. 2, Plate V.

5 – 8 FT.

Palisade of a temporary League Iroquois encampment. From Champlain's *Voyages* Vol. 2, Plate VI.

5 – 8 FT.

Palisade from Bressani's Map. *Novae Franciae Accurata Delineatio.*

15 – 20 FT.

Palisade of *Hochelaga* from Lescarbot's Map of *Nouvelle France.*

15 – 20 FT.

TYPICAL PALISADES

Such defense perimeters as these are usually found around Ontario Indian villages. The one discovered by George McMullen around the Boys site, however, was totally without precedence, and was thus a good psychic test case.

COPYRIGHT © 1971 BY THE HISTORICAL SITE BRANCH, ONTARIO MINISTRY OF NATURAL RESOURCES.

gives some idea of the size of the site with which one is dealing. The palisade was the 1973 digging season's most pressing problem. The second most important question was the location of the west wall of House I. The digging crew so far had turned up nothing but confusion: there were so many seemingly random post molds that no discernible line could be traced.

Archaeological digging is not simply going out and shoveling a hole in the ground. It is slow, maddening work in which the biggest digging implement is often a plastering trowel (sometimes it gets down to a dental tooth scraper). Two years spent searching for something without finding it, then, is more frustrating than the measurement of time would indicate. Reid was understandably discouraged.

Emerson knew the site well, it wasn't far from his home, and he and Reid had talked over a plan of excavation on more than one occasion. At Reid's request, they had begun to consider what George McMullen might be able to do. Finally Reid asked his former professor to arrange a visit by the psychic.

On May 19, 1973, Emerson and George drove out to Duffin's Creek, about four miles from Lake Ontario, to see what would happen. It was a warm Saturday. For the next two hours, and then for a total of three and a half hours in two sessions on the next day, George walked over what to Emerson looked like "a nondescript flat field covered with long grass, weeds, brush, a small stand of poplar and alder, and a big pile of gravel the government had dumped there back in the fifties. There was not much to look at, I can tell you." To George's eyes, however, there was considerably more to be seen.

After psychically orienting himself, he marched off unhesitatingly, with Emerson (tape recorder in hand) and Reid (with the excavation stakes and notebook) trailing behind. In minutes he had resolved the problems of years.

The palisade, he said, definitely existed, although much of it was now covered by the eight- to ten-foot-deep gravel pile. However, if Reid would dig to the east of the gravel, in a small ravine at the field edge, he would not only find it, but would also discover an entrance shaped like a "cattle gate." On the other side, the west side of the gravel, Reid would find that the fence reemerged and ran on for some distance. McMullen paced off the area, indicating a palisade that cut off the space between two ravines.

This both excited and disturbed Reid. For one thing, George had described a line of posts totally unlike any other palisade Reid had seen or read about: a mass of smaller posts shallowly set, creating a loose but relatively thick wall. For another, he said the palisade did not run all the way around the village because this group of Indians "were peaceful . . . they didn't need a stockade." In fact, McMullen said, it wasn't so much a defense line as "a snow fence" to keep the snows from sweeping down between the ravines and covering the village.

In answer to Reid's questions about the west wall of House I George was equally explicit. He said the wall Reid was searching for would be found running parallel to the medial hearths (fire pits) fifteen feet to the west. He then volunteered that the southern end of the house was fifty feet distant from the gravel deposit. The psychic also answered Reid's request for signs of burials by indicating three separate sites, one of which, he said, contained two people.

A week later McMullen returned for a third session at Boys and located what he described as a "different house," one that "was not used for living in," one he felt was probably a ceremonial structure. This second and totally unexpected house was located by McMullen in an area that, because of its heavy brush and gravel spillover, had never been excavated.

If he was correct, McMullen had, in those three sessions, solved most of Reid's location problems. But was he correct? Reid believed in Professor Emerson, and had come to believe in George. He decided to risk the time it would require to test this unorthodox guidance through excavation. It was a decision not lightly taken. The digging season in Ontario is not a long one and Reid, like all but a few archaeologists, faced chronic underfinancing. To follow this psychic information meant giving up other, more traditionally promising avenues of research, possibly even risking failure for the entire season—which might mean a threat to his master's degree. Yet here was an opportunity to do something that few scientists ever achieve at the beginning of their careers—to advance the cutting edge of their discipline by trying and validating a new technique. Reid decided it was worth the chance. It was a wise decision; the results proved extraordinary.

Reid decided to begin with the palisade since, if George was accurate there, the year would be a success no matter what else happened. And McMullen was very accurate. The Boys site had a pal-

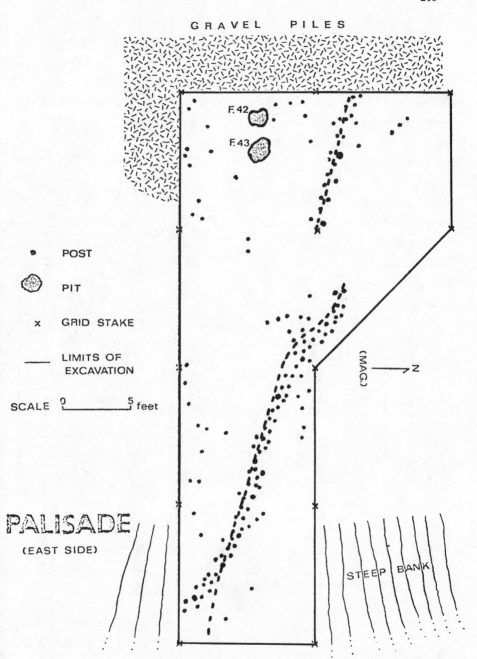

GRAVEL PILES

F. 42

F. 43

• POST

◯ PIT

x GRID STAKE

___ LIMITS OF
     EXCAVATION

SCALE 0 ⌐———⌐ 5 feet

(MAG.) ——→ Z

PALISADE
(EAST SIDE)

STEEP BANK

BOYS SITE PALISADE

The black circles represent post holes and moulds; the dashed line is
the demarcation staked out on psychic information.

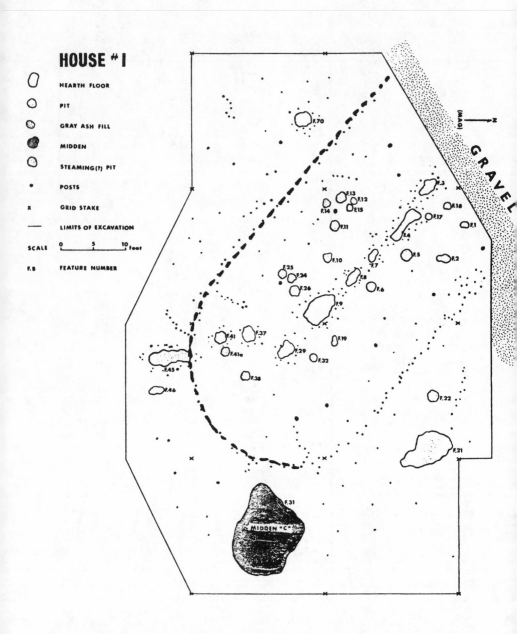

HOUSE #1

- 〇 **HEARTH FLOOR**
- 〇 **PIT**
- ◎ **GRAY ASH FILL**
- ● **MIDDEN**
- 〇 **STEAMING(?) PIT**
- • **POSTS**
- x **GRID STAKE**
- — **LIMITS OF EXCAVATION**
- SCALE 0 ____ 5 ____ 10 feet
- F.B **FEATURE NUMBER**

HOUSE I—WEST WALL
The dots are post holes and moulds; the dashed, George McMullen's line.

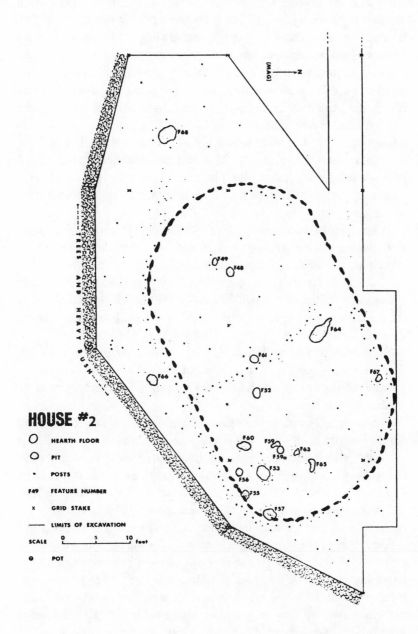

HOUSE #2

○	HEARTH FLOOR
○	PIT
•	POSTS
F49	FEATURE NUMBER
x	GRID STAKE
—	LIMITS OF EXCAVATION
SCALE	0 5 10 feet
◉	POT

HOUSE 2—BOYS SITE
This house was entirely discovered under the psychic guidance of
George McMullen.

isade and his psychic line exactly overlay a course that excavation showed to be correct for all but three feet at the lower east end. His description of the entranceway ("like a cattle gate") was equally impressive, and accurate in location to within less than twelve inches.

Furthermore, his reconstruction of its purpose, which was not for defense but "like a snow fence," was borne out, thus giving Reid an insight into the lives and nature of the people who had lived in this village. Parapsychologically, this particular commentary was also important because it met the requirement of a true triple-blind: no one knew the truth about McMullen's peculiar description of the palisade until it was excavated. Here was a case in which psychic information totally diverged from what was traditionally accepted about a subject.

When George described a fence made up of many small posts shallowly set in the ground, as opposed to a single line of deep-set stout posts, everything known about palisades said he was wrong. No such palisade had ever been found and there was no reason to believe one had ever existed. Yet Reid's fieldwork proved the psychic completely accurate. Here was "a palisade line unique in its construction."

Reid next turned to House I, and got equally exciting results. The west wall was just where McMullen said it would be, and the position of the end of the house was "within eighteen inches of its predicted location."

Work then began on clearing the underbrush and excavating where George said a nonresidential structure was to be found. His outline of this building, eventually known as House II, was if anything more accurate than the outline for House I, his dimensions and wall lines were exact. Also, as with the palisade, George volunteered more than simple location—he gave Reid insight into the building's use, which archaeological analysis proved to be accurate. As Reid concluded in his official report, "On the basis of a complete lack of [animal] remains, the division of the structure in two by a partition, the lack of sleeping platform supports, and the arrangement of the hearth and pit clusters, the preferred functional interpretation is that this is a ceremonial structure." Exactly what McMullen had indicated on that clear spring afternoon as he walked over a field where, as Professor Emerson said, "there wasn't much to look at, I can tell you."

Only in the question of burial locations did George *seem* to be off.

RECONSTRUCTION OF A LONGHOUSE FROM DESCRIPTIONS

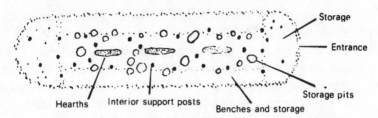

Idealized floor plan of a longhouse

0   10   20 feet

Reconstruction of longhouse from verbal descriptions and archaeology.

0   10   20 feet

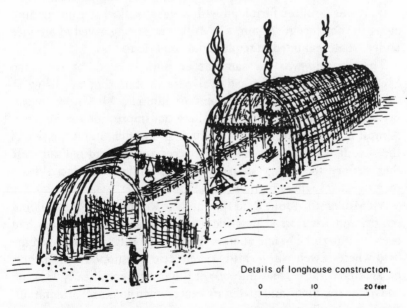

Details of longhouse construction.

0   10   20 feet

COPYRIGHT © 1971 BY THE HISTORICAL SITE BRANCH, ONTARIO MINISTRY OF NATURAL RE-
SOURCES.

At first this was attributed to his already known distaste for things involving the dead. However, more recent research, appears to dispute this initial judgement and, although the evidence is still conclusive, it seems to bear out George's accuracy. It is very possible that yet another assumption derived from orthodoxly obtained data will have to give way to more accurate psychic insights.

By the end of the summer McMullen's record for accuracy had so impressed Reid that he asked Emerson to arrange another test, this time at the Sewell site. Unlike the Boys village, which Reid had been working for some time, Sewell was only in the beginning stages. Nothing was known, for instance, about its size, whether there was a palisade, or even where any of the longhouses had been placed. All Reid had been able to do was locate two middens, already partially excavated by the time he requested George to visit.

After some delays the field test was arranged for the fall of 1973. Since for all practical purposes the digging season was almost over, Reid decided to concentrate his questions in two areas: How big was the site, and could George locate at least one longhouse and give its dimensions?

When McMullen and Emerson arrived at the Sewell location there was even less to see than at Boys. Sewell looked exactly like what it was—rolling farmland with a steep-banked stream running along the north edge. Again, McMullen was being asked to produce under what amounted to triple-blind conditions.

The scenario was the same as at Boys; indeed, by this time McMullen and Emerson had evolved a method they are using to this day. After psychically orienting himself, McMullen began briskly walking over the field, spilling out impressions as he went. Almost at once he dealt with Reid's first question about the size of the site. It was, he said, about ten acres and had no palisade. He then defined the village shape and limits while Reid placed metal digging stakes where he indicated.

McMullen then told Reid that he was working a canoe-building center, and went on to describe a series of structures he said had been built parallel to one another along a low ridge that bisected the field where Sewell was located. When Reid questioned him further, he answered the second purpose of the test by specifically locating one of the buildings and giving its exact dimensions and orientation. In all, McMullen spent about two hours on the site.

Since the weather was rapidly changing and snow was expected soon, Reid had his excavation crew immediately begin to dig a five-foot-wide test trench. Everyone entered into the work with good spirits and high expectations, remembering the success at Boys. But as the trench was dug across the low ridge where McMullen had placed the parallel houses, hope turned to despair. Where there should have been post molds or holes there was nothing.

Then the diggers began to notice faint traces of fire-reddened sand where George had indicated the structure would be found. When Reid researched the problem he discovered that a combination of sod farming and wind erosion had reduced the original height of the ridge and that what was now ridge top must have been, when the site was an active Indian village, the bottom of the post holes. Was the burnt sand all that remained of the posts that had been fired when the village was abandoned?

This supposition proved correct. As the trench was extended further east, post molds clearly demonstrated that McMullen's longhouse did indeed exist. As digging continued, the rest of the psychic predictions also proved to be accurate—although George's longitudinal orientation was about five degrees further south than it should have been.

By the time all this had been worked out, the modern world intruded and, along with the weather, brought the excavation to a halt. Archaeological sites are only fields to farmers, and this farmer wanted to disc his land before hard frost set in. The questions of palisades and canoes were shelved.

A bridge had been crossed, though, and, along with other Emerson students Tyyska and Garrad, Reid had helped move academic archaeology into new realms. By coupling the best archaeological techniques with psychic information derived *prior* to digging, he satisfied the requirements of a "hard" science. At Boys and Sewell, even more than at Ehwae or Etherita, psychic information had been used to generate hypotheses that were tested by actual fieldwork, showing that archaeological theories were "testable" in the sense that word is used in chemistry and physics.

It is only fitting that for the risks he took Paddy Reid was awarded the first graduate degree in archaeology ever given by a department of recognized stature for work in which the psychic played an overt role.

VII

INTO THE AMERICAN MAINSTREAM

*Weiant, Swanton, Long, and
the Mexico City AAA Meeting*

At the end of four months, after enduring rain, mud almost up to their hips, tarantulas, three different kinds of ticks, and "fat green worms" that dropped "like manna . . . usually down our necks," the joint National Geographic–Smithsonian Institution Expedition prepared to leave its Mexican dig.

Since before Christmas of 1938 until the spring of 1939 they had searched a two-mile-long area outside the village of Tres Zapotes, a hard day's ride on horseback then from the city of Vera Cruz. That season's work had produced over three tons of artifacts: It also produced one of the most famous single objects ever discovered in the Americas: the *Cabeza Colosal*. A giant enigmatic stone head six feet igh and eighteen feet in diameter, it weighed ten tons and was carved by an unknown people from a single massive block of basalt.

More important even than the head was a controversial broken stone tablet known as Stela C—since it was the third one discovered—which bore what some believe to be the oldest recorded date yet discovered in the New World, November 4, 291 B.C.*

Also included in the dig's findings was a statue of a priest, the only unbroken one uncovered by the expedition. This small figurine, although considered important because it was of the clear proofs that Tres Zapotes had been the meeting point of at least three cultures—Aztec, Maya, and Olmec—was overshadowed at the

---

*The Stela C date has been the subject of continuing controversy. Under what is known as the Spinden Correlation, the date is 291 B.C. However, more recent interpretations under the Goodman-Martinez-Thomas Correlation are currently accepted and these date the Stela at 31 B.C. The names assigned to the correlations are those of archaeologists who have attempted to first work out the Mayan dates and then to transfer them to the modern Western calendar.

time by the mammoth head and Stela C. Perhaps this happened because its full story has never been publicly revealed. It is a story as extraordinary as the man who discovered it.

Even in a discipline known for its distinctive individuals, Clarence Wolsey Weiant stands out. To begin with, archaeology was the last begun of three separate and successful careers Weiant pursued simultaneously. Indeed, he was in his thirties before he began to study archaeology and forty-one the winter that he went to Tres Zapotes.

His second career, in temporal order, was parapsychology. He had begun it in the early twenties by assisting Hereward Carrington in landmark experiments on thought-photography. Working first with a man recruited by Carrington and later with a woman photographer in New York, Josephine Rossi, Carrington, Weiant, and Weiant's sister, an x-ray technician, were able to demonstrate that it is possible, using some unknown psi factor, for human consciousness to cause images to form on fresh, unexposed photographic film. The woman photographer was so unconsciously adept at this that, as Weiant explains it, "She had to give up her photographic studio and lose her livelihood because she could not stop im-

WEIANT

It was to this small village of Tres Zapotes, about fifty miles from the Mexican city of Vera Cruz, that a joint Smithsonian Institution–National Geographic Society expedition came in 1939. No one, least of all he, could have anticipated that archaeologist Clarence Weiant would use a psychic to make some of the major finds.

WEIANT

This strange enigmatic head was one of the main reasons the Tres Zapotes site was selected for excavation. Discovered at the end of the nineteenth century, the head was almost wholly buried, but obviously of major importance. To this day its true history is shrouded in mystery; there is not even complete agreement among archaeologists as to which culture produced this extraordinary work of art. Its recovery was one of Dr. Weiant's first responsibilities. Here he stands next to it after the trenching was completed.

ages from appearing on commissions she had received. People did not take kindly to strange spiritlike pictures on their portraits."

This pioneer work in parapsychology predated most other research into this phenomenon by many years. A full study was not made until a Denver psychiatrist, Dr. Jule Eisenbud, began examining the psychic abilities of Ted Serios in the early 1960s. Ironically, just about the time Eisenbud was starting his research, Weiant was completing the last of four grants from the Parapsychology Foundation.

Before his parapsychological career began, Weiant had already become interested in one of the most controversial of the healing arts—chiropractic. In fact he was a doctor of chiropractic and a professor of the discipline at the Eastern Chiropractic Institute in New York.

How a chiropractor from New York with an interest in the psychic got involved in a dig recognized as one of the turning points in

Middle American archaeology—a dig that included almost every major Middle American archaeological specialist of the period and was funded by two of the most conservative institutions of America's scientific establishment—is attributable to only one thing: merit. C. W. Weiant was never a man to do things by halves.

Before he took up either chiropractic or parapsychology, Weiant was interested in archaeology, specifically Middle American archaeology. Through the years, no matter how deeply he was involved in his other two careers, his fascination with prehistory never waned. Although he often had to work at night to fulfill commitments to his students and patients, Weiant made up his mind to become a fully qualified academic archaeologist. He began by "persuading Columbia University to let me take courses as an undergraduate in anthropology." Four years later he became the first person to take a B.S. from that university with a major in anthropology. "At that time it was considered a graduate school subject."

Weiant then studied for his doctorate, in the course of which he impressed William Duncan Strong, one of the leaders in Columbia's anthropology department. Strong took Weiant to North Dakota on an Indian dig and, on the strength of his performance there, his overall scholarship, and the proficient Spanish he had taught himself, recommended him to Matthew W. Stirling, soon to be chief of the Smithsonian Institution's Bureau of American Ethnology. Stirling was one of the world's leading archaeologists specializing in Middle American archaeology, and he was looking for sites where the various prehistoric cultures of Mexico had met and intermingled. He felt Tres Zapotes might meet these requirements, and proposed to dig there, but needed a chief assistant. Strong suggested Weiant, despite the fact that his graduate student had not yet completed his doctorate; Stirling enthusiastically agreed.

Tres Zapotes became Weiant's doctoral fieldwork, and the results of the dig his Ph.D. dissertation. So meticulous was this report that both Strong and Stirling recommended it for publication by the Smithsonian. Thus in 1943 Weiant joined the very small and select group of modern archaeologists who can claim the honor of having their dissertations published by the Smithsonian. Weiant's work was also selected by the Johnson Reprint Corporation for inclusion in the series Landmarks in Anthropology. But his honored dissertation did not contain the real story of Tres Zapotes, the story of the laughing figure, the Zone of Burials, and Mound C.

Though forty years have passed, the incidents of that season are still clear to Weiant.

"I had started excavating the head. We knew it was there and digging it out was, of course, one of our main objectives. But I had also noticed that nearby there was a mound. I was anxious to do my job as well as possible and to keep one step ahead of Stirling so he wouldn't feel he had to keep planning my next move for me.

"Anyway, I took a certain number of men and had them dig trenches around that mound and one trench through—more nearly across it."

What his words do not reveal is what this meant in terms of actual labor. Even a picture cannot convey how big a task was involved, or how great a chance Weiant was taking. There were about fifty mounds—actually, they looked like vaguely unnatural hills—in the whole Tres Zapotes dig area. They ranged in height from a few feet to almost forty feet, and some were jungle-covered. Weiant knew it would be impossible to excavate all the mounds in a single season, so the decision as to which should be dug required serious consideration. He and Stirling had to make educated guesses based on firm archaeological principles. The mound near the head seemed a good choice.

The work progressed steadily but slowly in order not to miss or disturb the sequence of artifacts found. The men were restricted to the simplest hand tools—shovel and mattock—and often were down on their knees scooping out the heavy damp earth with cupped hands. Rain complicated the picture. Sometimes it was so heavy it took two days to clear trenches of rainwater so the digging could resume. Unfortunately, the decision "had not been a good one. We didn't have much luck, hardly anything turned up. It was pretty discouraging after all that work."

One evening while Weiant was considering the implications of this wrong turn a member of his work crew, Emilio Tegoma, asked to speak with him.

"He was an old man, the oldest man on the dig. From his historical remembrances, I would say he must have been well along in his eighties, and yet he dug right with the rest of the crew. Well, the old man saw that I was disappointed, and after we had quit for the day, he came up to me and said that tomorrow, if I would shift the digging to where he told me, I would have results. I would find what I was looking for."

How could an illiterate Mexican peasant in his eighties know where artifacts perhaps two millennia old were buried? Could he have "salted" them, Weiant wondered? But the vegetation was undisturbed, and besides, it was inconceivable that a man, or even a group of men, could "salt" an entire mound, sprinkling artifacts from a variety of cultures at different and historically correct depths throughout a wide area; and then restore the soil levels to their proper striations and order. Another archaeologist would have dismissed Emilio Tegoma as either a braggart or a fool, but Weiant, because of his parapsychological background, wondered if perhaps here, in this Middle American backwater, some other level of perception might not be at work.

Having already made one wrong choice, it was an enormous risk to gamble more precious time on what any rational archaeologist would say was little more than witchcraft. Nevertheless, in a matter of minutes Weiant reached his decision. He would follow the old Mexican's directions; he would move his crew to a new site.

The next morning, after he had performed his daily chore of checking in each member of the Tres Zapotes work force on his payroll log, Weiant took Tegoma aside. The old man again assured him that he could see things at a distance, things that were hidden, and that if Weiant would dig where he indicated, he would meet with

This Mexican workman, almost eighty years old, would become the key to Weiant's success at Tres Zapotes. Faced with a short digging season because of anticipated heavy rains, and the need to clear jungle before actual archaeological work could even be begun, Weiant was in the difficult position of having to choose his sites correctly if the expedition was to succeed. After Weiant's initial failure Tegoma came forward and told him that if the scientist would follow where he was led he would find what he sought. Instead of dismissing the man as a crank Weiant, who had already done extensive parapsychological research, agreed. This decision led to many of the major finds made that year.

WEIANT

WEIANT

After clearing the ground and beginning to dig, as shown, the elated Weiant soon learned that Tegoma's directions had provided everything he could ask for. Eventually this site, known as the Ranchito Mound, would provide insight into both the life customs and death practices of a vanished people.

success. Weiant agreed and Tegoma marched cheerfully off in a northeasterly direction. He had gone perhaps two-thirds of a mile along a flood plain that bordered the Arroyo de Hueyapa when he suddenly stopped. First he waved his hand over a cluster of four low mounds, indicating that this general area was the place, and then he pointed specifically to a low mound about ten feet high just on the edge of the slight ridge that overlooked the flood plain.

Here Weiant began to doubt his decision. He was quite a distance from the stone head that had been the original focus of his work. Nothing about this area, or this mound in particular, looked any more promising than the mound near the head that had proved a waste of time. That one had been selected on the basis of sound archaeological judgment and was at least an honorable failure. This one was being picked on the premise of nothing that could be defended scientifically. Yet he decided to dig.

"Within twenty minutes of the first shovelful, I knew the choice was the correct one, for in just those few moments the unbroken laughing figurine came out of the ground."

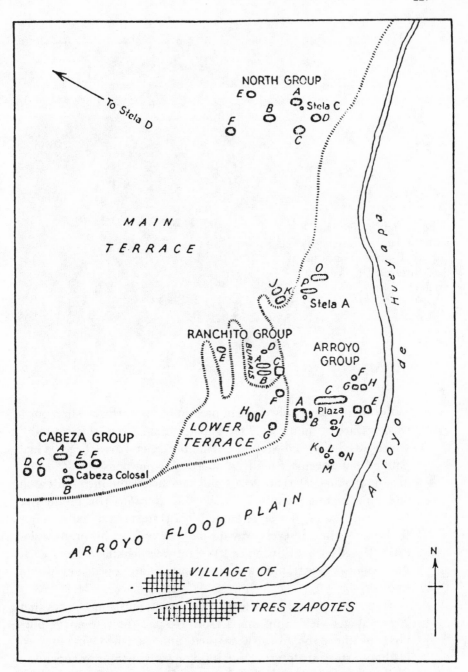

The general outline of the Tres Zapotes Site. The map is extracted from Dr. Weiant's doctoral dissertation published by the Smithsonian Institution.

The "laughing figurine" found in the mound where Emilio Tegoma had directed Dr. Weiant to dig.

Eventually, probably "ninety percent of the trichrome and polychrome sherds" found at Tres Zapotes came "from this mound," which was designated Mound C. And the general area Tegoma had pointed out became equally productive. Known as the Ranchito group, a joking reference to a small sun shelter the men built atop one of the highest mounds and dubbed *El Ranchito* (the diminutive for a ranch house of a large estate), it was the area of many major finds that winter. In every way the old man fulfilled his promise, locating exactly what Stirling and Weiant were looking for and what the National Geographic and the Smithsonian hoped they would find.

There were found all fifteen of the mysterious stone yokes that Tres Zapotes yielded up that season. No one knows the true purpose of these beautifully worked and polished fine-grained U-shaped stone sculptures. Nor is it known why some are open U's while others have a section that closes off the mouth of the U. The name apparently stems from a theory of the twenties that held the yokes were placed around the necks of persons intended to be sacri-

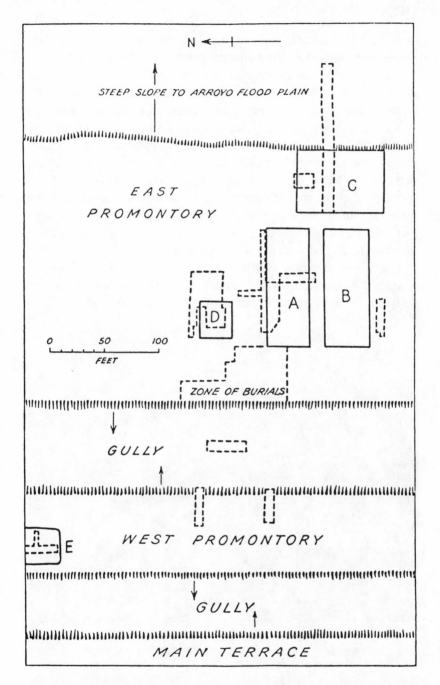

A general outline of the Ranchito Group. Dotted lines enclose excavated areas. This was the most productive part of the dig—except for
the great head and Stela C. The Ranchito Group was located when Dr.
Weiant followed the psychic directions of Emilo Tegoma.

ficed alive to the gods, perhaps to reduce their writhing when their chests were opened and their hearts removed, or to stop them from disturbing the ceremony's dignity by attempting an escape. No one knew for sure. Indeed, with the exception that this sacrificial theory is out of vogue today not much more is certain about the yokes than was known forty years ago. Some archaeologists now believe they played a role in a kind of religious ball game; but this is at best a theory, and in some ways even more improbable than the human-sacrifice one since it is hard to imagine how a man could move around a ball court with seventy pounds of stone secured to his waist. But understood or not, stone yokes have always been sought and prized by Meso-American archaeologists, and the excavation of fifteen of them at one site was a major discovery.

The Ranchito group also produced what was called the Zone of Burials, two layers of human remains. The first, only about a foot deep, consisted of cremated bodies; the second, at the four- to five-

WEIANT

Although limited to the most primitive tools, and restricted by archaeological protocol to a slow, tedious digging technique, Weiant and his workers at the Ranchito Mound were still able to unearth a considerable quantity of pottery, a burial area, and a number of strange U-shaped stone yokes. The site picked by the clairvoyant Mexican workman, Emilo Tegoma, would eventually prove to be one of the richest of the entire expedition.

foot-level, was "without signs of cremation." These were the only interments discovered at Tres Zapotes, and they played an important role in confirming Stirling's and Strong's original theory that the site was an area of multiple habitation and an intermingling of cultures.

Also discovered here was "a series of telescoped sections of clay tubing." Weiant was not certain of the function of the tubes but agreed with other archaeologists that they were religious in nature and designed for the escape of the soul. "One is tempted to speculate on the possibility that they were designed to provide a common exit for the souls of all of those whose fate it was to be buried in this cemetery."

Whatever else may have been in doubt, one conclusion was inescapable by the time Weiant and Stirling had finished the season's digging. With the exception of Stela C with its 291 B.C. date, and the *Cabeza Colosal,* which had been located by a villager clearing land in 1858, the Ranchito group and Mound C were the most productive efforts of the entire dig. (Even Stela C was an indirect result of, if not overt psychism, at least intuition. Stirling had made the find on January 16 when he played a hunch and had his crew turn over an unpromising chunk of stone. The seemingly ridiculous words of Emilio Tegoma once again demonstrated that a psychic informant and his information, when combined with a trained archaeologist's techniques, can efficiently produce results. As with the work of Reid, Pluzhnikov, and Scott Elliot, a psychically located excavation site produced within minutes discoveries that might otherwise have gone undetected or been unearthed only after hundreds, perhaps thousands, of hours of less rewarding work.

In fact, Tegoma's directions were so helpful that Weiant never needed to call on him again "since there was far more than I could hope to accomplish in a year in that first suggestion."

Weiant needed three years of archaeological laboratory research to sort out the various cultural influences and pottery types in order to make his interpretive analysis. When he finally sat down to write, he faced a question: What, if anything, should he say about Tegoma and his psychic perceptions? He had discussed the matter informally at the time with some of his colleagues, but that was a long way from giving such information an official imprimatur, especially that of the Smithsonian Institution.

On the one hand, he knew that Tegoma's directions "had made a

very significant contribution to my success." But "since I hadn't expected such a thing to happen, I had made no preparations for any
controls. Essentially, all I could report was my anecdotal account of
what took place—interesting, suggestive, but not a true scientific
test."

He also decided that mentioning the psychic would only embroil
in controversy the very significant archaeological insights gained at
Tres Zapotes. This was not a case of shirking; Weiant was no
stranger to the academic battlefield when he thought it warranted.
Sometimes he even sought such confrontations, as when he attended
an annual meeting of the American Anthropological Association
and presented a paper on the interrelationship between chiropractic
and anthropology. But be believed that controversy should serve a
positive purpose and be germane to the main issue. In the Tres Zapotes case he felt it would only be a disruptive digression. After
thinking the matter through one last time as he wrote his dissertation, Weiant concluded, "No one would pay any attention to
parapsychology combined with archaeology, even if it was conducted with the strictest controls that could be devised. Archaeologists simply weren't ready to concern themselves with an issue
like that."

He was unquestionably correct. There was no other archaeologist
in the United States with the various areas of expertise Clarence
Weiant possessed, and he knew of no other comparable experience
that might have provided him with supporting evidence. Bond had
been discredited. Language barred him from the work of Poniatowski in Poland. The Cayce material was hopelessly outside the archaeological mainstream, and in any case, it would be five years before confirmation came on Cayce's Essene commentaries and many
years more before even a preliminary correlation between the Cayce
readings and academic research was made.

Despite "the tremendous implications for archaeology in what
had happened to me at Tres Zapotes," Weiant chose to say nothing
about the matter. With only one exception he has remained silent
until now.

That exception came in 1961 at another American Anthropological Association annual meeting in Mexico City. As part of an
overall argument that it was time archaeology and anthropology
take note of the potential contributions of parapsychology, Weiant
made a fleeting reference to what had happened at Tres Zapotes.

The controversial Stela discovered by Matthew Stirling, on a hunch.

This decision to speak out was a deliberate act; Weiant felt that any controversy that might ensue was central to his main theme. He thought perhaps the mood of the discipline had changed. Considering what had happened in the almost twenty years since his dissertation was published, it was not an unreasonable inference.

During those years a body of literature had grown up that had both parapsychological and anthropological implications. Certainly, Weiant thought, others besides himself must now be considering what these two fields of research might give to each other. In light of what had happened to John Reed Swanton a few years earlier, however, Weiant's decision to speak out was a brave one.

J. R. Swanton was one of those very few scholars who in their own lifetimes become legends within their disciplines. He had trained under Boas at Columbia, preceded Matthew Stirling as chief of the Bureau of American Ethnology. On the occasion of his fortieth anniversary at the Smithsonian, the Institution published a special collection of essays illustrating his monumental influence on all phases of anthropology.

After a career of impeccable orthodoxy Swanton came out of retirement in 1952 to drop an anthropological bombshell. It took the form of an open letter to his fellows, and his words could hardly have been more straightforward. "A significant revolution which concerns us all is taking place quietly but surely in a related branch of science," Swanton said, and then stated, "It is not being met in an honest, a truly scientific manner."

He then argued earnestly for a study of the psychic by all anthropologists, of whatever subspecialty, telling how he himself had come to a point where "the thunderbolt has fallen." For Swanton, a clear choice now faced the discipline he had loved and served all his professional life. "Adhesion to current orthodoxy is always more profitable than dissent but the future belongs to dissenters. Prejudice and cowardice in the presence of the *status quo* are the twin enemies of progress at all times and (especially) of that 'dispassionate method' in which science consists."

This most uncontroversial and establishment-identified of men concluded with the unequivocal statement that to risk anything less than open-minded exploration of the psychic was to make science "a set of dogmas which the 'faithful' must accept or be damned."

His choice of words could not have been more pointed. Anthropology is a field of scientific inquiry that prides itself on recognizing and understanding how both individuals and whole societies are manipulated by all kinds of nonsense presented as rational thought.

Swanton died in 1959 so we will never know exactly what we hoped to accomplish unless this is revealed in some as yet unpublished personal papers. However, certain inferences seem reasonable. Swanton was not a stupid man nor, despite a modesty so great as to be almost a fault, one completely insensible of his position. So it seems likely that out of deep intellectual conviction he was willing to cast aside his usual restraint to assume the role of public advocate. Had this grand gesture dealt with some subject recognized as mainstream anthropology, Swanton's words would undoubtedly have had an immediate and significant impact. As it was, they were politely ignored. Good manners and respect for his position precluded criticism, so everyone acted as though the letter had never appeared.

Weiant had seen the letter. Although he had never met Swanton, he respected his work and was one of the small minority who not only sympathized with but understood what Swanton was saying. Observing that even a man of Swanton's stature could not get an objective hearing had convinced him that he must continue to remain silent. But Swanton's letter also gave him hope that things were changing.

The initial signs for what he was attempting were good. His request, addressed to Dr. Ignacio Bernal, director of the National Museum of Anthropology and History, who was serving as AAA program chairman, brought a quick and enthusiastic response: "I am delighted that you plan to present a paper on 'The Present Status of Parapsychology and Its Implications for Anthropology.' We are looking forward to seeing you at the December meeting."

When Weiant arrived at the Culture and Personality session, where his paper was the last of seven scheduled for that afternoon, the signs continued to be auspicious. As the day drew on, people drifted in until the room was full and they were standing along the walls. But, to Weiant, as important as the crowd size was the presence of the one person in the room whom he recognized, although her being there also made him nervous. From the dozens of papers being presented at other sessions that same afternoon, Margaret Mead had decided to come and hear Weiant's—the only one in the entire conference to deal with the psychic.

"I had no idea what stand she would take, but I had observed her in other sessions. Whenever a controversy arose, she always seemed [to be] able to make a statement that made any further discussion unnecessary."

As he spoke, Weiant was flattered by the close attention paid to his words, and buoyed by the applause that followed. His nervousness returned, though, when he saw Mead rise. To his surprise and relief, she not only did not attack him, she supported his entire thesis. "She said she had no objection to ESP research—indeed, she had cooperated with Gardner Murphy in his statistical research, but found playing with cards in the laboratory terribly boring. Then she went on to suggest exactly what I had hoped for: that it would be worthwhile for anthropologists to spend time in an area where sensitives are plentiful to find out what kinds of people are psychically sensitive, and why."

Weiant left the session, and the conference, feeling that perhaps the step required had finally been taken. He looked forward to research, papers, and seminars on the subject of parapsychology and anthropology. But in the months and years that followed, there was no such research. There were no papers. There were no seminars. Besides the interest of a few far-thinking anthropologists like Mead, there was nothing to show for his effort but the corridor story of an archaeologist's wife who said she had once seen a ghost in her apartment. For all their erudition, neither Weiant nor Swanton understood how a science deals with unwanted revolutionaries. First silence, then the flowing river response. Swanton's letter of 1952 and Weiant's paper of 1959 were like large black rocks widely spaced in a river—formidable in themselves, but too widely separated to make a barrier. The waters simply parted before them and closed after them, with only the slightest babble of sound.

So effective was this amnesia that when the second stage of the evolutionary process was reached some fifteen years later, the man who was to play the major role had only recently learned of Swanton's interest, and never heard of Weiant's paper at all. The roots of this third effort to bring the psychic and the anthropological together can be traced to 1970 and an anthropologist named Joe Long.

Long was in Mandeville, Jamaica, studying and comparing orthodox doctors and folk healers when "the coffin" appeared.

"It was the height of market day and both shops and street vendors had a lively trade going when the thing appeared. It was a three-wheeled open coffin apparently steering itself into the midst of the crowd. There were three live vultures perched at one end and a dead arm hung limply over the side. As if that weren't enough, a hollow voice issued from the coffin's interior repeatedly inquiring the location of one Jim Brown. Hundreds of people saw it—and heard the voice."

Long was not among them; he was in a nearby area doing other work. Arriving on the scene shortly after the event, he lost no time in questioning those who had seen, and heard, the incident.

"It was incredible. There were literally hundreds of people in that square and they all saw it, and heard the same words. More than that—and infinitely more important—they had all instantly *reacted* with behavior that showed they saw it. Within minutes the shops were empty, even of storekeepers. Everyone ran out to see the coffin and then just milled around, the way people do when they have seen something that has had a powerful effect on them."

Every person with whom Long talked told the same story (allowing for the minor differences that came from standing at different perspectives). They even broke off the narrative at the same point. "Apparently, it just drove itself down the street and around the corner. Nobody followed. You can understand why."

At first Long was inclined to think that since the Jamaicans had had time to talk the thing over, it "was a case in which one or two people have the hallucination and then the emotion of the moment somehow carries the others along." He changed his mind when he learned that "before there was any talking, they had all spontaneously reacted to the event." Long became convinced that "they truly believed it had actually happened—self-steering coffin, vultures, voice and all."

What he could not understand was how it could happen. To Long, who in those days "had read not a line about parapsychology except what appeared in the newspapers" and who knew "nothing about the boringly repetitious but scientifically important proofs arrived at in parapsychological laboratories, the answer was simple: "It could *not* have happened. I was perfectly well convinced that it hadn't. Indeed, I was rather dogmatic about that. There was nothing there. Things like that don't happen."

He was, however, honest enough to admit that "I didn't have a clue as to how to handle the matter. There was nothing in my anthropological training to prepare me for that coffin, and, had I seen it myself, I should doubtlessly have had myself committed."

Instead, he went back to his research on medicine and folk healers, eventually writing his doctoral dissertation on the subject. Through the years he said "not a word on the coffin. What was there to say?" But he never forgot the incident. "To this day, I can't explain it except to say there must have been some kind of unique mass telepathic hallucination. That's pretty weak, I realize, but how else to explain that several hundred people are in agreement about an event that cannot occur. As for a prank or purely physical explanation: If the CIA got all their geniuses together and developed the most diabolical mind-control device they could think of—well, it wouldn't equal that scene. And even if it could, why would they pick Mandeville, Jamaica, to try it out? None of it makes much sense—even now."

Perhaps because Long is one of those people compelled to make sense of things he found his anthropological research veering off the standard course and into alternative explanations. From this it was only a short step into parapsychology.

"Going over my field notes convinced me that I had witnessed a number of examples of psi phenomena, but instead of recognizing them for what they really were, I rejected that explanation because it did not fit into the model of scientific anthropology I had been taught." One can only wonder how many other anthropologists have had similar experiences that have been lost to science forever.

Unlike most of those others, Long decided to face these phenomena head on. Over the next three years, as his research progressed and he began producing what is now a long and distinguished list of papers, one central conviction grew. "Parapsychology and anthropology had much to offer each other, but because of intense specialization, there is very little crossover even within anthropological specialties let alone with another discipline. And if someone were seeking cross-discipline work, he would hardly begin with something as questionable—at least from an anthropologist's viewpoint—as psi research."

But Long put aside his prejudices and began doing interdisciplinary reading. Out of this research came an inner resolution: "Some effort had to be made to begin this cooperation." Since the effort had to start with someone, Long thought, it might as well start with him. But how to go about it?

At first he thought about publishing articles along these lines in the accepted anthropological professional journals, but he discarded this idea. Only a small fraction of articles submitted are ever printed, and in the "letters to the editor" sections of these journals only a limited exchange of thoughts could be achieved.

If articles were an unlikely prospect, Long reasoned, then the approach to take was to present a paper, or papers, at major meetings of the profession. Perhaps it is just as well that he had never heard of Weiant's 1959 effort. If he had he might have given up this avenue as having little likelihood of success.

Like all good career academics, Long is aware of how science politics works, and he knew how to plan a strategy. First he got himself on the program of the 1973 annual American Anthropological Association meeting. His topic was medical anthropology and parapsychology. He felt that by presenting a paper in his own field, he would get some idea of the profession's reception to the psychic without crucifying his career on a lost cause. The paper was responsible and well reasoned and the response was, if controversial, on the whole favorable. More important, in contrast to Weiant's experience, there was a call for more. Long decided to see if he could "put together an entire symposium for the next year's conference."

From the very beginning he recognized that unless he could "interest a group of sufficient prestige to present work" he would never get on the 1974 program. Consequently, before he said a word to anyone in the anthropological establishment, Long went about assembling a list of researchers whose "scholarly attainments were universally acknowledged, and generally not dependent on psychic research."

Long was also convinced from the beginning that this first symposium had to be as interdisciplinary as possible because "only in this way could I show anthropologists that responsible people with recognized achievements in many fields believed there was something to parapsychology and that such research had something of value to say to anthropology."

Like a man building a watch, he carefully put it all together and for his pains and good judgment was rewarded with success. By the time he approached Professor James Officer of the University of Arizona, who was the 1974 program chairman, he had a list any symposium chairman would be proud to present. With the exception of a few weak members, on whom Long felt he could afford to take a chance, it was an extraordinary group. Heading the list from anthropology was Professor Agehananda Bharati, chairman at Syracuse University. For his archaeologist Long enlisted Professor Emerson.

Although personally skeptical of psychic research, Officer and the other committee members were impressed by Long's work and panel selection. "We couldn't have called ourselves a science and come to any other decision," Officer would later say. Perhaps they also felt it was time for an open-minded discussion of the subject. The symposium was on for 1974.

As a political masterstroke, Long called it The Rhine-Swanton Symposium on Parapsychology and Anthropology.

Rhine, of course, was Dr. J. B. Rhine, the unchallenged father of what might be called the statistical school of parapsychology, the prevalent approach to psychical research in the United States. By getting Rhine's permission to use his name, Long publicly received

the blessing of the one psychical researcher even the most antago-
nistic anthropologist would be likely to know about. More than
that, he had the approval of a scientist whose work was so statisti-
cally verified—Rhine once published the complete record of eighty-
five thousand card calls using the Zener cards—that even Long's
most cynical colleagues would have to admit there must be *something* to psi.

Long next got permission from Swanton's son to use the 1952 letter and sent it out as part of his presymposium information package.

After much preparation Session 703 was scheduled to begin at 8
o'clock in the morning on November 23, 1974, in the Ambassadors
Room of the Maria Isabel Hotel. No one was prepared for what actually happened. Ironically, it would all take place, as it had with
Dr. Weiant, in Mexico City.

Before the meeting one thing had seemed certain: whatever occurred would be met with the reserved dignity that marks American
scientific symposia. It was not the seminar itself Long feared, but
the critiques that would follow, some of them months later. Scientists are not supposed to boil over—at least not in open seminars
where formal papers are presented to the entire discipline. Dis-
agreement is supposed to be polite, and an objective posture maintained at all times.

But as Long found out that day, if the subject is psychic phenomena, these rules do not apply. By 11:30 that Saturday morning, in
a room packed with anthropologists, one senior department chairman could be heard flailing away at another, "You're either lying
or cheating . . . I simply don't believe you . . . it can't happen . . . I
don't care what kind of evidence you've got."

The outburst had been building up all morning—indeed, all
week—and this exchange and the boos and hoots that followed
made Session 703 one of the most discussed seminars in a conference
that included over 225 symposia.

In a room designed to hold perhaps two hundred there were now
almost four hundred AAA members. They lined every wall and
filled every chair. It was something of a shock, after days of proper
and professional soporific monologues, to hear this exchange. The
morning had begun calmly enough with Jule Eisenbud, whose
thought-photography and psychoanalytic approach made him an
early choice by Long for the panel. There followed such other schol-

ars as Professor Robert Van de Castle, past president of the Parapsychology Association (an affiliate of the American Academy for the Advancement of Science) and faculty member at the University of Virginia.

In terms of real contributions to parapsychology, the most impressive and important figure there was Dr. Evan Harris Walker, a quantum physicist. Walker's research provided a theoretical mathematical validation of psi phenomena, and his work should have ended all the conceptual objections anthropologists had to either psychokinesis (moving objects with the mind) or ESP. After Walker made his presentation, Long thought the discussion could proceed to what psi *meant* and could be used for, rather than whether it existed. Unfortunately, other than the speaker himself there was probably not another person in the room who understood much more than Walker's introductory anecdote, a fact Walker quickly guessed. Hence he skipped most of the slides that were to accompany his talk, slides made up of the symbols that are the language of theoretical mathematics.

That they did not understand Walker did not matter to the audience; this talk was not what they had come for. They wanted a confrontation between anthropologists. As the morning progressed, this came to mean specifically Bharati taking on Emerson. After Emerson talked about George's work on Iroquois sites and the argillite carving, the crowd got what it was waiting for, though no one expected it to end with Bharati screaming and waving his hands in the air.

Bharati, a former Hindu monk, was, despite his heritage, now fervently convinced that the mystical/psychic traditions of the East or any other culture could not yield hidden or "privileged information" because "such information does not exist." He claimed that "psychokinesis is fraudulent—all of it!" Emerson, on the other hand, who had spent most of his life as the most *un*metaphysical of men, now believed such information not only existed but that mystical/psychic research was the way to get at it.

Bharati based his position on the peculiarly anthropological argument of *emic* versus *etic*. Of these two terms, first developed by linguistic anthropology, of which Bharati is a theorist, *etic* originally meant the range of sounds the human larynx can produce, and *emic* this range compared with the way a specific culture chooses and arranges those sounds from the total range available. In rough coinage, however, the words had come to have a different meaning. In the vernacular, *etic* now stands for objective "absolute" truth, while *emic* means subjective observations.

The contradiction between Bharati's emotional outburst and his paper, which called for objective *(etic)* standards to be applied to any research involving parapsychology and anthropology, was not lost on the audience. One anthropologist leaned over and in a stage whisper said to her companion, "How much more *emic* can you get than 'I simply won't believe it?'"

Eisenbud, after giving his own presentation, had been quietly watching the exchange from the back of the room. Asked to give a critique of what he had heard, he walked to the lectern and said tartly, "I take special umbrage, of course, at Dr. Bharati's statement that all psychokinetic phenomena are fraudulent . . . a flat broadside, a blanket statement. Now this is not *etic,* and this is not *emic."* Then drawing on the terminology of the physician he is, and looking around the crowded hall, Eisenbud said with a faint smile, "It shares some of the *emic* characteristics [but] it is sheer *emetic . . .* We find that some thinkers and some investigators rationalize what they are doing in terms of hard-headed super criticism, or 'scientific critique,' when what they are doing is puking. When they can't stand certain data, they puke . . . and it comes out as a paper that gets into the philosophy of science."

Lost in all this heat was the fact that of all the papers Bharati could have chosen to attack, he had singled out psychic archaeology in general, and Emerson's stories of the Iroquois work and the argillite carving as his special target. Archaeology, traditionally one of the more unscientific of anthropology's children, had, by the addition of a psychic component, suddenly become the one area that could provide clear, indisputable, testable information. Shamanism may be subject to several interpretations, but whether an artifact is where a psychic says it is, is identified as described, and is positioned as psychically perceived is not. The psychic is either right or wrong. Bharati—and everyone else—instinctively recognized this. Every question from the floor was addressed to Emerson. Only *his* paper (really more of a fatherly talk than a research treatise) was discussed.

Yet despite the fact that most of the audience, and obviously many on the panel, had cast Bharati as the reactionary spoiler, both his presentation and his later virulent attacks were important. This senior academic, respected scholar, and departmental chairman had become a symbol of the second phase in a discipline's reaction, the phase in which unwanted evolutionaries are perceived as revolutionaries. The large black rocks were becoming more numerous and now blocked the traditional stream flow. The result was an angry boiling over, a dangerous rapids in which no clear flow was apparent. To the established point of view, psychic archaeology was now a threat.

Bharati chose to play his reactionary role to the hilt. Coming to the lectern for his presentation, he ostentatiously took from his briefcase an enormous magnifying glass and waved it about saying, "Excuse this contraption . . . I must use it because I lost most of my vision looking for valid psychical phenomena in India." In his anger and excitement he said things he shouldn't have, but one must respect his willingness to do battle for what he believed.

Still, Session 703 did take place—an entire seminar on parapsychology and anthropology in a mainstream academic setting. And at least one segment of anthropology could move, if only tentatively, from inexact social science to experimentally validatable data.

Recognition of Long's achievement was not delayed. He was asked by the respected Scarecrow Press to edit an academic book on parapsychology and anthropology; Van de Castle was asked to do a chapter in a handbook on anthropology. Others also began to write. Academic books and articles make citations for other scholars to draw upon, and citations are the sure sign that open and responsible mainstream discussion is beginning. A second seminar was held in 1975; significantly, someone else chaired it, another was held in Houston in 1977 with the University of Virginia's David Barker as chairman, and Long will return to the chair in 1978 for a meeting in Los Angeles. At San Diego State University, Professor Philip Staniford, has begun explorations using psychic information in the field of cultural anthropology while, at Florida Technological University, in Orlando, David E. Jones and Ronald L. Wallace have successfully carried out several projects, including one to develop an ethnography of Folsum Man, using psychic respondents. Most revealing of all a proper, referred-to journal, *Phoenix: New Directions for Man*, was begun in 1977 with the express aim of academically exploring the formerly unexplorable (if one wished to preserve one's career) realm of the psychic in the framework of anthropology—and, already, it has a backlog of professional papers from anthropologists who wish to enter the lists. There is nothing like discipline-wide acceptance yet, but the critical first steps have been taken toward considering the psychic as a real tool in the anthropological armamentarium.

Joe Long, a quiet, almost shy medical anthropologist, had accomplished what no one else had been able to achieve. Clarence Weiant, who at seventy-eight is still practicing all three of his professions, although no longer actively digging, could not be there. But Weiant would later hear about Long's success and smile. John Reed Swanton, wherever he may be, must have felt a sense of vindication.

VIII
KUHN, CONTEXT, AND REVOLUTION

Thomas Kuhn

Right after World War II, a young graduate student at Harvard was nearing the end of his doctoral dissertation in theoretical physics when, to indulge an interest in the history of science, he became involved in an experimental course designed to present physical science to the non-scientist. To his "complete surprise," Thomas S. Kuhn found that "exposure to out-of-date scientific theory and practice," which had been discussed by way of leading up to current research, "radically undermined some of my basic conceptions about the nature of science and the reasons for it's special success."

So powerful was the impact of this course, in fact, that it caused Kuhn to fundamentally alter his career plans. After receiving his doctorate, in 1949, he shifted his research first from physics to the history of science and, then finding the standard approaches in this field inadequate, he branched out from the orthodoxy of "relatively straightforward historical problems . . . to more philosophical concerns." From there he began to develop what he had been unable to find: An explanation of the real purpose of science and the dynamic by which it changes.

Now in his fifties and M. Taylor Pine Professor of Philosophy and History of Science at Princeton University, Kuhn has never written a word about archaeology, psychic or otherwise, but his research and his theories are crucial to this story. For Thomas Kuhn's work produces the one thing that has so far been missing—context. Without it, the research of Ossowiecki, Bond, Emerson, and the others is only an anthology of psychic searches conducted over the past seventy years in countries scattered throughout the world. Each effort is but an isolated occurrence having little meaning beyond its own peculiar circumstances.

Facts are only part of what is needed to understand both the real nature of psychic research and the implications it holds for science. The other portion of the solution is found in context, which requires shifting focus from the details of specific research to the largely unconsidered principles that lie at the heart of modern science. Not an easy task, but a necessary rite of passage for anyone interested in going beyond the superficial. Kuhn is important because he is one of the few men who can serve as a guide for the trip.

His speculations and theories are articulated in a number of articles and books, but his extended essay known as *The Structure of Scientific Revolutions* (first published in 1967) is perhaps the most brilliant and original modern work in the history and philosophy of science. Here Kuhn's presentation is so elegant, simple, and profound that virtually every university offering a course in this field uses the work as a text. Even courses in the physical and social sciences employ it as assigned reading. Kuhn is also important to an examination of archaeology because so many archaeologists talk about him and have attempted to adopt his perspective for their particular discipline.

To begin to understand what Kuhn is saying, we must first deprogram ourselves of the myth and folklore with which laymen (and most scientists) are burdened. Perhaps the most fundamental of these myths is the assumption that science has, by the gradual accumulation of information over the centuries, consciously and purposefully moved toward the basic "truth" about the universe and everything in it.

This is a comforting thought but almost certainly wrong, as Kuhn demonstrates by studying the great scientific revolutions, including the Copernican, Newtonian, and Einsteinian. His evidence makes it clear that, for most scientists, this is neither their true premise nor their goal. Further, he states that even if scientists did have this as their aim, such an end could not be reached by the plodding, gradual accumulation of knowledge down through the years.

To Kuhn, an open-minded examination of history reveals a very different story. From his perspective, "The developmental process has been an evolution *from* primitive beginnings—a process whose successive stages are characterized by an increasingly detailed and refined understanding of nature. But nothing . . . makes it a process of evolution *toward* anything. Does it really help to imagine that there is some one full, objective, true account of nature and that the proper measure of scientific achievement is the extent to which it brings us closer to that ultimate goal? . . . The entire process may have occurred as we now suppose biological evolution did without benefit of a set goal, a permanent fixed scientific truth of which each stage in the development of scientific knowledge is [an improved] exemplar."

While it is possible that Kuhn's perceptions about ultimate principles are not so much wrong as self-limited, stemming from his own conditioning to the current scientific view of what is real, all other portions of his theory are not only valid but almost weirdly lucid. In his own field he is as revolutionary as the now famous revolutionaries who are the subjects of his research; like them, he has made observations that seem so profoundly correct it is difficult to understand why they were not perceived before.

Kuhn begins laying his groundwork by describing science and scientists as a special community dedicated to solving certain very restricted and self-created problems. He then presents what is perhaps his masterstroke—the construct of the paradigm. According to the dictionary, a paradigm is, among other things, "an example or model." To Kuhn, it is much much more.

A paradigm by his definition is a series of "universally recognized scientific achievements [in a given field] that *for a time* provide model *problems and solutions* to a community of practitioners" [emphasis added]. Kuhn postulates that for the scientists who use it, a paradigm is a world view. Its boundaries outline for them both what the universe contains and, equally important, what it does not contain. Further, the paradigm theories explain how this universe operates.

Having said that some specific truth is not science's goal, Kuhn then argues with great persuasiveness that the real purpose is simply puzzle solving, and that "in its normal state . . . . *a scientific community is an immensely efficient instrument for solving the problems or puzzles that its paradigm defines*" [emphasis added].

This vaunted efficiency would be impossible, he explains, without set boundaries, since without an agreed-upon framework, no observation has any greater importance or weight than any other. Without this relativity, science *per se* is impossible; the very narrowness of view is what makes depth achievable. As Kuhn points out, the narrowness increases as a science matures, and manifests itself in increased subspecialization; one is not simply a chemist but a molecular chemist. It should be obvious then that "one of the reasons why normal science seems to progress so rapidly is that its practitioners concentrate on problems that only their own lack of ingenuity should keep them from solving . . . intrinsic value is no criterion for a puzzle, the assured existence of a solution is."

This efficiency in puzzle solving underlies what Kuhn calls "normal science." To most of us, however, this is simply science, since it comprises all but an infinitesimally small amount of the research going on in laboratories and at field sites in the world.

This research, as Kuhn describes it, has only three aims, all of them within the agreed-upon paradigm. It seeks the "determination of significant facts, matching facts with theory, and articulation of theory." Kuhn never calls it this but, here is the source of the Myth of Gradualism, the conviction that scientific progress is the result of laying one research brick atop another. Obviously normal science is accumulative, but is it also innovative? Is its goal Copernican leaps, insights that will change the course of history? Kuhn says the answer is No! Normal science, he insists, is specifically *not* interested in the very thing it is popularly supposed to be obsessed with doing.

Revolutionary advances and normal science, it is now evident, are incompatible because to seek the discovery of new phenomena unaccounted for by the paradigm, or to attempt the breaking of new theoretical ground, would threaten the paradigm, which, obviously, is almost a synomym for the word *science.* A researcher engaged in threatening activities would almost by definition be practicing antiscience and be a nonscientist.

Indeed, as Kuhn points out, while it is possible to perform scientifically such tasks as measuring accurately or experimenting and recording results carefully without a paradigm, the sum total of such practices, however faithful they may be to high scientific standards, is not necessarily science. There is a critical difference and paradigm is at its core.

There is also a human cost in obtaining "progress" and "efficiency." The efficient solution of problems, as we have seen, requires an agreed-upon limit to what is attempted. To reach such an agreement—paradigm—demands a special kind of education, one that does not teach the student how to seek the "truth" but instead conditions the aspirant, by stages of initiation, into a commonly shared body of experience. The fledgling scientist concludes such an education only after demonstrating through examinations and papers that he has learned what work, and only what work, is supported by his group's world view. As Kuhn puts it, "One of the things a scientific community acquires with a paradigm is a criterion for choosing problems that, while the paradigm is taken for granted, can be assumed to have solutions. To a great extent these are the only problems that the community will admit as scientific or encourage its members to undertake. Other problems, including many that had previously been standard, are rejected as metaphysical, or the concern of another discipline or sometimes as just too problematical to be worth the time."

Such inculcation of shared perception should not be disparaged, however, for it allows the paradigm to be acknowledged as a sort of common law. And without a common law, the community would be condemned to rehash its basic role and principles with each generation of researchers—a most inefficient practice. Only when everyone agrees what game is being played can the rules governing it be assumed. Only then can the players do what all players always want to do: play the game rather than discuss it.

But achieving this agreement levies a certain tariff and requires the acceptance of some fundamental distortions, not the least of which is a false presentation of the past.

It may be offensive to historians, and it is usually hotly denied by scientists (when they think about the issue at all), but the truth is Western science has no past, excepting perhaps the careers of the teachers of those now practicing. What is even harder to admit is that there is no real need for the past. This is especially true if a science's paradigm has changed, or is about to. Past research, particularly if it operated under different rules, is unscience by definition. The only thing the past has to offer, then, is those few laws or rules that have survived the vicissitudes of progress; and these can be presented in their most condensed form since the context in which the researcher who formulated them worked, or the philosophy that motivated him, is of no interest or help to a present-day investigator.

Another factor no one seems to have paid much attention to is how a scientist communicates his research to others. From its post-classical roots until well into the twentieth century scientists usually presented their major findings in books issued to the world, and it is a popular myth that they still do so. The fact is, however, that the days of Darwin's *The Origin of the Species* or Newton's *Philoso-*

phiae Naturalis Principia Mathematica are over. As the paradigm-achieved sciences matured, one sign of their maturity was the development of jargon that might be considered a sacred language.

Before the individual paradigms had solidified to the point they have today, a scientist was also a philosopher. Indeed, until well into the nineteenth century many researchers styled themselves natural philosophers; the word *scientist* is the product of the latest scientific epoch. Early theorists in electricity such as Ben Franklin, for instance, called themselves electricians if they did not style themselves natural philosophers. As a philosopher, it is possible to talk with the world at large. As a scientist, it is not. Not only is the language that is used almost impossible to understand for anyone outside the paradigm, even when the material itself would not be, but talking to those outside the paradigm group is considered unscientific since it means addressing ideas to nonscientists.

Despite popular myth, since the twentieth century, ideas and propositions in the paradigm-achieved sciences (the "hard" sciences) have been communicated to peers not by books but through papers, seminars, and professional journals. It should come as no surprise, by the way, that paradigm-aspiring groups, which include all the social sciences, feel an emotional need to mimic this format of presentation; indeed, they tend to be excessively "scientific." Such disciplines as psychology, sociology, and certain specialties in anthropology have created a sacred language so dense that even simple thoughts are sometimes incomprehensible to other members of the group.

As Kuhn notes, a scientist's standing in the "hard" sciences is as likely to be diminished as helped by publishing a book, particularly if it is accessible to the general public or if it dwells unduly on the past. Just as there is an unwritten taboo against going to political authority to enhance a controversial position, so it is bad form to "go public." The fact that in the social sciences, books debating philosophy and publicly proposing new theories are still being written is an indication that here the paradigm-achieving process continues.

All this does not mean that books have no place in science, for they most definitely do. If the book is no longer the primary vehicle for the presentation of original work, it has another equally critical task. The book, in the form of the textbook, is currently the main processing mechanism used to condition aspiring scientists. It is essentially pedagogical propaganda, and for this reason textbooks are molded to a very specific pattern. They report only the research that supports the paradigm and its normal science techniques; rarely are alternative explanations of reality and the research that produced those explanations presented. These volumes deal with the past in only a slightly more charitable manner; it is usual practice, for instance, for textbooks on archaeology and physical anthropology to explain, in two to three pages, the complex developments that led to the acceptance of evolutionary theory. If history is summarily dealt with, past researchers get even shorter shrift. When they are mentioned at all it is with no sense of context and usually only from a distorted hero-worshipping point of view. Earlier paradigms that might have been applied to much of the same data get almost no coverage. In fact, the entire concept of paradigms is omitted.

The rationale is obvious: Why glorify what is perceived as unscience, or men who are now unscientists?

In essence, as Kuhn describes it, "Textbooks . . . begin by truncating the scientist's sense of his discipline's history and then proceed to supply a substitute for what they have eliminated. From such references both students and professionals come to feel like participants in a long-standing historical tradition. Yet the textbook-derived tradition in which scientists come to sense their participation is one that, in fact, never existed." It may be pleasant and good for morale that paleontologists, for instance, trace their professional genealogy back through the twentieth-century Kenya-born Englishman Louis S. B. Leakey, to the eighteenth-century Frenchman Georges Cuvier, to the fifteenth-century Italian Leonardo da Vinci, to the sixth-century B.C. Greek, Archelaus, assuming an unbroken continuum of research. But this is a fiction made possible only by distorted hindsight. In truth, these men operated either under no paradigm or under radically different paradigms. The only valid continuum is that they each represent an attempt to solve similar puzzles in the context of their own age.

As he undergoes this educational process, the aspiring scientist not only learns a false tradition but also tends to lose some of his empathy and ethical and philosophical overview of life. And all too frequently he also develops what in some cases is an extreme antagonism toward anything not consistent with his newly acquired perception of the universe.

The loss of empathy occurs, in part, because the language the new scientist learns to speak isolates him from all but those inside his group. The more proficient his mastery of the group's self-appointed tasks, the greater his isolation, resultant lack of objective perspective, and loss of sympathy for those who have not undergone the same conditioning. Shop talk becomes almost the only conversation possible. The degradation of his ethical and philosophic overview is also a concomitant of paradigm allegiance.

Since normal science, the servant of paradigm, is not concerned with values, at least in a societal context, but is aimed at problem solving, any task that can be successfully completed is inherently worth doing—whatever it may cost the culture at large. This the curse of rampant technology that devours the earth's assets.

The antagonism paradigm researchers feel toward anything inconsistent with, or threatening to, the paradigm stems from both a sense of personal ego threat and a perception of danger to the group with which one has cast one's lot (in a sense, the same thing).

Finally, there is the most subtle and long-range cost of all. Since science is concerned only with what can be empirically proved, it is self-evident that the metaphysical or nonempirical portions of life are considered a counterproductive line of inquiry. The scientist, unless he gains sufficient insight to join a very special group within a group—the extraordinary innovators—is almost by definition tied to a three-dimensional reality.

Articulating these concepts of paradigm and normal science is only half of Kuhn's gift to the philosophy of science and, indirectly, to our discussion of psychic archaeology. Having pulled aside the curtain of myth, he does not leave a void, but carefully fills the empty hall that lies beyond. With careful and incisive reasoning Kuhn explains how a science first attains paradigm; then how normal science, which is its result, eventually brings on the very change it does not seek; the nature of the crisis that that change represents; and finally, what occurs in the postparadigm resolution phase of a science's development. Along the way he deals with the differences and similarities in the paradigms of the various sciences, and the relationship that exists between members of a paradigm group and society as a whole.

From this discussion, although Kuhn himself does not make the distinction, it becomes clear that there is a difference between a science and a discipline, though it is only poorly understood. A group may call themselves archaeologists or chemists, and society may come to acknowledge both their name and their mission, but this acceptance does not make their practice a science. To become a discipline is a social phenomenon, not a scientific one. Only one thing makes a discipline a science, and that is paradigm. What we have called paradigm-achieved science is a tautology. It should more properly be paradigm-achieved or paradigm-aspiring discipline.

Also, because a paradigm is a world view, anyone outside of the paradigm-aspiring or paradigm-attained discipline is a layman. An M.D. is no more a member than is a plumber of the discipline of quantum physics. Through social conditioning, however, our culture tends falsely to see science as a monolithic whole, and "non-professionals" as a fragmented collection of unrelated trades.

After a group has clustered over a period of time, certain theories begin to draw adherents and schools (of thought) are formed. This is in a sense the second step on the road to paradigm. What makes completion of the task so difficult, Kuhn points out, is the fact that in a preparadigm group "all facts that could possibly pertain to the development of a given science are likely to seem equally relevant. As a result, early fact gathering is a far more nearly random activity than it will be once a framework of paradigm is established." Achieving paradigm is the accomplishment that produces values.

Gradually, this school phase begins to give way to yet a third stage of development when one school "gains status" by being more successful in solving what the discipline has set up as its most acutely pressing task of the moment. This does not mean that this school's theory and techniques are more "truthful" or that they can solve all problems. It only means that the school is more efficient and successful at solving the critical problems in question. Indeed, since by definition a paradigm is a set of boundaries, the victorious school and its theories are only designed to solve a selected list of puzzles.

Understanding this point is complicated by another myth that clouds the judgment of those who view the process. Most of us believe that one set of data admits to only one "true" interpretation. However, a researcher trained in, say, the paradigm of chemistry and another trained in high-energy nuclear physics might look at the same set of data and deduce from it very different and yet equally workable world views.

Once a view has proved successful, the school it represents draws adherents from the other schools until a kind of critical mass is achieved. At this point one set of theories predominates and becomes the entire discipline's paradigm. Obviously, though, not all members of a discipline are willing to accept the dictates of the dominant school; some have a vested interest in their alternative theories. What happens to them? Theirs, it seems, is a rather ruthless fate. If they persist in clinging to their now "unscientific" views, they are drummed out of the community, "which thereafter ignores their work." The only other alternative, and for emotional reasons it is not often taken, is to relinquish identification with one's former discipline group and align with some other discipline whose paradigm is not inconsistent with one's own. It is a rather cold comfort.

Having achieved paradigm, a group is now in the fourth stage; it begins to practice normal science. At this plateau, "The scientific enterprise as a whole does from time to time prove useful, opens up new territory, displays order, and tests long-accepted belief. Nevertheless, the *individual* engaged on a normal research problem *is almost never doing any one of these things* [emphases Kuhn]." He finds himself instead working from a different motivation, the desire to demonstrate that he is capable of solving a problem within the paradigm that no one has ever solved before, or has not solved as elegantly. "On most occasions any particular field of specialization offers nothing else to do, a fact that makes it no less fascinating to the proper sort of addict. . . . Scientists normally [do not] aim to invent new theories, and they are often intolerant of those invented by others."

As we have already noted, Kuhn outlines only three classes of problems for normal science: "determination of significant fact, matching of facts with theory, and articulation of theory." Where, then, does an Einstein, a Newton, and, in a slightly different way, a Freud come from? The group within a group mentioned earlier. And how does this extraordinary researcher's work, which is genuinely radical and not simply an extension of normal science, get into the mainstream? The answer is one of Kuhn's subtlest perceptions: The seed of innovation lies within the dynamic of normal science. Although Kuhn does not give it a name as such, there is a kind of Metamorphosis Mechanism contained within the very being of a paradigm.

Since it is by nature narrow and rigid—and this should not be construed as a pejorative description because the vast bulk of research could be practiced in no other way—normal science always produces anomalies in the course of its work, and as it proceeds inevitably to reach its boundaries, the encounters with anomalies increase. The reason is simple: before paradigm is achieved, clearly nothing can be anomalous; after paradigm, a great deal will be; and as the limits of paradigm are reached, what lies beyond is that much closer.

Normal science, however, abhors anomalies since they are not tailored to the tidy scheme by which it defines the universe. At first, then, anomalies are ignored on the assumption that later normal science research will deal with them when either instrumentation or theory articulation or both are improved. If this does not happen, an attempt is made to extend the endangered theory in the hope that an extension of the paradigm's accepted propositions will bring the anomalies back into the fold. For a while, better instrumentation or theory extension does eliminate most of the anomalies by making them conform; some, though, will not conform, no matter how artful the experiment or ingenious the development of the original premise.

If this should occur, the anomalies are left in a state of limbo. Everyone knows they are out there lurking on the edges of the paradigm like hungry wolves around a fort. But unless they do staggering damage, most problems can still be contained within the paradigm, and so, for a time at least, normal science continues, and the paradigm provides a reasonably secure framework.

However, as research continues to get closer to the edge of the "known," it pushes so intensely, and with such specific focus, that this exploration produces just the opposite effect from that desired. Not only does it fail to strengthen the paradigm, which was its original purpose, but it produces still more anomalies. These begin to cluster until so many exist that not only theory but the paradigm itself is called into question. When this happens, the science enters a fifth phase, a state of crisis from which there is no turning back.

There is extraordinary resistance to this fifth stage. Science hates crisis even more than anomalies. Researchers delay retooling as long as they can, since this is expensive and involves much aggravation. When it does become irresistible, several significant events take place.

First, the perception of the universe the paradigm has both espoused and represented begins to go out of focus. As this happens, the rigid restrictions that have dominated normal science begin to relax because researchers in the community become less dogmatic and secure in what dogmatism remains. This insecurity is reflected in the papers and seminars that normally reinforce the community's perception of itself and what it can and cannot do. Books again appear. Debates on the philosophic foundation of the community take place—an activity that is almost nonexistent in the normal science phase since philosophy is taken to be a given.

Most of all, crisis allows the reexamination of problems that were formerly assumed to be either solved or unscientific. To do this, what Kuhn calls "extraordinary research" is begun. This research, as in the preparadigm phase, begins to cause fragmentation ·and then a reassertion of schools.

When this stage is reached, two segments of the community become its critical practitioners: the most senior and the most junior. The latter are important because they will probably be the ones to engage in the extraordinary, indeed, revolutionary, research that will relieve crisis. They have been in the community the shortest length of time, have the smallest vested interest in the past way of doing business, and are most open to alternate perceptions. They may also, although there is not much unquestionable evidence for this, be at a biological creative peak intuitively as well as physically and intellectually.

The seniors are important for very different reasons. The fact that extraordinary research can articulate a new paradigm does not mean that it has solved all the puzzles that its formulation represents. By definition it cannot since that more mundane task lies within the domain of normal not extraordinary science. Consequently, although juniors may make the breakthroughs, it is the graybeards around whom the emerging crisis schools will form. Because there are few answers, and only new puzzles, practitioners within the community align themselves with new theories not only on the basis of intellectual scientific merit but also (and this is almost never admitted even when it is recognized) on faith.

Because seniors are respected and securely placed within the profession, their association with with one of the new schools carries great weight, even though the research that created the school may have come from a junior. The great bulk of the community, the middle group, responds intellectually to the juniors only after it is made emotionally secure through allegiance to that research by seniors. Emerson, in Canada, is a classic example.

One other source can produce the revolutionary innovators, and for archaeology this group has been particularly important. Occasionally, researchers from another paradigm group find themselves attracted to puzzles that have significance for the first group. Because they are not fully conditioned to the paradigm outside this field, and have less vested interest in its maintainence, these investigators function very much like juniors. Furthermore, they have a great mastery of research skills. Extraordinary advances are often the result of this interdisciplinary contact.

Regardless of whether the innovators are juniors or investigators from other paradigms, however, the final result is the same. Gradually, as in the preparadigm days, one school emerges supreme, the world is redefined, a new paradigm is established, extraordinary research is suppressed as "unscientific," and normal science can begin "the mopping up operations [that] are what engage most scientists throughout their careers." Revolution is over and the cycle begins again. And although to an outsider it may appear that things are much the same (and they are in the sense that the same words in most cases are still used and many of the old solutions are still valid), there has been the most fundamental change possible. The world of that scientific community has profoundly altered because what the universe is, and how it operates, is radically different from before.

Several questions hang in the mind at this point. One of the more obvious is: Why doesn't anyone ever talk about these revolutions as changes in paradigms? The answer, in part, is that no one has quite had Kuhn's vision. Equally important, the nature of scientific history and education requires that all the teaching books be rewritten. When the paradigm changes, so does the world view; revolutions are therefore invisible except as highly distorted hero-worshipping of certain past researchers. They are presented not as revolutionaries who tore the world apart, but simply as men and women whose vision made the science's knowledge move more rapidly forward—*but still in the same channel.*

One crucial question remains unanswered: Whence comes the inspiration for the extraordinary research and the insight into what its results mean? On this point Kuhn is almost silent. He does note that it represents a change in *gestalt,* a change in "beingness." "Normal science," he says "ultimately leads only to the recognition of anomalies and to crises. And these are terminated not by deliberation and interpretation, but by a relatively sudden and unstructured event like a *gestalt* switch. Scientists then often speak of the 'scales falling from the eyes' or of the 'lightning flash' that 'inundates' a previously obscure puzzle, enabling its components to be seen in a new way that for the first time permits its solution."

Kuhn is also willing—since the evidence is so great that it cannot be denied—to invoke the inspiration of dreams, although how this actually works he does not venture to say. In fact, he makes only one speculation on the nonintellectual aspect of puzzle solving (and it will prove important in a later chapter). He notes, "No ordinary sense of the term 'interpretation' fits these flashes of intuition through which a new paradigm is born. *Though such intuitions depend upon the experience, both anomalous and congruent, gained with the old paradigm, they are not logically or piecemeal linked to particular items of that experience as an interpretation would be* [emphasis added]."

Before leaving Kuhn's paradigm theories two final points bear examination. Although Kuhn himself does not explicitly discuss either of them, one at least is the logical extension of his work to the next level. More than that, these points are critical to understanding how the psychic relates to archaeology specifically and, more generally, to science as a whole.

Clearly, while each science—again, a discipline that has achieved paradigm—has a world view distinctly its own, chemists varying slightly from biologists and so forth, there is also what might be called a metaparadigm. For although each science has apartness, it also has certain core perceptions it holds in common with all the other disciplines that recognize one another as having attained paradigm level.

Just as a paradigm discipline tolerates a certain leeway among competing schools, and schools of thought tolerate certain variations among individuals, so individual paradigm attained disciplines are allowed a measure of latitude. But at each level membership in the greater whole implies agreement on several critical assumptions. In the case of the current metaparadigm—which, because it is the scientific expression of materialism, will be called the Grand Material Metaparadigm—there are at least four of these critical assumptions. They are: (1) the mind is the result of physiological processes governed by bioelectrical postulates; (2) each consciousness is a discrete entity; (3) organic evolution moves toward no specific goal but simply flows according to Darwinian survivalism; and (4) there is only one time-space continuum and it provides for only one reality.

Western science can be practiced because it accepts these world perceptions; without them, it would be impossible, or at least would be something else. Essentially, all sciences that accept the limitations of a metaparadigm are, in aggregate, that metaparadigm's normal science.

Under the rules, then, by which the metaparadigm's normal science is put into practice, although specific technique may vary from discipline to discipline, it is always presumed that: (1) the researcher and the experiment can be isolated from affecting each other except in controlled and understood ways; and (2) since the experiment exists in a time-space continuum, the conditions under which it is carried out can be duplicated and the experiment replicated by any other researcher if it is valid.

Although it is usually argued that the difference between the "hard" sciences and the "social" sciences is demarcated along this line of experimental replication, this may be simply an attempt by practitioners of the Grand Material Metaparadigm to deal with the anomalies produced by the "social" sciences by demeaning them as unsciences. The real problem is a basic and irresoluble discontinuity between world views. This conclusion is borne out by the fact that all but the most rabid paradigm-oriented practitioners of the "social" sciences will concede that despite decades of attempting it, the "social" sciences have yet to reach paradigm, or at least one that is acceptable to the Grand Material Metaparadigm. Why they have not, and if they ever will, remains to be seen.

One final observation: It should now be clear that there is an entire hierarchy of science within Kuhn's model, one that begins with the individual researcher; goes on to the school (sometimes literally the institution with which the researcher is affiliated); then to a discipline; then a paradigm-achieved discipline (or science); finally, a multiscience community made up of the disciplines that have achieved paradigm and share in a metaparadigm. Within this hierarchy the key figures are those individuals who produce extraordinary research not by force of intellect or will alone, although these are important, but because they have had psychic or intuitional insights at the same time that there was a crisis. Prematurity usually leads to martyrdom if the intuitional researcher pushes too hard.

However, when intuition and crisis are correctly juxtaposed, what makes these key figures revolutionaries is not only the nature of their work but also the source of their success. They are set apart because their information derives from a source Western scientific paradigms so far have not been able to deal with. At the deepest level the process by which the information is obtained is as revolutionary as the information itself, although the resistance to the psychic component in their research is never articulated as such.

The rite of passage ends here. But with the legacy of insight that it leaves, it is now possible to understand what Poniatowski, Emerson, and the others have to give to archaeology—and most important of all, what their work means to all of science and the future.

IX

PREJUDICE, PAIN, AND PARADIGM

The subject of the interview had been cordial, even gracious, and in most areas of the afternoon's talk interested to the point of animation. But now he was clearly nervous and unhappy over the turn the conversation had taken, although he knew in advance that the psychic would come up. As one sign of his agitation he shifted from looking straight across the desk to surveying his book-lined office, seeming to call on the absent authors of those works to come to the aid of a colleague forced onto the front lines. Then he reached into his shirt pocket, pulled out a pack of cigarettes, and consulted a small card stuck into the cellophane. It had little black pencil marks, one for each cigarette smoked that week. The professor smiled ruefully. "I'm trying to stop . . . joined Smoke-Enders . . . you're not helping any. Listen, can we take a break? I need to go to the john . . . and I want a smoke." He put the card back without marking it.

By the time the break was over, this very decent and quite intelligent archaeological scholar had become a politician. The ground rules were changed; he would only continue if his name was never used.

"Archaeology is in trouble, for Chrissakes. We're not a science, and I'm not sure we ever will be. Certainly, we don't have a paradigm, at least not as Kuhn defines it, if that's what you mean. But we still do a lot of good. We answer what I think are some pretty important questions about what man is and how he operates. You know the purpose of archaeology is not just the past. It's really about the present, even the future.

"Can you imagine what would happen to us, to our work, if we embraced the psychic? How do you think the other sciences, the other departments of this university, would feel about me if I did? Well, you can multiply that a thousandfold . . . or however many members there are in the American Anthropological Association.

"I don't give a damn if the psychic answers every question we ever asked—and you could prove it. I don't want to talk about it. Frankly, I don't care whether it works or not. It can't work, science says it can't. So even if it did, who would believe me?

"To admit the psychic could be a real technique would mean turning the whole world upside down. Would you be ready to take that job on? I wouldn't."

There was an awkward silence. What more was there to say? Politeness found a few words, but the interview was over.

It did not matter that he had asked that his name not be used; in a sense, it even helped. His was, after all, the opinion of the majority, and his stipulated anonymity made him a spokesman. His sentiments and the anguish with which he expressed them were common enough, and not just to archaeology. All the sciences or disciplines that operate, or strive to operate, under the assumptions of the Grand Material Metaparadigm take the same stance. That is one reason why in the West at least, psychics have rarely been used for anything beyond the statistical tests of the laboratory.

The problem is not that they cannot provide accurate information upon which successful research can be based. It has already been demonstrated in one field, archaeology, that they can. The problem is that obtaining such information is theoretically impossible under the metaparadigm, and even if its possibility were admitted, actual pragmatic use of psychic information costs too much. The cost is damage to the paradigm, and that outweighs any possible contribution the psychic might make. Since archaeology is particularly vulnerable to the temptation of the psychic, many of its practitioners overreact to the suggestion of using psychics.

There are several reasons for this vulnerability, but they all begin with the fact that although it is perceived by nonarchaeologists as a monolithic discipline, archaeology is in reality not one but several disciplines, not all of them compatible. It is even possible to take degrees from different departments in the same university—*all of them in archaeology.*

Each avenue of archaeology has its own professional societies and journals, all aimed at what it perceives as the professional community. But these societies, their journals, and the archaeology each of them espouses have little or nothing to do with one another. It is a little like being an osteopath as opposed to an allopathic M.D. The purpose of both groups is healing, both study much the same curriculum and, in practice, use many of the same techniques. But the two groups stem from different roots and therefore have almost nothing to say to each other. It is the same within archaeology, and consequently the popular vision of the archaeologist as the discoverer of pyramids and cavemen is both right and hopelessly muddled.

The oldest form of archaeology is an extension of history, almost a technique of history, whether it be in the specialty of architecture, painting, or pottery. It was one of the earliest of the Western disciplines to develop, and it traces its roots, falsely, back to the Greeks and Romans, correctly, back to the emergence of humanism in Italy and the Italian Renaissance. The most significant meaning of its great age is that this branch of archaeology developed under a different metaparadigm than the Grand Material world view that is in force today.

Historical archaeology* began and flourished during a period in which the Christian Church was not only a religious but a political and scientific institution as well. As such, Catholicism provided the very first metaparadigm within which most of the Western sciences were formed. The world view of this metaparadigm was the Bible, its dominant school the Roman Catholic Church, and its central theory Creation as described in the First Book of Moses. The most important rules of paradigm theory were that any form of evolution was impossible; all life was "preformed," immutably shaped to an appearance it would hold for all time; and man himself was the result of a Special Act of Creation. These conventions, because of their source, will be called the Genesis Metaparadigm, or just Genesis.

Although striving toward Genesis began as early as the fifth century A.D. when the Church began to consolidate as an ecclesiastical and secular authority, it would be almost a millennium before the metaparadigm was firmly established. The real impetus came in the years immediately following 1453. Christianity very early on had split into two main paths, Eastern and Western, with the Eastern centered in Constantinople. Because it was more tolerant, it was in this city that the last stagnant, but still deep, pool of classical knowledge survived.

*Not to be confused with the subspecialty historic sites archaeology, which concerns itself with such reconstructions as Colonial Williamsburg in Virginia.

When the Turks finally overran that lonely citadel in 1453, scholarly survivors fled to the great universities of Italy, Germany, France, and Scandinavia, taking with them an intellectual flame that was to light up all of Europe. This influx from the East, in fact, was responsible for bringing science back to life in Europe. When it combined with the rising nationalism that followed the breakup of the Roman Empire, a desire to know the past, both classical and local, arose.

For the first time in hundreds of years the literature, culture, and art of Greece and Rome, which had been closed books, were opened and study of their riches became a major avocation of the wealthy. An entire class of scholars grew up whose name provides a word that survives with a slightly pejorative flavor to this day—the *dilettanti,* those who "delighted" in art, history, ancient texts, architecture of the past, and the social customs of the Greco-Roman era. They studied everything, in fact, except the correct but proscribed classical speculations by such thinkers as Democritus, Herodotus, and Archelaus concerning evolution and the nature of prehistoric cultures, including a theory about a Copper, Bronze, and Iron Age sequence.

In spite of these limitations their efforts are still important. The *dilettanti* not only established collections that in time became the foundations for many of the world's great museums, but they developed a tradition of interdisciplinarism and acceptance of research outside the strict academic community that was eventually to bring to archaeology some of its greatest intellectual yields.

Most important of all, these antiquarians, as the first archaeologists, began to consider cultural differences and cultural interrelationships along a temporal continuum, as well as cultural differences at the same time in different geographic locales. These principles, of only secondary importance to historical archaeology, would carry over and become fundamental to the archaeology and cultural anthropology that developed later. In fact, this archaeology of literature cultures, which developed under the Genesis *gestalt,* has provided much of what we know about Egypt, Greece, the Roman Empire, and the Holy Land. For most people, with the exceptions of the New World civilizations such as the Aztec, Maya, and Inca, and the vague category "caveman," this *is* archaeology. Although the metaparadigm has changed, this branch of the discipline still survives and makes major contributions. The recent discovery of a

2000 B.C. Semitic civilization in Syria by two Italians, Paolo Mat-
thae and Giovanni Pettinato, is an example of that excellence.

But if much developed out of Genesis, much more was suppressed, and archaeology, because it held such potential for mis-
chief, suffered from restrictions more than most disciplines. The result, like Chinese foot-binding, was a grotesque distortion of natural
growth and the strange bifurcation that afflicts archaeology to this
day.

Although anomalous artifacts and remains probably were discovered as early as the fifteenth century by investigators potentially
able to explain them, as late as the seventeenth, James Ussher, Irish
scholar and Archbishop of Armagh, could calculate back through
the begats that litter the Old Testament to arrive at 4004 B.C. as the
year of creation, and have the conclusion accepted as fact by most
scientists of the day. Even later, in an attempt to buttress and further refine this date, no less an academic than Dr. John Lightfoot,
vice-chancellor at Cambridge, stated that his calculations proved not
only that Creation had occurred in 4004 B.C., but that the event was
consummated precisely at nine o'clock in the morning on the twen-
ty-third day of October!

This conception of time also made the most obvious source of prehistoric information, fossils, useless. Despite the surprisingly modern theory of fossilization developed around 1499 by a middle-aged
Leonardo da Vinci, who at the time was turning them up as he
dredged and repaired Milan's San Marco Canal, for most scientists
fossils were almost anything but mineralized casts of ancient plants
and animals. Some researchers held they were the products of a cosmic *vis plastica* (plastic force); others felt that they were the result of
mysterious emanations from the sun, moon, and stars; still others
were convinced that they were the remains of giants, elves, and fairies. Even more bizarre was the theory that fossils were the offspring
of "seminal gusts" that had fallen on the rocks in which the fossils
were found. The source of the "gusts" never seems to have been
specified.

When it became impossible to continue to explain away fossils because they were too obviously what they actually were, and when
paradigm researchers were forced to admit that geologic time extended considerably further back than Archbishop Ussher's few
thousand years, crisis seemed imminent. As might be expected, before this could happen there were attempts at theory extension.

The most prominent of these was the eighteenth-century development of the Catastrophe Theory, first propounded by Georges Louis Leclerc, Count Buffon, a brilliant eccentric and contradictory Frenchman who paradoxically seems to have been a closet revolutionist, and brought to its final refinement by a German-born Frenchman, Georges Cuvier. This paradigm defense held that there had been multiple catastrophes (as many as twenty-seven, according to one school), of which the Biblical Flood was simply the last, although the only one recorded, because man was created after the twenty-sixth. With each catastrophe almost all the creatures and plants on earth were destroyed. Fossils were thus finally admitted to be the remains of life, but life destroyed by divine fiat—for wickedness, it was sometimes held. When no living creature could serve as the modern counterpart of a fossil, the explanation was obvious: that animal or plant had been unusually wicked and was not recreated.

The Catastrophe Theory, of course, subjected the Genesis Metaparadigm to considerable battering. Yet the explanation remained logical in terms of the observational technology of the time. More importantly, it allowed normal science to cling to the emotionally sacred core of the Genesis *gestalt:* Man had no prehistoric past; he was the result of a Special Creation; evolution was impossible. In spite of the fact that the Catastrophists, as they were known, had the earth in an almost constant violent upheaval, science *wanted* to believe, and so, somewhat shakily, Genesis endured for another half-century.

Finally, however, so many anomalies accumulated (including incontrovertible evidence of early man mixed with prehistoric creatures) and so many advances from related fields (such as geology) brought pressure on the Catastrophists' fragile compromise position that the Genesis Metaparadigm was forced into crisis. When that happened, extraordinary research was allowed as well as philosophical debate. These two trends found their most famous figure in Charles Robert Darwin (not, strangely, in Alfred Russel Wallace, Darwin's contemporary who had arrived at the same conclusions). The theory is now known as evolution and the book that expressed it was *The Origin of the Species by Means of Natural Selection.*

Not surprisingly, neither Darwin nor Wallace was the original man to propose this theory, nor even to offer proof to support it. The first person to expound on the issue in postclassical times appears to have been Sir Walter Ralegh, as he killed time in the Tower of London between 1603 and 1618, before he himself was killed. Sir Walter had been all over the world and he was a keen observer. As he reflected on his experiences in the course of writing his five-volume *The Historie of the World,* it occurred to him that there were simply too many different kinds of animals. For years he had staked his life on his ability to estimate exactly how much cargo a ship could carry for a voyage, and he was sure Noah's forty cubits would not have been enough to do the job. Only one solution, he felt, could objectively explain this: a given number of original creatures had mutated. Since this proposition contained the germ of evolution, to many of the scientists who were his contemporaries, Sir Walter's theories seemed as good a reason for execution as his pirating and conflict with King James.

Charles Darwin's own grandfather Erasmus Darwin had become involved in the problem, and his speculations were surprisingly similar to his grandson's finished product. But Erasmus was nothing if not eccentric, and chose to publish his views in the form of an extended poem known as *Zoonomia.* Since didactic verse was not a favored mode of scientific exposition, sadly, most people ignored it.

It was necessary for theory and crisis to reach a point of confluence. Ironically, if Charles Darwin had not been such a shy and deeply religious man, concerned lest his work harm Genesis, he might, like Ralegh, today rate nothing more than a footnote in the saga of evolution and the Genesis Metaparadigm revolution. Darwin had written the basis of *The Origin of the Species* in 1838 but withheld it from publication until 1859. Had it been published when written, Darwin also would have been premature. As it was, his publication, plus the Englishman Charles Lyell's explanation of geologic principles and the Austrian monk Father Gregor Johann Mendel's theories on genetics, joined to bring about the Genesis Metaparadigm's collapse.

By 1905, with the appearance of the first four of Albert Einstein's revolutionary discoveries, Newtonian physics also underwent crisis and resolution, the physical sciences drew abreast of the natural ones, and the process of metaparadigm establishment was complete. The Grand Material world view now held unquestioned sovereignty. Its premises, the reader will recall, are: (1) the mind is a result of physiological processes; (2) each consciousness is a discrete entity; (3) organic evolution has no specific goal; (4) there is only one time-space continuum, providing for only one reality. Its rules concerning the separateness of researcher and experiment and the Covenant of Replicability became the only acceptable basis for science. All else was at best unscience and at worst mysticism or quackery.

Today it is easy to laugh indulgently at educated people who seriously entertained the idea that "seminal gusts" were the source of fossils, or accepted twenty-seven creations, or believed that the world began on a cool October morning in 4004 B.C. We should remember, however, that under the philosophic assumptions at the core of the Genesis Metaparadigm all of these theories were not only perfectly plausible but also logical and fully supported by the normal science of the time. More to the point, research derived from the philosophy of materialism is not inherently more "truthful" than research whose underpinnings had been supplied by the Christian philosophy: it simply appears to be more "truthful" today because no other "scientific" interpretation is recognized at this time.

It is also easy to dismiss condescendingly Genesis's researchers as less dedicated because their paradigm is no longer considered "science," or to assume that their work was less rigorous because of this. Many examples could be used to refute these beliefs (at heart yet another collection of science myths), but Georges Cuvier is perhaps the best choice since he is sometimes accorded the honorific "Father of Paleontology" and this is now one of the primary disciplines of evolutionary study. Cuvier considered theories of evolution and prehistoric man to be heretical fantasies, but this did not preclude his being able to painstakingly and accurately reconstructing an entire skeleton from a few fossilized bones. Indeed, he was known to contemporaries as "The Pope of Bones" and his work is still influential in this field.

He was also no less zealous than, say, a Food and Drug Administration physician pursuing alternative cancer therapies when he perceived someone dallying with the unscience of prehistory. With the full support of "science," he did not hesitate to destroy the careers of those who would not recant. He coolly and guiltlessly drove his earlier patron, the gentle Abbé Jean Baptiste de Lamarck, to what sounds very much like a nervous breakdown for espousing such theories.

As for Cuvier's conviction of the scientific truth of his beliefs, there are few modern researchers willing to go to the lengths he did to defend his hypotheses. At the very beginning of his career Cuvier had developed what he called the Law of Correlation, which, in general terms, stated that all creatures develop the form and function of each organ of their body in relationship to every other organ, and that knowing one organ makes it possible to surmise the rest. An animal with vegetarian teeth, for instance, can be assumed to have horns and hoofs, while an animal with carnivorous teeth will have ankle bones and claws.

This concept makes perfectly good sense in the light of day, but as the German scholar Herbert Wendt reports; in the dead of an eighteenth century night with no comforting bedside lamp to snap on, when something crawls through your window and stands silhouetted against the moonlight, it isn't so easy to retain that sense of conviction. When that figure calls out in a cold and evil voice, "Wake up, thou man of catastrophes ... I am the Devil. I have come to devour you," there are not many who could, roused from a dead sleep (as Cuvier was by a disguised student) maintain sufficient scientific objectivity to observe carefully the intruder, and on the strength of the law conclude that since the Devil has horns and hooves he must be a vegetarian, make this observation and, without a further word, roll over and go back to sleep.

So, the earlier researchers, in conviction and motivation, were no different from those working today. And thus the fate of Genesis serves as an antidote to the malignant humor of scientific smugness and an object lesson for future comparison when examining the contributions of psychic data to archaeology, or any other discipline.

Not that the collapse of Genesis did not cause fundamental change: it did, and no area of research experienced this as much as the disciplines dealing with man and nature. Suddenly humankind had a past; evolution of the forms of life was not only possible but a key paradigm assumption; and *Homo sapiens* stood at the head of the class, not by virtue of Special Creation but as a result of that evolution. The foot-bindings were off, and between Darwin's *Origin of the Species* in 1859 and Einstein's theory of relativity in 1905, the study of man's newly opened past coalesced into the discipline of anthropology and its daughter movement, the field of anthropological archaeology.

This new expression retained many of the techniques developed by historic archaeology under the old metaparadigm, but for all that there was a fundamental difference not only in its philosophic basis but also in the task that the new archaeology set for itself: to study, through objects, ruins, and remains, man's cultural and social past; to order the past chronologically, then reconstruct it; and finally, to explain the *gestalt* (beingness) of the people under observation. The concern was now with people rather than with their objects.

At first the differences between this goal and the study of objects as an end in itself seemed only an obvious and unavoidable response to new conditions. The earlier form of archaeology had developed as a technique of history because there was no need for it alone to explain what anything meant, or why it had developed in just a certain way, or even what it was used for. Since this archaeology was limited by metaparadigm to cultures that went back in time only so far, a written record was usually extant. Although this was more apt to be stone hieroglyphics than books, if reconstruction was attempted, this material rather than the objects themselves was the first, easiest, and most efficient source of information. Archaeology simply had to locate it, dig it up with a sense of context, and preserve it.

But this relief from the responsibility for reconstruction was ultimately detrimental. It meant that no world view or philosophy specific to historic archaeology had to be developed, and without such a foundation there could never be a paradigm, no matter how sophisticated the techniques employed in research. It is a situation that still exists.

Although they rarely said so publicly, it had always rankled those who studied the past that, no matter how vehemently they made the claim, their field was not considered a science by those disciplines that recognized one another as having achieved paradigm. Now the new ground rules of anthropological archaeology not only allowed investigation without time limit into the past, which was exciting, but the nature of the problem—older cultures have no written record—meant that reconstruction would be an inescapable part of their work. It had to be extracted from the artifacts, remains, or ruins themselves, and this meant that, for the first time, archaeology was in the emotionally satisfying position of being independent of other disciplines. It could make its own way, develop and test its own hypotheses, just like any other research field. It could even develop a philosophy, achieve paradigm, and become a science—if only it could provide testable reconstructions and explanations!

At first not much effort was made to reconstruct, let alone explain. The discipline was just forming, sorting itself out, classifying and dealing with the anomalies that had been accumulating for so long under Genesis. There were strange mounds in Ohio, vast and largely uncharted ruins in the Americas, prehistoric skeletons and cave art in Europe, and an entire list of other projects that had either been ignored or explained in some other manner. There was not much concern about locating things since there were more than enough sites to go around. The problem was learning to dig them in a professionally competent manner, once it was determined what that should be. One of the best sources to determine this was the third President of the United States, Thomas Jefferson, whose excavations had been carried out at the end of the eighteenth century.

In his thirties Jefferson reached an ebb in his career. He was ending a second and unhappy term as the Virginia Commonwealth's governor, had recently endured the ignominy by being forced to flee the capital, Williamsburg, when the British overran the commonwealth, and seemed to have little choice but to retire from public life. In spite of this turmoil Jefferson found time to examine the mystery of the American Indians and their place in the New World.

Before 1781 he decided that the answer to their present might lie buried in a past that could only be rediscovered by excavation. He surveyed a number of possible sites and finally settled on one. With a work crew probably made up of slaves, he rode over to the banks of the Rivana River some few miles from his home Monticello. Above the Rivana's principal fork he began a careful layer-by-layer excavation of an Indian burial mound. As he dug, he made a complete record of the context in which bones and artifacts were found, noting their relationship with one another, the geologic strata, and the positioning.

Jefferson saw digging as only one phase of the solution. He was convinced that artifacts should be sought not just because they were collectible objects of beauty, but also because they provided a method of obtaining chronology as well as material that could be used to synthesize a cultural interpretation. It was his interpolation by analogy from present back to past, his comparison of excavated remains with then current tribes, his comparison of tribal languages, and his theories about the effects of nutrition and environment on individuals that mark Jefferson as the first modern archaeologist and a modern cultural and linguistic anthropologist. This minor dig, then, on the banks of a small Virginia river comprises the first anthropological archaeology in the world. But like the work of da Vinci, it was hopelessly ahead of its time.

Jefferson's approach to archaeological excavation was so premature it was to languish almost forgotten for close to a hundred years. Only with the end of the nineteenth century, when the change in metaparadigm was well under way, would his work be rediscovered and refined. And not until the 1920s could the problem of reconstruction, that mandatory first step toward explanation, begin to be resolved.

The first great effort at reconstruction was made by V. Gordon Childe, an Australian who at the time was Abercromby Professor of Prehistoric Archaeology at the University of Edinburgh. His story is worth going into because Childe was more than an innovator whose theories profoundly influenced the discipline for more than a generation; his experience illustrates what was going on in orthodox archaeology as its anomalous cousin, psychic archaeology, was developing. What happened to Childe has happened to many other researchers in this century, although Childe is perhaps the clearest example.

In his work it is possible to see both the brilliance of archaeological investigation and the inherent shortcomings of traditional archaeological methods that that brilliance has never been able to overcome. His orthodox work serves as a counterpoint to the psychic location work of Scott Elliot and the Russian rod-walkers, or the reconstructions of Ossowiecki and McMullen.

Childe is one of archaeology's legends; in any age or science he would be a character worth noting. In a discipline whose hallmark is digging, he dug hardly at all, preferring to be the community's Mycroft Holmes. Colleagues brought him problems, or he simply surveyed the field from his armchair and synthesized the solutions others had not seen. The early 1920s found him busily at work in just such a manner as he looked out over Europe from his vantage point in Scotland and observed that while the archaeologists in each nation were trying to make reconstructions in their own special areas of interest, no effort was being made to pull the discoveries into a coherent whole. To Childe, this seemed a critical task, so he set about doing it.

He chose as his point of entry a dating technique recently postulated by the Englishman William M. Flinders Petrie. Although primarily an Egyptologist, and hence a literate-culture archaeologist, Petrie's interests lapped over into the anthropological side of the house. He had worked out a method of dating finds and comparing cultures based on pottery and independent of the written record. Childe decided to use Petrie's method for his synthesis since pottery, once it had been introduced in an area, was almost sure to turn up no matter what kind of site was being dug. It was one of archaeology's universals.

After assembling the pottery collections he would use, Childe knew he had geographic distribution. The problem he faced was in arriving at a time relationship between cultures at the same stage of development but separated by distance. When the same type of pottery appeared in different parts of the world, Childe wondered, what did that signify?

Petrie had theorized, on the basis of an unproved but logical assumption, that similar pottery in any culture followed the same relative sequence; glazed pottery, for instance, came after unglazed pottery because it represented greater technical sophistication. Childe accepted this and took it a step further.

He began by making another assumption: similar pottery was found in a certain sequence because civilization began at one place and then flowed out from that area until it encompassed all the known world. But how did the flow run? Since anthropological archaeology was a child of Western culture, like its mother, it had always looked back to Greece and the Near East. There were many reasons for this: habit, recent history, religious conviction, and accessibility. There was also the archaeological fact that the oldest known recorded dates were Egyptian. With this emotional and intellectual bias, Childe made another assumption. Civilization, he reasoned, flowed East to West, from the Near East to Europe.

This made marvelous sense. There was also a certain elegance in believing that civilization had spread like the rings caused by dropping a pebble into a still pool, successive ripples representing waves of culture moving out from the center. It was such a satisfying reconstruction that twenty years after being proved wrong, schoolchildren are still being taught this as unquestioned truth.

Childe published his theories in 1925. Revealingly, he chose a book as his medium, entitling it *The Dawn of European Civilization*. For decades its views were to the the final word on this subject. The only problem was that as normal archaeological science progressed and attempted to develop both new techniques and improved technology, instead of buttressing this established position it produced anomalies. Then, too, questions arose about what role the Far East might have played, or at exactly what dates civilizations had attained certain cultural levels. Metal technology was a prime example. There were also many questions about the cultures of Middle, Central, and South America: reconstructions of the Maya, Toltec, Inca, and Aztec civilizations, begun with such assurance, were now beginning to resemble suppositions wrapped in assumptions.

Theories were soon tagged speculations and considered suspect, which put archaeology in an untenable position since theories were the road to reconstruction. Without them it was impossible for archaeology to be anything more than a discipline dedicated to the collection of things. Childe survived, partly because there was as yet no absolute case against him and partly because dismissing his theories left a chaotic void. It was one thing to have problems in such psychologically peripheral areas as Thailand or Peru (peripheral only because archaeology reflected European political consciousness), but to attack Childe was to attack the heart. It was better not to talk too much about the matter of civilization and Europe.

By 1940 the situation had deteriorated so much that Harvard professor Clyde Kluckhohn felt moved to write (about Middle American archaeology, but by extension archaeology as a whole), "To begin with, I should like to record an overwhelming impression that many students in this field are but slightly reformed antiquarians . . . the industry of workers . . . is most impressive as is their technical proficiency and the scrupulous documentation in their publications, but one is not carried away by the luxuriance of their ideas."

Eight years later another professor, and former Kluckhohn student, Walter W. Taylor, was so upset with his discipline that in *A Study of Archeology* (sic) he noted that "the task will be to analyze what the archaeologists say they have been doing and what they have actually done and then to see how these two bodies of fact compare." After making a meticulous attempt to provide such a comparison, Taylor was forced to conclude, "It is the blunt fact that archaeology, as it is currently practiced . . . is neither historiography nor anthropology (that is, the study of man)."

Taylor went on to suggest a new interdisciplinary synthesis he felt would solve the problem. But as confused as the situation was, it was not quite desperate enough to give up the old ways: Taylor's reception by his fellows was distinctly chilly; he was premature. A small minority listened, however, and their number grew.

Then, between 1945 and 1959, a researcher from another science got interested in archaeology, a field in which he had no vested interest, and ultimately, although indirectly, he deprived the community of the succor of Childe's reconstruction. The scientist was Willard Frank Libby, a professor of chemistry and specialist in nuclear research at the University of Chicago. His contribution to archaeology was the now famous radiocarbon-14 dating technique, for which he won the Nobel Prize in 1960.

By measuring the level of deterioration of carbon-14, a substance that is present in all organic matter and that decays at a known rate, Libby gave to archaeology an entirely new method for reconstructing chronology. The discipline, dazzled by the attention and technology from one of the most esoteric of the "hard" sciences, and certainly in need of what it had to offer, thankfully and immediately accepted Libby's gift.

At first it appeared that Childe's theories would be buttressed by the new method. But as the work progressed, an anomaly developed in the one place where none should have occurred—the unimpeachable Egyptian dates. Carbon-14 produced a chronology hundreds of years closer to the present than the hieroglyphs indicated. What had happened? After a rather nervous time it was proposed that the discrepancy was due to inaccuracies that resulted when the Egyptian calendar was matched to the modern one. Either that, or in addition to it, the samples upon which the carbon-14 Egyptian dates were based had become contaminated with more modern organic matter as they sat on museum shelves. Everyone relaxed, congratulations went round—it seemed technology was going to allow the firm validation of a reconstruction. Childe, now dead, had a new claim on life. A few years later, however, a young professor at the University of Arizona, Charles W. Ferguson, got interested in the puzzle and began to wonder if the explanation for the discrepancy was as accurate as everyone seemed to think.

Ferguson was an expert in dendrochronology, dating by counting tree rings. The idea that these clear-cut growth layers could be used to calculate dates was an old one. Leonardo da Vinci, just as he had with fossils, seems to have been the first European investigator to make the observation. But it was not until A. E. Douglas, also at the University of Arizona, began to study the subject that it developed into a useful and accepted tool. Ferguson was carrying on Douglas's work, and in conjunction with several researchers in California, using the bristlecone pine, he had so refined the method that he could date as far back as 9000 B.C., accurate to a single year. This was a far more definite method than C-14, which has a plus or minus factor of up to two hundred years.

The challenge of comparing C-14 dates with his tree-ring chronology was too obvious for Ferguson to ignore, and so he took it up. Soon he concluded that one of the basic assumptions of radiocarbon dating, that the world's atmosphere always contained the same amount of C-14, was in fact wrong. He believed it had changed, perhaps because of man's increased use of combustion in one form or another as his world became technically more sophisticated.

Ferguson based his case on what at first seemed an equally untested premise: that when the weather changed in California (affecting the growth of tree rings), it also changed throughout the world. Such an assumption made his conclusions as vulnerable as those derived from carbon-14 dating. However, when he began his Egyptian tests a very important agreement appeared; Ferguson's dates agreed with the hieroglyphic dates! His case therefore seemed proved and archaeologists accepted his correction factor for C-14 results. That seemed to be it, until the correction factor was applied to European dates and a ghastly anomaly appeared.

Childe had based his theories not only on pottery but also on the mysterious stone monoliths known as dolmens. These peculiar stone structures were, he felt, a sure proof that civilization had flowed West, since they could be found from Greece to England. With Ferguson's tree-ring correction factored in, however, Stonehenge was clearly hundreds of years *older* than any such stonework in the Near East. This conclusion was fortified by the trade pottery found in the area, which was in turn keyed in to Egypt. Then copper implements discovered in Hungary were dated at 4000 B.C., far earlier than similar Near Eastern metalwork. Suddenly nothing was secure.

The issue was further confused by the discoveries of Pisit Charoenwongsa, curator of the National Museum in Bangkok, Thailand, and Chester Gorman, of the University of Pennsylvania. Following leads produced in the 1960s, they discovered Bronze Age implements that were as old as 3600 B.C. near the small Thai farming village of Ban Chiang. This find predates similar discoveries in China by one thousand years, and in the Near East by six hundred. It also means that a Thai copper period must have existed even earlier, or as Gorman puts it, "These people had an understanding of metallurgy that seems to have been unparalleled in any other area in the world at that time." Until the next discovery.

The situation was further complicated by Gerald S. Hawkins's spectacular, and still controversial research, slightly earlier, which showed that dolmens in general, and Stonehenge in particular, were the work of prehistoric builders possessing highly developed astronomical and mathematical abilities. The traditional archaeological attempts at reconstruction become very tortured indeed. Even ancient man was undergoing revaluation, in considerable measure because of the work of the Peabody Museum's Alexander Marshack, whose study of calendar bones appeared to prove than even "cavemen" were capable of calendrical observations and had the math to record them.

When one also includes such anomalies as the pyramid mathematics discussed by Peter Tompkins and Professor Livio Stecchini in Tompkins's book *Secret of the Great Pyramid* and the growing academic interest in Atlantis, the comfortable assumptions about the past seem muddled beyond archaeology's powers of clarification. Ignoring these anomalies as wildly controversial has been the traditional archaeological response, but even when Atlantis and other such material is eliminated, there is still a great deal unresolved, unreconstructed, or unexplained. Even without considering anomalies at all, there are difficulties that cannot be swept away, one of the worst being the problem of location.

If little attention was paid in the first decades of this century to reconstruction, even less was focused on location. Despite the outcome of earlier controversies, such as Heinrich Schliemann's disputed discovery of Troy by following Homer's ancient poems, the question of location hardly arose. There were simply too many sites, too much to dig; the problem was not where to dig, but what to dig first.

By the 1920s, though, site location had become a subject of concern. Most of the known major excavations of interest had been begun, and what was already excavated raised questions that only additional digging could answer. Puzzles about exactly how two cultures had overlapped is an example.

The development by the English archaeologist R. J. C. Atkinson of aerial surveying, which he perfected in cooperation with the RAF by utilizing World War I aerial reconnaissance techniques, at first promised the answer to location problems. And many *were* solved, but many more remained. There were also solutions to be found in the growing understanding of what kind of building, burial, and refuse patterns to expect when excavating a given culture. It became possible, for instance, to say that if a burial ground was uncovered, it could be assumed that village ruins must be nearby. For a while it seemed that normal science would solve all future locations problems, and certainly it has helped.

However, these new solutions together did not cover every situation, and worse, sometimes created additional problems. For example, in the study of prehistoric man the problem got worse as the more obvious sites were worked out and new ones were sought. Then there was the question of cultures that were not known to exist, and so could never be consciously searched for. The first and most basic of all archaeological tasks—location—is as perplexing today as ever.

It could hardly be otherwise because the location of a site, when all that may remain are skeletons and some stone tools, is a proposition of almost incalculable difficulty. Some idea of exactly what is being attempted can be gained by comparing the archaeologist's difficulties with those of the United States Navy, which constantly seeks to pinpoint Russian ships, especially submarines. Using satellites, underwater sensors, and all the other hardware that only billions can buy, it is still an imperfectly performed operation. Even when the ship is identified and the waters in which it is sailing known, sometimes down to a relatively small area, location success is far from certain. And if the ship is a submarine cruising a hundred feet or more below the surface, the odds are even slimmer.

Imagine, then, trying to locate a five thousand- or ten thousand-year-old site buried beneath many feet of earth. Small wonder that location remains largely a serendipitous occurence, as a survey of recent major finds illustrates: Chinese peasants digging a village well discover an "army" of perhaps six thousand life-size ceramic soldiers and horses, thus opening whole new chapters on the 221 B.C. dynasty of Shih Huang-ti. A young graduate student stumbles in Thailand and unearths a vase that begins the search that results in the discovery of a 3600 B.C. Bronze Age culture. A bulldozer operator cutting a roadway along the Rio Grande in New Mexico turns up signs of man in the New World thousands of years before he was supposed to have been in the area.

The problem of location has also had a deleterious effect on archaeology's efforts to achieve reconstruction, and that in turn has contributed to making the next step—explanation—difficult if not impossible in many cases. To understand a culture it is necessary to know that the things one is discovering represent a meaningful cross section of that society. If the site is not an old one and represents an urbanized cluster, or is known perhaps through historic records, then there is greater assurance that the material coming out of the ground is a reasonable cross section. However, if the site is very old, the culture very primitive or nomadic, or if the site is only a temporary camp, then the challenge is much greater. A real cross section under these conditions is virtually impossible from one site and only problematical from several.

To get some idea of the complexity of the challenge, project yourself fifteen thousand years into the future. Among the remains of the high-technology cities of which Western cultures are so proud would be toilet bowls, possibly some forms of plastic—and nuclear waste. Electronics, television sets, telephones, antibiotics—in fact, many of the essentials of sophisticated modern life—would have vanished. It is interesting to speculate what kind of reconstruction would be made of this surviving selection of artifacts. Toilet bowls might be seen as household shrines; plastic, some kind of sacred material; nuclear waste, perhaps a sacrificial mechanism.

Even before reconstruction could be attempted, how would future archaeologists know where to look? Could Manhattan be reconstructed fifteen thousand years from now through the discovery of Appalachian West Virginia? And what would the investigators make of "contemporaneous" Colonial Williamsburg and nearby modern Norfolk. This futurist illustration is, admittedly, almost certainly inaccurate because some written records might survive and future archaeologists might have techniques we could not begin to imagine, but still it illustrates the enormity of archaeology's task as it struggles to become a science.

The sad truth is that anthropological archaeology, using modern technology, became in many instances little more than a psychotherapeutic journey into history via objects. The artifacts and natural remnants performed for a culture the same function as dreams and symbols for a single human psyche. Thus few, if any, absolute interpretations were possible. Indeed, archaeological reconstruction was one step removed from the answers it sought because it viewed only objects produced and not the minds that produced them. These serious qualifications did not keep certain researchers from flights of fancy in which access to an entire culture was postulated via a random collection, or even a single artifact. More responsible, or at least less fanciful, archaeologists recognized that despite radiocarbon dating, thermoluminescence, dendrochronology, pollen analysis, and a host of other technical advances, in spite of the excavation of thousands of sites and discovery of tens of thousands of artifacts, the essential information remained locked away. Archaeology was only what the Museum of Northern Arizona called it—"a shadow" discipline.

The realization of this, even if only at the subconscious level, is what drove Kluckhohn in 1940 and Taylor in 1948 to propose the conjunctive approach. This same awareness in 1958 led Gordon R. Willey and Philip Phillips of Harvard to state, "Archaeology is anthropology or it is nothing."

This, plus the recognition that archaeology has rarely been able to locate sites at will, or advance testable reconstruction hypotheses of the sites it did discover, finally brought the discipline to a crisis. After over a century of hoping that technology would provide the key to paradigm, it became evident that it was leading instead into a maze of new puzzles.

By the 1950s a new movement developed: the path lay, it argued, in assuming that culture was evolutionary and followed certain laws. If one assumed that all humans operate from certain universal urges, that they need to eat, to copulate, rear families, and that all cultures grow through a series of universal and predictable stages and respond always in the same way to specific stimuli, then perhaps it would be possible to *prove* reconstruction. While not going quite this far, Willey and Phillips, for instance, did feel that general regularities did exist. This trend finally culminated, however, in a presentation by one of the great mavericks of modern archaeology, Lewis R. Binford. Entitled *Archaeology as Anthropology*, it formed the basis for what has become known as "the new archaeology."

There was only one problem: all the evolutionary laws were *assumed*. The state of "beingness," the consciousness or *gestalt* of a culture was presumed to be the same in all cultures. Unfortunately, as studies of Australian aborigines and the teachings of the Shoshone medicine man Rolling Thunder have shown, this assumption is highly questionable. Cultural evolutionary theory is logical and does seem to work when applied to physical needs or biological urges (people do eat, make love, and rear children in every culture of every race), but beyond these overt material phenomena there is a barrier that cultural evolutionists cannot leap.

Why did the Maya build an empire in an area devoid of the resources necessary to their culture? What was the purpose of those strange runway-like markings found on the Peruvian pampa? What did a Chinese priest feel about his society? No matter how carefully chronicled, the detritus of a culture can never bring it back to life. Climatology may reveal that the rains stopped coming and so a tribe was forced to find new territory, which, in turn, accounted for an Indian war. But it cannot reveal how the people felt about all this, or how they explained the failure of the rains. The relevant information about the past, the data needed to answer these questions, is simply not available through traditional archaeological methods.

This unpleasant but inescapable conclusion made it clear by the1970s that paradigm achievement was as far from the discipline's grasp as ever. In fact, Willey, writing in conjunction with another Harvard colleague, Jeremy A. Sabloff, felt compelled in 1973 to say, "We do not have, and we are unlikely to get . . . empirical laws that deductive explanations demand." The only hope they saw was a change in the paradigm ground rules, which they acknowledged would probably not be acceptable to other sciences.

How much this sense of scientific inadequacy and frustration actually touched the working life of average archaeologists is difficult to assess. Certainly many practitioners do not think very much about theory and have no desire to alter the status quo; complacency is the hallmark of any majority under most circumstances. But by 1974 the sense of blockage and feeling of crisis had become sufficiently intense for serious discussion at seminars around the country of what at first sounds like a suicide pact—a moratorium, an end to all digging, until archaeology developed a better approach to its chosen task. One that made the location of sites selective and efficient, and the interpretation of what was brought out of the ground understandable, useful, and relevant. One that offered testable reconstruction, even explanation.

The most convincing sign that even the silent middle group had been touched was an open letter written in 1974 by Charles

McGimsey III, in his official capacity as president of the Society of American Archaeology. McGimsey's point was simple. Archaeology, he said, must stop taking an attitude of "keep your heads in the pits, boys" and start coming up with "innovative approaches."

The worth of practitioners of psychic archaeology is clear. It is the anomaly to normal science that, now that the Metamorphosis Mechanism has been activated by crisis, points the way out of chaos to the security of resolution and, finally, to the achievement of paradigm.

Bond's major problem was not a prickly personality but prematurity. He was using a method with which the metaparadigm even now cannot easily cope, and which the raw young discipline of archaeology in 1918 could not possibly have assimilated, and so he was crushed. The tragedy of Poniatowski and Ossowiecki is even crueler, the loss more agonizing. They too were premature, but because of Poland's social structure and their place within it, they were protected from the cruder criticisms Bond had to endure. Had it not been for the war, Poland might well have led the way in the application of such information, and psychic archaeology might have entered the mainstream years before Long's 1974 Mexican success.

It is understandable why the archaeological community chose to ignore the venerated Swanton rather than attack him. To attack him at the end of his life would have been to attack, if only indirectly, everything he had said, and his earlier work had provided much of the foundation for archaeology's self-image. Weiant risked more than Swanton because he was not a senior in the same sense, and his career could easily have been destroyed. Only his earlier outstanding work (which included the Tres Zapotes dig, whose psychic component was unknown), his financial and emotional independence, and his freedom from such leverage points as academic tenure made his professional survival possible.

Long's achievement in organizing the Rhine-Swanton Seminar is a reflection of both his abilities and the relaxation of archaeology's world view brought about by crisis. In the same light, Emerson becomes the transitional man that Poniatowski might have been: a senior around whom others could cluster for security as they blazed a new trail. (Pluzhnikov may serve the same function in Russia.) General Scott Elliot, Evan Harris Walker, and others are the "men outside," the researchers who are not career archaeologists either because they are dilettantes or because they belong to other disciplines. Like Willard Libby in the twentieth century with his radiocarbon dating, or Charles Lyell in the nineteenth with his theories on geology, these psychic researchers offer answers to problems that archaeology cannot itself supply because of the limitations of its present vision.

With Kuhn's sure hand as a guide it is now possible to discern not only the role these psychic researchers have played for the past seventy years, but the probable contribution they will make in archaeology's search to become a science. The key lies in McGimsey's call for "innovative methods," a phrase that sounds remarkably like Kuhn's "extraordinary research."

Indeed, it hardly seems a disputable point that a psychically sensitive person, working in conjunction with a trained investigator, offers archaeology at least the opportunity to resolve its crisis. In spite of the rancorous arguments—which are sure to continue for a time—there is now sufficient evidence that a new way lies open. Here at the very-least are solutions to problems of excavating a valid cross section, or selectively locating the single site necessary. By utilizing the sensitive's information systematically, it is possible to reverse the traditional process—that is, instead of locating a site and *only then* framing hypotheses, one can develop a hypothesis based on psychically obtained data *prior* to digging. The dig itself then becomes a test allowing verification at four levels before reconstruction is even considered: (1) testable information on site location; (2) testable information on surface geography and subsurface geology; (3) testable information on what artifacts and remains will be found and at what depth; (4) testable information on the specific positioning of those finds; e.g., "the broken blue edge points straight down."

Not all the pioneers concerned themselves with every level of this ascending hierarchy of psychic data—Emerson, for instance, knew where the site was but needed information on houses and artifacts; Weiant did not ask for more than location itself—but the general framework for all four levels has been forged and awaits tailoring to new situations. Using this psychic format, these researchers were finally able to effect the scientific "if this . . . then this" presentation that has, for so long, eluded all other approaches to archaeology.

More than that, their work offers not only reconstruction but in many cases explanation. The question is: Can the reconstruction, let alone the explanation, be accepted as accurate? In a sense there is no final answer to this query. To be completely accurate, a reconstruction would have to take as long as the original it seeks to represent, involve all the same people, and duplicate the environment; practically, a *re*construction is impossible. And an explanation about what it all meant to the people who lived it would require a totally successful psychoanalysis of the entire population. Ultimately, then, these two puzzles can never be fully solved, only approximated. However, the approximation achieved by an Ossowiecki or a George McMullen—*after* he has provided other related information that has been successfully validated at up to *four* separate stages, each of which has then been evaluated by trained researchers, compared with any existing historic records, and subjected to all the applicable standard techniques—logically must be considered more believable than a traditional reconstruction that is developed *after* digging based on a random assemblage of material serendipitously discovered. The psychic method, properly applied, *employs all the steps traditionally used,* and then adds verification hurdles orthodoxy could not even attempt.

Once its complexities are understood, this psychic work points the way toward the development of an intuitional model, which could be applied in *any* field to almost *any* research problem.

If so, why is resistence to it so high? The answer is that within the terms of the Grand Material Metaparadigm psychic information is impossible. Where could it come from? How could it have been preserved, through millennia in some cases? Where could Cayce or Ossowiecki have possibly gone in the accepted time-space continuum to get information either unknown or contradictory to the knowledge of the day until excavation proved it to be correct? How, for instance, did Cayce know there were women in the Essene community at Qumran?

In another area the implications of the psychic are even scarier to anyone aspiring to the title *scientist*. Psychics come equipped with several different levels of ability; some guess cards, some are psychometrists, some move things, to name only a few of the complex and little understood skills possible in this field. But for all their variety they break fairly neatly into two main categories: those who do and those who do not discuss the nature of man.

Beyond certain very limited applications (unfortunately for science), those who can provide information of use in applied research almost invariably fall into the first category: they talk about man. Worse yet, they tend to regard him, if not as a spiritual being, at least as one whose true identity and consciousness exist independent of the world science thinks of as reality.

The major exceptions are dowsers, which is unquestionably one reason scientists in the Soviet Union were allowed to attempt research with dowsing in the archaeological field. One can only smile in wonder at the reactions of Marxist materialists to, say, Bond's Band of Watchers calmly mixing accurate and testable archaeological location and reconstruction data with descriptions of the transcendental nature of consciousness or statements concering reincarnation as a significant factor in understanding man's history and development. If this is impossible for Russians to accept, the truth is it is hardly less difficult for Europeans or Americans. It is no coincidence that the majority of parapsychologists have chosen as their field of research such things as card calling, dice rolling, and dreams. They will do almost anything rather than openly deal with clairvoyant readings on the order of Cayce or even McMullen.

Although it may seem a contradiction, parapsychologists also are striving for paradigm and so, even in a discipline that by its very nature appears antiscience, they do their best to conform to accepted standards. This is one reason why there is so little theorizing about the source of dynamics of psychic energy. To push too hard in that direction is to come into unavoidable confrontation with man's nature. This more than anything else has split academic parapsychologists off from "those interested in the psychic," even if "those interested" have Nobel Prizes in other fields.

It is not really surprising that it should be this way: these resisting men and women, like everyone else, are children of their culture. With the exception of a few rare individuals, researchers of every era have shied away from the loneliness of purgatory, the isolation that comes from nonconformity—whether it be to religious principles in the eighteenth century or to scientific principles in the twentieth. To be academically associated with psychic research is daring enough. The only defense against ridicule and ostracism is to publish reports so burdened with statistical analysis that they are almost unreadable. Better a pedant than a pariah.

This explains in part why parapsychologists have not flocked to what looks like the ideal test conditions offered them by archaeology. Archaeologists do not have to concern themselves with the source of information, only whether it is accurate or not, and this has allowed them to avoid questions on the nature of psychic sources. The parapsychologist who wishes to do such practical research is in no such protected position. By self-definition his discipline is dedicated to examining the phenomena themselves. And the phenomena that produce information for use in archaeology—except for dowsing—almost always include observations on man. It is possible to avoid theorizing in dangerous areas when dice are being selected, but far more difficult when dealing with location, reconstruction, and explanation of man's past. Consequently, although the four stages of psychic archaeology's test hierarchy—particularly the fourth stage, positioning—offer unusually potent experimental rigor, to date *not one* member of the "official" parapsychological community whose primary training is parapsychology has ventured onto the playing field. Only *physicists,* such as Harold Puthoff, Russell Targ, and Edwin May of the Standford Research Institute, and a very few others seem at all interested in doing so.

It is clearly understood (although rarely articulated in these terms) that dealing with psychic abilities capable of providing information on man's motives and his works almost inevitably leads to the disquieting perception that man is a being of transcendental consciousness. Such an idea is about as threatening to the metaparadigm as could be imagined. It presents an obstacle so formidable that even knowing that psychic information might be accurate, and could resolve problems that have been plaguing investigators for decades, has not made the data attractive enough for parapsychologists to use.

It is all the more amazing, then, that archaeology *has* taken this step, since it is a discipline whose only operational premise and claim to admittance to the councils of science has been adherence to the metaparadigm. It is like watching an orphan with a painful past reject the only foster parent who has been kind to it—with no assured promise that there will ever be another home. Not only must archaeologists reject the metaparadigm, but they must also be willing to serve as pointmen for an attack on that home. It would be hard to conceive of a more insecure psychological posture.

It has caused real anguish for them, and the fact that it may have been inevitable does not give relief. There never was a solution to archaeology's problem within the Grand Material Metaparadigm. The study of man means more than just his body or the outward physical symbols of his "beingness," but that was all that normal science deemed admissible. Consequently, archaeology, psychiatry, and all the other disciplines whose answers lie beyond these limits were always condemned to fail at paradigm achievement. Under materialism, they could never be sciences, but must always remain arts.

Now the choice facing archaeology is being made. Crisis and evolutionary survivalism have finally given much of science the courage to overcome the psychic obstacles. It began in such paradigm-achieved disciplines as physics, then spread to the *relatively* safe area of plant research, and finally to the distinctly threatening inquiries of archaeology.

For the archaeologists there is a reward: one happy and (one hopes) painless benefit. Under the psychically based paradigm the two unnaturally separated branches of archaeology can become one since even the written record is not complete nor have all historical questions been answered. At the level of explanation, they must become one.

Psychic archaeology resolves some old difficulties only to create new ones in their stead. The pioneers, from Bond to Pluzhnikov to Reid, have done what pioneers always do; they have framed the general answer, but by no means perfected all the steps to reach it. It is time to examine both the complexities and the promise that psychic data, practically applied, represent not only for archaeologists but for all of us.

X

COMPLEXITIES, PROMISE, AND IMPLICATIONS

"Capella St. Edgar. Abbas Beere fecit hanc capellam. St. Edgar . . ."

It is seventy years since these Latin words written down by two men in an English office led to the first disciplined application of psychic information in archaeology. The acceptance of this technique since that first experiment in 1907 has been as slow as all such revolutionary advances. Now, however, since its admission as a discussion topic at the meetings of the American Anthropological Association, since its open advocacy by both Russian and Canadian scientists, psychic archaeology has begun to move into the international research mainstream.

This is not to say that many middle-group researchers will suddenly become willing or able to accept what it has to offer. It may take yet another generation for full acceptance. As physicist Max K. E. L. Planck, who won the Nobel Prize in 1918 for the formulation of the quantum theory, once observed, "A new scientific truth does not triumph by convincing its opponents and making them see the light, but rather because its opponents eventually die and a new generation grows up that's familiar with it." Resistance to psychic archaeology thus is only a holding action; no matter how many years are required for *full* integration into archaeology, no open-minded scientist can any longer sustain the position that using the psychic as a practical testable technique for exploring the past is impossible.

Today psychic archaeology can handle the doubters. It has been demonstrated that this technique can be conducted in a manner that meets, or even exceeds, the best standards attained by strictly orthodox research. Time is also on its side, as are the anomalies research

290

has produced. Within perhaps as little as a decade there is good reason to believe that a metaparadigm crisis will be precipitated; and the resolution of that breakdown can only strengthen practical psychic research, psychic archaeology included. What this new subdiscipline cannot withstand, however, and what it will have a much harder time dealing with, is the deadly embrace of overzealous or half-trained true believers. These people first promise more than the approach can deliver, then compound that misrepresentation by performing their research sloppily, and finally top the devastation by making outlandish claims for anything their excavations do discover.

Just as science once turned away from the study of life energies pulsing in fields around all living bodies because of the overstated claims and poor research that frequently accompanied such statements, so psychic archaeology could be similarly undercut. It has taken most of this century for life-energy field research to come back to active consideration, and the lesson is an obvious one. The psychic technique is an invaluable new tool, but it is no more a panacea than any other single technique. Almost three-quarters of a century of science history makes it clear that psychic archaeology works best when its own special demands are met, and the results are then combined with the most advanced methods developed by more traditional approaches.

But overcoming the challenge of the zealots does not mean an end to psychic archaeology's troubles. The approach is also vulnerable because it seems so easy: Ask a question. Get an answer. Dig! At first glance the only training even a generally qualified archaeologist needs is the ability to talk. Reality is a little different; psychic archaeology is no more likely to become a universal skill in its discipline than brain surgery is in medicine. But because in format it is just two people engaged in correspondence or conversation, it gives an appearance very much at variance with its real essence. Unless this is understood, money will be spent and effort expended by archaeologists who are not specialists in this technique; and when the dig finds nothing, failure will not be ascribed (correctly) to researcher inadequacy but instead to that easy target—the psychic. It won't take many such failures before those who do not want to accept the psychic will feel they are justified in their prejudice. Although this will not alter the inevitable, it will unnecessarily slow it down. The bulwark against these assaults by both true believers and casual users is the same: a thorough understanding of the complexities posed by the practical application of psychic material, in archaeology or any other field.

To begin with, practical work with sensitives is very different from standard parapsychological laboratory research; it even studies a qualitatively different psychic episode. In the lab the goal has been to accumulate vast amounts of statistical data designed to substantiate the theory that psi talents exist. This has traditionally been done by studying psychic minievents—the calling of Zener cards being the classic example. These tiny psychic episodes were chosen because they had easily controlled variables, the most limited research goals, the ability to be repeated over and over with each event taking only a few seconds—thus creating the numbers needed for statistical analysis—and the least number of disturbing philosophical implications. This is almost another world from the metaevents involved in practical psi research. Metapsychics involves relatively few episodes, almost no repetition, and most important of all, seeks out the reconstructive and explanatory information on the nature and history of man that causes such trauma to the laboratory researcher. Minievents are the realm of the parapsychologist whose interest is psi validation. Their proof comes from repetition. Metaevents to date have been the concern of nonparapsychologist specialists who see the psychic not as an end in itself but merely as a stage toward achieving a larger goal. For them, proof lies not in the psychic incident but in the subsequent fieldwork that puts the sensitive's information to a practical test.

The laboratory researcher, for whom there is no later practical testing (the psychic response being an end in itself), designs his experiments under two imperatives: eliminate fraud and isolate what is going on. After establishing that the psychic is not a fraud (not always an easy task), the researcher has to determine whether, for instance, the sensitive is displaying a precognitive ability when he guesses the card or has simply read the researcher's mind.

To the nonparapsychologist scientist interested in practical application of psychic data, both these considerations quickly dispose of themselves. Prefieldwork testing (described in Appendix I) quickly eliminates fraud as a possibility; material is either where the psychic predicts it is or it is not; reconstructions are either accurate and in accord with information previously hidden from the sensitive or they are not. As to the question of which psi talent is at work, this is largely irrelevant in practical research.

The most important difference between laboratory and practical research has been that the practical research team *assumes* the existence of psi. This frees them to move beyond the restrictions of confirming minipsychical research, and there is every justification and an excellent precedent for such a working stance. No one knows to this day exactly what gravity is. All that can be discussed are its observable and predictable effects. Yet this has not stopped every astronomer since the seventeenth century from plotting the movements of even unseeable celestial bodies based on the assumption that gravity exists. The astronomer not only relies on gravity's effects, he proves the validity of this premise by accurately calculating movements that his "fieldwork" later confirms. In very much the same way the successful psychic archaeologist assumes that although psi causes are unknown, the reality of psi effects may be gauged. How he does this, and exactly what complexities he faces, can be best described by the analogy, first begun the chapter on Bond's work.

The psychic archaeologist is like a submarine detection officer aboard a surface ship. Both are trying to scan a vast "space" for something they are not sure is there, and they have to work in two environmental mediums as they do this; something they achieve only through the intervention of an intermediating mechanism (the psychic or the ship's electronic apparatus). Worse, they and their target are moving at different "speeds" and there is much "static" cluttering up their sensor reception.

As the detection officer searches for his quarry, he never hesitates to meet the demands of the sensor array of underwater microphones and radiation receptors, which his ship is pulling at the end of a cable like an underwater kite. Nor does he slight the requirements of his computer and its print-out displays, which provide him with a presentation of the information the sensors have picked up. He will, in fact, meticulously check to see that his submerged platforms are in good order, and that his transmission and receiver cables are without flaws. Even in the tropics he will concern himself more with the cooling and dehumidifying of his computer spaces than with his own quarters. He knows and accepts that computers operate only within very narrow temperature and humidity parameters.

To talk with that computer, he will, without balking, master its binary mathematical language, so foreign to everyday conversation, or see that someone else on his team knows it. He will also go to any lengths to see that the programs entered into his computer's memory banks are the best his team is capable of. The detection officer is fully aware that failure to achieve proficiency will surely invoke fulfillment of perhaps the oldest computer adage: "Garbage in—garbage out." If his program is faulty, or his mastery of the language imperfect, every other phase of the operation, from flawless controls to clear reception, will be invalidated. He will additionally see to it that his print-outs are properly calibrated so that the resulting profile provided him by a combination of sensor data and computer program is accurately depicted. He will even learn and tolerate an "eccentricity factor," realizing that nothing so complicated as the systems he is working with can ever be completely predictable.

Anyone seeking to do practical psychic research must also make some accommodations and realize that he is working with a biosensor and biocomputer, one "designed" to probe an immaterial realm with a sensitivity, sophistication, and potential quite literally—for the moment, at least—beyond his comprehension. To be successful, as all the researchers from Bond on soon discovered, six major factors have to be considered. In substance, they are very much like those faced by the detection officer of the analogy. First is the condition of the channel, and second, that of the researcher; third is the development of a mutually understood language; fourth, the formulation of the question (or program); fifth, their joint involvement in the transmission of questions and reception of answers; and sixth, the analysis of the psychically obtained data. Even when, as with General Scott Elliot, the archaeologist and the psychic channel are one, these same points are valid. To ignore these six requirements is like asking the detection officer to put his equipment in an exposed open boat transiting Africa's Cape of Good Hope in a thunderstorm.

The condition of the psychic is the most obvious of the six requirements: if the psychic fails to perform in practical work, then there is no experiment (the negative conclusion valued by laboratory scientists is of no use to field researchers). The psychic's well-being is so obviously important that it would seem that no one could dispute it. How can any researcher who fails to create optimum conditions hope for a successful outcome?

In most nonpractical parapsychological research, though, this is not how things have worked out because until very recently—certainly no further back than the past five years—with very few exceptions psi researchers did not (many *still* do not) believe they were an influencing factor in their own experiments. Werner Heisenberg, a successor to Dr. Planck and also a Nobel winner (1932, for developing Planck's theories into quantum mechanics and for research in hydrogen), observed some twenty years ago that the experimenter is not just an observer but a force affecting the experiment's outcome; the implications of his observation are still largely unconsidered.

As recently as 1974 Evan Harris Walker, in a discussion at the AAA Rhine-Swanton Seminar conducted by Professor Long at Mexico City, felt moved to note that "while everyone in physics knows about the Observer Effect, not many in physics concern themselves with what it really means. . . . And the record is even worse in other disciplines." This situation is predictable, since the very idea that the experimenter is part of his research problem is not only offensive but threatening to a scientist working under the assumptions of the Grand Material world view.

For the minipsychic events being studied in the laboratory this noninvolvement conceit was only one of a gamut of factors working against a successful outcome. It was probably no greater an inhibiting influence than the deleterious psychological effects produced by conducting experiments in rooms—or worse, cubicles—whose decor was as sterile as the inside of an icebox; or the negative, if not actively hostile, attitude held by many of the individuals conducting the research. That the hostility was usually not personal but rather the conditioned stance of the properly skeptical paradigm researcher made little difference. Hostility and negative anticipation were so prevalent that, as sensitive Ingo Swann remarked, when these conditions were added to the mind-numbingly dull multithousand repetitions, "it . . . only succeeded in grinding the diamond into a dust pile while trying to capture the sparkle."

Swann's comments are particularly apt because he is one of the few sensitives who, in spite of all that is baleful in parapsychological experimental protocol, has been able to perform almost unbelievable experiments, including altering the decay of a magnetic field established in a quark detector and magnetometer at Stanford University's Varian Hall of Physics. This apparatus, encased in an aluminum container inside a copper container, inside a super-conducting shield, buried "some five feet beneath the floor in concrete," was, by careful "design and thorough testing," supposed to be "impenetrable to outside influences." Swann stood over it, in the presence of several researchers, and with his parasensory perceptions correctly described equipment inside he had never seen and then visualized the effects he wished to happen. For an impossible ten seconds the magnetic field altered—just as had been requested.

Swann is not alone in his feelings, but he is among the most articulate of the current generation of psychics. "I simply grew tired," he said "of putting up with experiments designed from the point of view of everyone but the person who was expected to produce results, the psychic. When I went to work out at the Stanford Research Institute I insisted that I would only participate if I had a say in what we were going to do and how we were going to do it." It is this taking the matter beyond feelings to rebellion that makes him a distinct exception. Most psychics submit to conditions almost perversely structured to work against their flickering and usually poorly controlled psychic talents. This has caused the greatest tragedy of parapsychology—the stamping out of fledgling psychic talent because of the pressures built into the experiment themselves. Known as the Decline Effect it is the one constant to thread its way, year after year, through the literature of American academic parapsychology.

Anthropologist Margaret Mead is something of a fairy godmother to psychic research. It was she who years ago defended the work of Professor Gardner Murphy when he was attacked by his fellow psychologists; it was she who encouraged Professor Weiant in 1959; and it was Dr. Mead's powerful plea that, after two previous rejections, led to the affiliation of the Parapsychological Association with the American Academy for the Advancement of Science, the beating heart of the scientific establishment. Today, after having held high office in the AAAS, as well as remaining the Curator Emerita of Ethnology at the American Museum of Natural History, she continues in the role of protector. She has always had an interest in sensitives; one of her early papers was a study of a family friend who was one. To her, the reason for this tragedy of psychic loss is a clear one:

"The trouble with this whole field . . . they either want to prove that it is true, or that it isn't true . . . They already have their con-

clusions. ... They don't want to find out exactly what is there. ...
It is this kind of thing that I regard as totally unscientific. You have
to realize that in culture after culture the gifted sensitive always
doubts himself. You know I advocated, and I still am advocating,
. . . that the sensitives are a special type of people . . . and they occur
with about the same frequency in every culture whether they are
picked up or not. ... The seeming disparity between cultures is accounted for by whether the culture does pick them up or not. That's
why you seem to get a lot of sensitives in places like the Kentucky
mountains, or the Scottish mountains ... because *the culture ex-
pects them to be there; recognizes them when they do occur; and
teaches them how not to be destroyed* [emphasis added]."

The first requirement for practical work, then, is to make sure the
psychic is in good physical condition—since clearly no one does very
good work when overtired or run down—and that he also knows he
is genuinely liked by, and important to, the researcher and his experiment. It is no coincidence that the great metapsychics of history
either developed a particularly close relationship with one person,
often an academic of stature, such as Croiset with Tenhaeff or
McMullen with Emerson, or else they were individuals of great personal charm who drew to them supportive friends, as did Ossowiecki and Cayce. The sensitive, no matter what form his talent
takes, is above all else a *person,* and like all people he possesses an
ego and feelings.

For this reason the emotional climate in which he must operate is
just as important to the psychic's biosensor—which is to say, himself (particularly the part of himself psychiatry would label the unconscious)—as humidity and temperature are to the computer environment. The greatest single complexity in achieving successful
practical research is the psychic's humanity. And because in any experiment there are sure to be at least two humans, this issue has a
major corollary: the basic question of how people communicate.

There are two phases to this complexity, both are critical to the
success of practical psychic research, and each is dominated by the
same overriding consideration. It is not just the intellectual content
of the words spoken, but the attitudes—indeed, the ideals—held by
the psychic and the researcher that affect the experiment's outcome.
To begin with, there is the nature of communication at times other
than when the research is actually under way.

In the past decade psychologists and others have begun to look at how we really convey our messages. Their research makes it clear that if the other person does follow our meaning, it is perhaps as much because of our eye movements, facial expression, cadence of speech, and body language as it is because of our words. These signals are sent not by our conscious intellect but by our deep-seated often unconscious attitudes. Moreover, when communication fails, or at least appears to fail, it is not necessarily because of the choice of words so much as the disparity between the words and the attitudes behind them. Nonintellectual signals may be subtle but they are powerful, sometimes even strong enough to override the speaker's words. Putting aside for the moment any psychic perception that may come into play, it is obvious that if the researcher's words say "I have confidence and faith in you and your abilities" while all other signals transmit "I think you are a fraud," successful psychic research, particularly practical research, is doomed.

The situation becomes even more delicately balanced in the second phase of psychic/researcher contact—communication during the experiment with the source itself. In this portion, although much of the nonverbal, physical transmission is lost, apparently unperceived by the psychic source, researcher intention is even more clearly transmitted than before. Motive and attitude take on an even greater, a primary, significance, (a situation which holds true even if psychic and researcher only communicate via correspondence). How this happens is not clear—it is apparently a genuine psychic communication—but the evidence of the psychic archaeologists is unequivocal and is supported by other psi research as well. In psychic contacts unspoken feelings shout!

Swann, for example, began noticing that "sometimes it was as if there was a barrier between me and what I was trying to find out. At first I thought the problem might be coming from me, perhaps I was more tired or bored than I realized. But I found that even when I was very excited about what we were doing and enthusiastic about the outcome, the barrier would still sometimes be there. . . . Obviously it had to be coming from another source. I began asking those overseeing the experiments how they felt about what we were doing. Then I got more sophisticated and began giving the testers tests. If it turned out they expected a negative outcome, or were bored or irritated—even if by something totally divorced from our work—why, I just told them they shouldn't participate that day. The 'attitude-of-the-researcher'—that's something all too many scientists don't like to look at, particularly if it is brought up by somebody like the psychic who isn't supposed to know anything at all about such matters."

Parapsychologists are doubly threatened individuals. Most began their careers as psychologists in a paradigm-aspiring (and thus insecure) discipline. Joining the even more tenuously established sister field of parapsychology compounded their problem; nothing is further from the physical-science core of the American research establishment than the psychic. Perhaps these scholastic tensions are the reason that most of the practical research done with psychic material has not been done by parapsychologists but by researchers already secure in other fields.

For example, Professor Poniatowski, an ethnologist, clearly understood the impact of the researcher's attitude, the Observer Effect, and the question of the surroundings in which the experiments took place—as he realized so much else that is only now being accepted. He worked very hard to create the correct environment for Stefan Ossowiecki, inevitably conducting the work in the most emotionally secure place possible—Ossowiecki's own study. To deal with the Observer Effect, he always made an effort to see that the light conversations on subjects unrelated to archaeology or the experiment, which Ossowiecki wished to have quietly going on around him as he began, were kept up. The reason for this is obvious:

"Such conversation distracts the conversants from thinking about the guide object, which in turn makes it easier for Mr. Ossowiecki to concentrate on it. . . . When Mr. Ossowiecki is in the trance he can perceive the thoughts of those present with regard to the experiment and therefore their thoughts, when focused on the guide object, make his concentration difficult. He preferred a neutral theme for conversation."

To deal with researcher attitude, Poniatowski and Ossowiecki made sure that only those they felt had the right ideals took part in their work. The importance of these attitude-control mechanisms is illustrated by their one obvious failure, the twenty-seventh experiment. Two psychologists had asked to observe and Poniatowski acceded to their request. As he later wrote, they were more interested in the psychic phenomenon than in prehistory. Throughout the experiment, despite Poniatowski's careful presession explanation of what was required, they attempted "by mental suggestion" to influence what Ossowiecki saw and said. Their success was archaeology's loss. Poniatowski and Ossowiecki both decided that, because of their influence, his contact with "the Kentian culture [was] . . . a bad experiment."

Frederick Bligh Bond's automatic writing technique also recognized the importance of his and Captain Bartlett's attitudes; they avoided experiments when they felt that emotions such as hostility and anger would interfere. And they were equally meticulous in taking into consideration the need for a "neutral" atmosphere. This is the explanation for Bond's reading aloud from light literature during the course of each session.

Poniatowski and Bond are not alone in reporting the importance of such influences in psychic archaeology. Aron Abrahamsen, by profession an electrical engineer and by experience a senior member of the scientific team to study the feasibility of sending a manned expedition to the moon, as well as advanced studies of the space platform, space shuttle, and recoverable booster vehicles, began developing clairvoyant abilities in his forties. Today he works full time as a consultant in this field, operating on much the same lines as Edgar Cayce. Like Swann, he is an educated man and articulate in explaining what he has learned about the psychic process after thousands of individual readings, research readings for scientists, and an abortive dig in the San Francisco Peaks outside of Flagstaff, Arizona. For this excavation he provided a number of readings, and actual fieldwork proved that his subsurface geologic predictions—for example, what soil strata would be encountered at a specific depth—were uncannily accurate. However, some of his other predictions proved far less correct and this led him into speculation about what makes for "a good contact."

"The attitude of those asking questions is crucial. You must consider just what need motivates the questions asked of a psychic. Is it idle curiosity? Is it a desire to use the information to gain power over others . . . or to make yourself famous or rich? These attitudes are one of the first things a person should examine before consulting myself or, I believe, any other psychic. If the need is great—a person is in deep inner turmoil over something, say, and has done everything they can before turning to the psychic—why, then, the answer is usually succinct and to the point. They stand a good chance for a clear and accurate answer. If the need, however, is only to satisfy curiosity, or the motive is selfish, or they have *not* done everything they can at their end . . . well, if that is the case, the answer may be long or clouded. It may even take several readings, and they may never really understand what has been given. Certainly, all psychics make mistakes. But I have come to believe that not all the reasons for error lie exclusively with the psychic. Ideals and purposes are critical in dealing with the psychic realms."

Abrahamsen also raises a rarely discussed yet apparently critical factor: that the researchers do "everything they can at their end." This factor has never been carefully considered since to do so would be to accept a premise rejected by the Grand Material Metaparadigm. While many researchers believe correctly that the psychic ideally should have no knowledge of the subject under experiment, since such knowledge might only pollute or complicate the research program as unnecessary data foul a computer setup (by possibly forcing the psychic to view the problem, he has been given analytically, left brain? rather than intuitively, right brain?), the obverse consideration of researcher preparation rates little concern. Yet without exception, all the outstanding individuals using psychic archaeology were already significant authorities in this field. Bond may have had only a limited formal education, but he was widely respected for his work in ecclesiastical architecture and archaeology. Poniatowski was one of Poland's leading ethnologists; Emerson is one of the world's experts on Canadian Indians. Even General Scott Elliot made sure he had learned his archaeology by the time he began dowsing for sites. The one body of psychic information that has had the greatest difficulties, the Edgar Cayce readings, undoubtedly faces these problems because no anthropological professional ever used the Cayce channel. It is not enough to be a parapsychologist, nor is it simply a matter of academic stature (Bond, Garrad, and Scott Elliot were all amateurs). But the evidence of success does indicate that there is a correlation between intellectual excellence in a field and getting high-quality psychic data, particularly when intense interest and good researcher attitude are also present. It may be romantic to think of huddling with a psychic and then going off to discover a lost city—no previous preparation required—but it does not happen that way.

Even when attitude is correct, and expertise attained, the requirements for success have still not all been met. There remains the question of language, literally the words chosen, the formulation of the question—the equivalent of the computer's program. Hugh Lynn Cayce remembers with great vividness an incident that illustrates both this point and the matter of attitude.

At the height of the Depression a group of individuals associated with Edgar Cayce attempted to locate a treasure by using Cayce's psychic information. There followed, in the course of several readings, a highly detailed narrative of a Civil War skirmish between a Confederate patrol and a Union troop laden with gold coins that President Lincoln was sending to pay Federal volunteers. The complex discourse Cayce offered was specific down to individual names.

Although the incident was truly obscure, and Cayce's reconstruction all the more impressive for that, with the help of military historians the group of treasure seekers was able to ascertain that the psychic's words were breathtakingly accurate. Indeed, the only variation between Cayce's account and the war chronicles was the historians' conviction that the Confederates had made off with the coins, whereas Cayce said the Federals had buried them in two iron laundry tubs and pushed their campfires over the site to hide the newly turned earth.

With everything but the disposition of the gold confirmed, the group decided to make the effort to recover the funds, and with the help of hired workers, dug about seven holes each the size of a railway boxcar. As they dug, fruitlessly, their frustration increased. They went back for additional readings on the treasure's location. Significantly, this is what they called it in their questions—a treasure. Despite guidance, they never found any gold coins. They did, however, find a quantity of buried artifacts, the pathetic garbage of war: belt buckels, harness tabs, metal buttons, and rifle actions. All of this confirmed that there had indeed been a battle at this location—which had the effect of further tantalizing the treasure hunters. They brought the items back to Edgar Cayce. To their surprise, he took the discovery of what they viewed as mere souvenirs very seriously and in a quaint little ceremony carried them out past the driveway, prayed over them, and then built a fire and threw them into it. That was the end of the treasure hunt.

Almost forty years later Hugh Lynn sat in his office and talked about the entire event, the first time he had done so in quite a long while:

"At the time, I thought it was a question of ideals. The readings had warned us repeatedly that we would not find the treasure unless our ideals and purposes were correct—that is, there must be nothing selfish or oriented toward personal profit in our motivation. And even then we could see that although our heads, our intellects, may have believed we had no desire for this wealth . . . well, in our hearts I think we all knew we were not so sure. It's hard to know yourself that thoroughly, to have sufficient control to work with those deep-seated attitudes.

"I still believe ideals played a major role; indeed, they may have had implications we still don't fully understand. But at a very simple level, if we were unclear about our real purposes, we were even more imprecise in the language of our questions.

"At the time there were several very powerful experiences on the site itself . . . things that I suppose some people would call hauntings. I know that sounds strange in this pragmatic age but it was almost as if that location was a time trap . . . as if at some level this incident was still present tense . . . not so much happening again and again . . . but a frozen continuing moment. One of these events was so terrifying that it, as much as anything else, caused us to stop.

"Also there was the question of Dad's behavior when we showed him what we had dug up . . . the praying over the stuff and then burning it in a fire. At the time when I asked him he told me that he had done that because there were some people trapped there by the violence of their emotions . . . things they were focused on with their whole being, even up to the moment when they died. He said what he had done freed them. I realize now I only partly understood what he meant by all that.

"You know, my father was not a medium. The source of his information was not people we would think of as dead. But on certain rare occasions there were signs that sometimes the channel . . . it's almost as if it got clogged by such influences . . . mediumistic interference, as it were.

"Looking back on the incident of the Civil War treasure, I wonder . . . we kept asking for treasure. Now, although we meant gold coins, that isn't always what we said. We said 'treasure.' And if you stop and think about it, what would be a treasure to someone trapped in time? Not gold coins! What good would gold be? But perhaps . . . just perhaps, a treasure would be the thing that released you. Considering the entire episode, I believe we asked a question, and because our ideals were less than they should have been, we influenced the channel . . . and the information became mediumistically tinged. This was further compounded by imprecise language. It started out all right, I really do think that treasure is still there. Maybe because we didn't understand exactly how this all worked, we split our target. That is, part of the information was about gold coins . . . and part of it was about something else. If you consider it from that point of view, I think we got an accurate answer. We *were* given accurate directions to a treasure! And we proved it by digging up that battlefield junk. Certainly, everything else about what Dad said checked out. One thing I am sure of, though . . . after years of studying it, I am convinced that psychic research—particularly when it has practical applications—is *much* more complicated than it appears."

The formulation of questions, then, can be an unseen trap undermining success even when the researcher's attitude is positive and open, when he has expertise, and when the sensitive has proven ability in the area being studied. Unfortunately, language skills are not a possession of most people. Ironically, with the paradigm demanding jargon words, neologisms, and the semantic grotesqueries of scientific language, the very people who have the expertise to ask a question often lack the skills to word it. Nor does the problem end there, since language is the primary medium for not only the question but also the answer. Just as the researcher must learn to deal with his own pecularities, so must he learn to understand exactly what linguistic eccentricities the psychic is heir to.

As usual, Poniatowski was one of the first to make this observation. To him, "Every instrument of research, even the most sensitive measuring apparatus, has observation errors and is dependable only within certain limits. It is therefore important to understand that the clairvoyant as a research tool is dependable only within limits and may be used for scientific purposes only when we know all of his observational errors."

With this as his point of view, it is not surprising to discover that when Ossowiecki spoke of "iron decorations" in one experiment, tens of thousands of years before the technology existed to produce them, Poniatowski did not dismiss the statement as a "miss" but inquired further. Having taken the time to learn that his psychic partner often reported in his descriptions not actual materials but his subjective impressions based on appearances, he stopped Ossowiecki and asked him to elaborate. Was it really iron, he asked, with no implied rebuke that would alert the sensitive to his mistake

(Ossowiecki knew nothing about the correlation between culture stages and chronology). Immediately Ossowiecki amended his earlier words, saying that no it was not "iron" but was some other material that looked like "iron." Because he understood the nature of the psychic's observations and his use of language, Poniatowski was able to clear up what on the surface appeared to be a bad mistake, and also gained insights into the appearance and texture of prehistoric "jewelry," insights that would not otherwise have been his.

Emerson also discovered that "I needed to learn George's language and the fact that he sometimes lacked words for what he was seeing." As an example, at a Canadian Archaeological Association meeting a friend of Emerson's asked McMullen if he would give a reading on a skull and mandible (lower jaw); he agreed. This, Emerson explained, "was quite literally the first human skeletal material George had ever touched. He knew nothing more than any other untrained person would about the sort of things that would interest a physical anthropologist who was studying a strange skull."

However, as McMullen began to talk, he focused on exactly that kind of information—only his words were not adequate to the task, at least superficially. He spoke of "fang teeth" and "maulers." Had Emerson not been present, and had he not been knowledgeable about McMullen's linguistic limitations, it is very likely that what was said would have seemed to those listening little more than an entertaining series of malapropisms. But with many hours of jointly shared experience behind them, Emerson and McMullen knew how to talk to each other, and Emerson was able to establish that "fang teeth" were canines and "maulers" molars. "His diagnosis was interesting: subhuman, Mongoloid, and grain-eating." As Emerson said later, George's performance impressed the woman who had supplied the material, confirmed things she had suspected, and gave her food for further thought.

On the basis of proven research done in psychic archaeology, it would appear that those psychics who receive their information as if they are looking first at a film, which becomes "live action" (Ossowiecki, Croiset, Conway, and, to some extent, Abrahamsen), have considerable difficulty in passing on to the querying researchers all that they are seeing. Also, since these psychics are usually conscious, or only in a very light trance state, as Professor Valkhoff observed, "The impressions pass through [the] . . . subconscious mind and may have been slightly altered, or even confused with memories and other impressions." Perhaps because he is awake, this type of psychic seems to have no greater command of language than his conscious vocabulary would indicate. Even when greater language skills are evident, suggesting a "higher" intelligence has come into play, the more probable explanation is that we all hear and store far more words than we normally use. An interesting line of inquiry would be to study the results of exercises *jointly* undertaken by both team members, researchers and psychics, to develop a large shared vocabulary. This does not mean sensitives should be trained as specialists—as archaeologists, for instance. Such training would negate most controls because the psychic would then be in contact with material that would precondition and bias his perceptions. (The one exception to this is dowsing, since this technique at best can provide only a rudimentary reconstruction or explanation.) In general, the goal should be learning language, not theory.

Waking or semitrance "film" psi perception is by no means the only psychic format useful in practical experiments, as pioneer psychic archaeologists have demonstrated. At least one variation does seem to permit access to a vocabulary that even unconscious awareness cannot explain. This is the Edgar Cayce manifestation, perhaps the rarest form psychic talent takes. Although Cayce definitely was in a deep trancelike state, it was not the same state as that of a medium. He was, however, truly unconscious, did not know what he had said during a reading, and had to read his words like everyone else to learn what had transpired. But there is some evidence that toward the end of his life, in 1945, this changed and he did, or at least could, remember. This would seem to suggest that unconsciousness was an involuntarily assumed response to avoid being burdened all day long with other people's traumas. The reason for this change cannot be stated with certainty but one possible explanation both covers the facts and reveals something very important about metapsychics. Like Ossowiecki, who was almost his exact contemporary, whatever Cayce's motives may have been at the beginning of his talent's manifestation, he ended life less a psychic than a pilgrim. He too had learned that money and the psychic do not mix, not because money is inherently bad but because it brings to bear on the psychic frightful pressures to perform no matter what. Cayce, like Ossowiecki, had also come to realize that in its proper context psychic ability is a byproduct and not a final goal, and that using it only for service is the greatest protection the sensitive can have against the neurotic djinns that seem to plague most psychics. Having reached this state of perception, Cayce's unconscious perhaps no longer needed what might have been selective amnesia.

Whatever the explanation, since he was unconscious, the conscious biases Cayce held seemed to play a relatively small role. Indeed, personal waking belief and psychic convictions were initially diametrically opposed (on reincarnation, for instance). And although Cayce sometimes said he saw things during readings, it is not really clear exactly what his subjective experience was and whether it occurred in all episodes. What is certain is that Cayce did not suffer the language limitation of waking psychics. On more than one occasion he gave readings in foreign languages—when awake, he could not have said a word in any language but English. Even more regularly he used medical, scientific, and engineering terms, sometimes stopping, still in trance, to spell them out, as if he could see that his secretary, Gladys Davis, did not know how to do so correctly. However, if his vocabulary seemed unlimited, that did not assure lucidity. McMullen's "fangs" and "maulers" are the essence of clarity compared to a Cayce reading: in his sleep state the Kentuckian's syntax was so bizarre as to constitute an English-related but foreign language. The language reminds one of a Picasso painting of the forties in which all sides of a woman's face are portrayed, but on a two-dimensional surface. Similarly, Cayce seemed to be stuffing a complex and multifaceted observation into an inadequate verbal vehicle, and thus his sentences often had five or even ten dependent clauses. He frequently made the same point over and over again, each time using a slightly different phrasing, often ending a sentence with "See . . . do you see." Clearly, as the study of the Cayce material by archaeologists and physicians has proved repeatedly, when working with this kind of channel, one of the first priorities is a mutually understood language.

A third variation of the language problem, and perhaps the simplest, is found in dowsing. In a sense the dowser is truly a biomechanical sensor and, more than any other psychic skill, dowsing resembles a computer operation. The dowser has only two modes: yes or no. The rods or pendulum move to express only positive or negative reactions; otherwise they are still, neutral. It is an exact analogy to a computer's binary language. The simplicity of the response, however, does not make the formulation-of-the-question part of the exercise any easier. As every water dowser soon learns, it is not enough to ask to locate water (the questions are either said aloud or held in focus in the mind). To state the question that way is to risk digging a well down to polluted or mineral-saturated water. The question must be some variant of "I am seeking *potable* water." There must follow questions about depth and flow, and whether the water will dry up in the summer.

If attitudes, expertise, and language are resolved, there still remain major hurdles to successful research, and one of the most difficult is gauging psychic bias. Here again, although for different reasons, the psychic's humanity has made the problem one that all but a few exceptional people have avoided. Most researchers prefer to think of sensitives as subjects—indeed, they are often referred to that way or, even more demeaningly, by only their first names. The logic for this is ostensibly to protect the psychic's identity—as one would protect a patient with a social disease. It is difficult, however, to read parapsychological literature or talk with researchers without realizing that this is not the only reason. Dehumanizing the sensitive by calling him a subject or by using just his first name is a protective mechanism, one that allows the parapsychologist to avoid seeing the psychic as an equal, a full partner in the research going on. This may work for minievents but for two reasons it is deadly to the practical research of metapsychic events in fields such as archaeology. First, it fails the requirement of joint respect and acknowledgment. The researcher may be the senior partner but he must publicly state *and truly believe* that it is a partnership. Emerson's avowal, "my friend," dismissed by some in Mexico City as "too homey . . . too unscientific," is a critical factor in his success. Until the psychic is known as a whole person, and not just an odd talent, it is difficult if not impossible to understand fully and appreciate his biases—which is to say, the parameters of his skill or his limitation in research. Emerson, for instance, learned over the months he talked with George McMullen that "because of experiences in his childhood, George does not care much for burials or things involving churches or religious events." This was not the sort of thing a person would mention casually, nor would it make much of an impression in a straight field test, or even in most psychological evaluations. McMullen was not afraid of death, nor unwilling to talk of dying. He was not unbalanced in the matter, but he did not seek out psychically or otherwise things involving death. "Not really such an abnormal position, you know," Emerson noted. "Who in Western society besides undertakers does find death or dying of itself a subject of interest?" But he began to realize that McMullen and other sensitives with whom he worked might unconsciously be filtering out impressions, perhaps not even seeing them psychically, or altering them. This is a bias very different from the language hurdle.

This realization taught the professor that he needed to pay close attention to what he asked his psychic friends to do, always keeping in mind that "they were people just like me or any other scientist, and just as we have our hangups and pet ideas, we can hardly expect the psychic half of the team to be any different." Emerson came to learn, as Bond and Poniatowski had, that the best approach to practical psychic research was the team concept. Getting multiple impressions by different people of the same event or site tended to cancel out such bias limitations. There is also clear evidence in every psychic archaeology project from Bond onward of a learning curve. The more the team works together, the better quality information it gets.

It is also necessary to deal with a category of difficulties common to all forms of psychics, and one that is particularly critical in archaeology. Abrahamsen, for example, has proven something Poniatowski suspected but was unable to test because of the war: questions of measurement and chronology need careful attention. "It is very hard to measure from psychic perception since I may be seeing all sides of a thing at once. Is it a foot or a yard; sometimes it is hard to tell."

There is even some evidence that indicates sensitives often have a kind of psychic astigmatism. They may, for example, always measure long horizontally and short vertically. Since the distortion appears to be consistent, it can be adjusted for in the same way that a spectacle lens adjusts for visual inaccuracies. Clearly, it will be important for future archaeologists to learn by prefieldwork testing whether a psychic has such a perceptual problem and to adjust for it before attempting an actual excavation. Once on site, it also means that very careful survey points will have to be established so that questions can be phrased with maximum accuracy. Failure to calibrate, query, and measure carefully could mislead an archaeologist untrained in psychic work to think, after unsuccessfully digging, that he had inaccurate information when in truth it was simply uncalibrated.

The other difficulty of this type is even more complex and less understood. Suppose a psychic describes artifact material at a certain depth, but upon digging it isn't there. Was he wrong? Not necessarily. The psychic may have seen it, and it may have been there—at *some point* in the site's chronology, at the time psychically perceived. Soil conditions may have subsequently caused the bones or fabric to decay so that, in the present, it is no longer there.

Perhaps psychics also have uniform chronology variation. Unfortunately, past researchers, ever including Poniatowski, do not seem to have considered this and, thus, their reports do not provide much insight into the problem. Two things do seem clear. First, chronology should be established by specific reference, obtained through questions, to such things as pollen or carbon-14 or soil strata; and some kind of sequence should be established through orthodox procedures. Second, failure to consider these questions of measurement and chronology could easily result in information being dismissed as worthless which, had the psychic researcher done his work properly, could have produced extraordinary results.

Bias and limitations, do not stop at the researcher and the sensitive; they extend to the psychic source. Sometimes, as when one of Bond's monks answered, "I do not know about these things," the limitation is admitted. And since by definition psychic information comes from suprasensible realms, source limitation is the most difficult complexity a practical research team has to face. It may not matter to the applied researcher where the data come from, but still he must learn the source parameters within which his psychic partner works if their efforts are to have the best chance of success.

After decades of psychic archaeological research there remain more questions than answers about the dynamics of the metapsychical event. There are still enigmas blooming in a field of mystery, but certain basic outlines have emerged. Although in an ultimate philosophic sense all sources may be one source, at the applied level each of the basic metapsychic groupings seems to have, within itself, the same type of source, and each of these has its own special problems for the researcher to master. Exactly what all these factors are is still so poorly understood that they cannot be completely defined, let alone resolved.

One critical factor, however, is clear: The more personalized the source (that is, the more the sensitive appears to be getting information from a person either named or unnamed), the greater the possibility of error. This does not mean that such a channel is worthless or usually wrong, but the accumulated record does caution the researcher always to bear the nature of the source in mind.

Emerson, for instance, discovered that when McMullen was called on to answer questions about physical anthropology, "he appeared to be getting the aid of what could conceivably have been a dead specialist in this field." In the professor's opinion, "When information comes from such a source—and that is not always easy to tell—then George doesn't really seem to have access to anything more than a good physical anthro man would know. It is still an incredible psychic feat and George will know things he never normally knows. But that doesn't mean that we end up knowing something we never knew before." This should not be mistaken for mediumism, however. No one is speaking through McMullen or other sensitives with similar skills; they never surrender control to any personality or power. Rather, it is as if these psychic individuals go into another room, have a conversation, and then report its conclusions back to a researcher who could not be present—as McMullen did for Emerson when he contacted the six Indians who said they represented two hundred years of history on that site.

The Canadian archaeologist even sees a benefit to this kind of source, since it seems that McMullen can get help from "experts" *as he is actually viewing a scene in the past.* "It is as if a living specialist could go piggyback, back into time with George . . . and that can certainly add to what you know." The benefit Emerson enjoys is not a qualitative advance in a field but quantitative access to a fuller picture than a similar specialist would have if he were working today only from remnants. Sources working at the personality level also have a problem suggested by Bond's Watchers but never fully explored by him or anyone else; it could be called the Limitation of Standard Knowledge—what an average person, contemporary with the period being studied, would know about a given topic.

When the source of information is, say, a Roman soldier, although he may be excellent at location, his reconstructive knowledge could well be very inaccurate. This could be interpreted by those hearing the information as a "miss" when it is actually a result of source limitation. A Roman soldier simply didn't know very much about the complexities of government policy or social practices; even the names of those making decisions probably were unknown to him. But this would not stop such a source from stating that he *did* know, and then advancing in unequivocal words pure drivel. This limitation is not surprising; the same response pattern can still be heard in off-base bars today. And the Limitation of Standard Knowledge is hardly restricted to the military, past or present; nor does professionalism in one field assure competence in any other.

For the practical researcher, the pitfall is that all genuinely psychic information is not by definition accurate! It is the responsibility of the researcher to recognize this limitation and to adjust for it. This is not always easy. The fact that a psychic may have access to many personalities may make him appear smarter and more certain than he normally is; it may even provide some valid information. But, by analogy, if six people sit in a room and each has total awareness of what the other five know, there is still no assurance of accuracy. Equally significant, there is not even any real increase in new knowledge; there is simply better communication. This does not mean that there is no benefit to be derived from such a network of contacts, only that it is important to identify correctly what that benefit is. While there may be no absolute increase in information, there may be an increase in insight. New inferences based on the same body of information are not only possible, but should be cultivated as a genuine tool of the multihookup provided by this type of psychic source. And where more than one psychic is employed, the effect is increased geometrically. Also, since the number of inputs available is larger, mistakes due to any single source's Limitation of Standard Knowledge are at least partially accounted for, in the same way that a committee of researchers physically sitting in a room tend to buttress one another's weaknesses.

In the case of a contact that appears to draw its information from impersonalized racial memory or collective unconscious, the difficulty is somewhat different. Here, since there is no personified source on the psychic side, one of two things is likely to happen. Either the sensitive will misinterpret the record, or if he is a time traveler for reasons already described, his perceptions may be no better than those of an eyewitness to a bank holdup. With a source such as Cayce's there is yet another twist. Cayce seemed to contact some kind of record; at least he said he did. But if it was racial memory it was in some way different from that providing information to other psychics. In his case the major source limitation lay in its very limitlessness. The range of Cayce's contact was so vast that it caused tortured syntax and obfuscating language, the central restriction to the practical usage of his readings.

This long list of complexities, involving the psychic, his source, and the researcher, would seem so formidable as to preclude any successful research were it not for the persuasive results obtained by practitioners of psychic archaeology working in fourteen countries over the past seventy years. And they operated largely unknown to one another, isolated from the help of shared experience—pioneers reinventing the wheel generation after generation, each working through the same problems with nothing more than trial and error to guide him. But it need not always be this way. Now there will be papers, seminars, communication, all the accoutrements of mainstream investigation. More important, the uncertain nature of the research itself is changing. Recent studies in biofeedback training, left brain—right brain investigators, the Western adaptation of yoga techniques, even the use of electroacupuncture, have suddenly begun to play a role. All this seems destined to make psychic channels stronger and psychic communications clearer and of a more consistent and higher quality. As a result the future of metapsychical experimentation lies not so much in the traditional approaches of parapsychology as in the fields of physiology, neurology, physics, and electrical engineering. Such research is hardly in the toddling stage, yet it has already shown a promise almost beyond prediction.

The psychic archaeological team of tomorrow, an interdisciplinary group composed of the traditional archaeological specialists as well as, possibly, biofeedback technicians, a staff parapsychologist, and, of course, a team of psychics—preferably each with a different skill and type of channel—will obtain its data for fieldwork, reconstruction, and explanation using methods whose sophistication will be as incomprehensible to us as the world of electricity was to nineteenth-century man. This psychic subdiscipline may not become either a universal skill or a cure-all, but it will without question radically alter our understanding of the past, and even the most superficial evaluation makes it clear that archaeology itself will never be the same.

Perhaps now the question of Atlantis, so violently but inconclusively debated since Plato first introduced the subject to Western science in his dialogues *Timaeus* and *Critias,* can finally be resolved. The Maya hieroglyphs can at last be deciphered; the reconstruction of almost every culture made more complete. All the closed doors to our past can be opened, at least a little. And the black sheep mysteries of strange earthworks, peculiar stone monuments, and odd dating sequences may finally be brought into the fold of context through reasonable explanation. With that promise, and psychic archaeology's past record, can the argument about whether psi even exists still be taken seriously? Can any complexity be allowed to stand in the way?

Its contributions will also go a long way toward making archaeology a testable science for the first time in its history. Perversely this attainment, the very thing that should assure archaeology's paradigm achievement, will cause its continued denial. It is impossible to reach paradigm status using a method whose very existence is refuted under the terms of the metaparadigm, whose tenets are the basis for all science. What is far more likely to happen, as a result of the tests provided by psychic archaeology, can best be understood by moving up a level from consideration of this single profession's future to that of Western science as a whole.

It is apparent that a metaparadigm crisis is approaching, although its development is still in the early stages. The roots of this upheaval do not lie in archaeology, although archaeology will have its role to play, but in quantum physics, which provides the theoretical basis for a new world-view. As physicist Dr. Fritjof Capra notes in his book *The Tao of Physics,* here is not only "intuition, but also . . . experiments of great precision and sophistication, and . . . rigorous and consistent mathematical formalism." This is the encompassing structure archaeology itself could never provide.

Psychic archaeology, though, has made and will continue to make a real contribution to the nourishment of this new metaparadigm entity because its research generates the anomalies needed to bring on crisis. It is even possible that this psychic subdiscipline will bring the issue to critical mass. Certainly, the discovery of the "hall of records" based on the combined commentary of Edgar Cayce and George McMullen and other sensitives could provide such a catalyst. Even without that, this new approach has done much to demonstrate that the time and space framework so crucial to the Grand Material world-view is by no means as absolute a construct as most paradigm scientists believe. And it has done this through the location of sites, the prediction of subsurface geology and artifacts, and the accurate reconstruction of paleolithic cultures—all far more comprehensible to most people than the arcane world of higher mathematics.

The exact form this new metaparadigm will take is as problematical as the timing that will bring it into being. But it is certain that the Unified Metaparadigm (for that is a reasonable name) will no longer maintain the artificial distinctions between consciousness and the physical, time and space, or researcher and experiment. Rather than dealing with compartments, it will see continuum, a spectrum running from physical reality to suprasensible realms—as visible light shades into X-rays and beyond.

There is yet a higher level involved, one that supersedes any consideration of specific information from one or even a series of experiments in psychic archaeology or any other discipline. Science as we know it began under the Genesis Metaparadigm, then, with Darwin and Einstein as the catalysts, moved on to the Grand Material World-View. Now the metamorphosis into the Unified Metaparadigm seems inevitable. This is surely not the end of it, but no matter what changes, what permutations and alignments science goes through, its goal is still the same. Historically, each metaparadigm has been more comprehensive in its explanations of things than the world-view it replaced, and more specific and refined in the techniques it used to reach those explanations. Yet each stage has produced not so much the discovery of new facts as the rediscovery of old cosmological truths. We have come almost full cycle from our beginnings. The scientist and the mystic blend together seeking the same answer, the solution to the only quest that has ever really mattered: Who are we? Why are we here?

I
PREFIELDWORK TESTING

The first step in any experiment in psychic archaeology is obviously assembling the team of psychics and researchers.

The selection of the latter requires little comment beyond noting that since psychic data are involved, the orthodox archaeological component of the work must be as good or better than prevailing standards. Only this offers protection from misdirected but potentially damaging criticism. The ideal team, then, should have the best possible representatives of all the specialties normally involved in a dig of the type planned.

To these people can be added a parapsychologist and a statistician, who will move the experiment into the multidiscipline catagory. Their concern is whether the psychic hypotheses check out during the fieldwork, and what can be concluded from a statistical analysis of the successes and failures of these hypotheses.

Researchers, however, are only half of a team of equals. And it is in the selection of the other half, the psychics, that the experiment is most vulnerable. Perhaps because practical psychic research has just begun to receive serious and open consideration, many people, including scientists of notable attainment in their own specialties, tend to assume that an individual with highly developed psychic skills in one area must have equal skills in any area. These same researchers would never think of asking a physical anthropologist to do work in linguistics, despite the fact that specialists in both fields are nominally called anthropologists.

The first step in selecting psychic teammates is to find out who can really perform the work at hand.

Surprisingly, as Professor Emerson's work indicates, people who seem to have no other psychic ability can psychometrize. For this reason, no one who has an interest in trying should be discouraged from doing so, a conclusion that has very recently received a boost.

At the Stanford Research Institute, senior consultants Harold E. Puthoff and Russell Targ have been testing what they call Remote Viewing. After checking literally almost anyone who walked through the door, they came to believe that everyone, once he relaxes, can move from what sounds like just fantasizing to seeing in his mind's eye the surroundings of a target person who has traveled to a location that could not consciously be known.

From this research, and the Emerson work, it would appear that psychometry is a movement along the time line, and Remote Viewing is movement through space, although very possibly these are different facets of the same continuum. Exactly how this works in either case is unknown, but from an applied research point of view, it doesn't matter. What counts is finding individuals who can do it, and the best of them may not have been previously considered psychics at all. Since movement in both time and space is important to psychic archaeology, what follows is a simple test capable of quickly determining who has such capabilities.

GOALS

The experiment has three major goals:
1. The location of a site on the basis of nothing more than maps and photographs.
2. The location of artifact material either at a distance or on site.
3. The psychic reconstruction of background material concerning artifact material either buried at the site or physically in hand.

EXPERIMENT PRÉCIS

Essentially the experiment is based on first psychometrizing a series of objects the respondent holds in his hands (Phase One), followed by (Phase Two) a two-part section in which he is first asked to locate a rope grid from a distance and then identify and position objects within that grid. This is followed by an on-site inspection of the grid, at which time the psychic is again asked for location and positioning of target objects. Finally, in both segments of Phase Two, the respondent is asked to give a psychometric reconstruction while the object remains buried in the ground.

MATERIALS NEEDED

PHASE ONE

Three artifacts—potsherds, axe heads, and arrow points—are best. However, any man-made object of any era can be used. What is important is that its history be accurately known.

PHASE TWO

1. One length of colored ¼-inch nylon line 36 feet long.
2. Four lengths of colored ¼-inch nylon line each 9 feet long.
3. Sixteen wooden surveyor stakes.
4. Thirteen signs 12 inches square each.
   a. Nine signs should each bear *one* letter of the alphabet, beginning with A and ending with I.
   b. Four signs should bear the letters A, C, I, and G (one each sign).
   c. Each sign should be light-colored enough so that the letters clearly stand out.

d. Each letter should be a capital at least 4 inches high.

e. Thirteen wooden stakes, somewhat longer than surveyor stakes.

f. To these longer stakes are attached thirteen sign squares.

g. One mallet (for driving in stakes).

h. One small shovel, entrenching tool, or backpacker's spade (for digging small holes into which artifacts are placed).

i. One medium-sized brown paper grocery bag into which have been placed twelve artifacts similar to those mentioned in the Materials list of Phase One.

j. A piece of twine with which to seal the bag.

k. One 3x5 card and a sealable envelope. On the card should be drawn a miniature of the grid (see questionnaire).

l. Topographic map (see below, Selection of Site).

m. Twelve small opaque bags (into each of which one target artifact has been placed, and the bag then sealed).

SELECTION OF SITE

1. The site selected should be in a place protected enough to allow the grid (see Preparation of Site) to stand for whatever period of time is necessary to complete the experiment. Since very little actual space is needed, a backyard, the edge of a farmer's field, an unused piece of land on a campus are all possibilities to consider.

2. The main considerations should be that the grid is some distance from buildings (to avoid confusing the psychics), so that they are unlikely to run across the grid accidentally or be told by someone that it is there.

3. The site should also be *unknown* to the person asking the psychics the questions. Indeed, *this individual should not even know who does know the grid's location.* (This is a double-blind protection against telepathic leakage.)

4. The site should be on land shown on a topographic or county map. It is critical that the map offer sufficient detail to allow the psychics to mark precisely the grid's location. It should not be so detailed a map, however, that the psychic has no chance of failure.

PREPARATION OF SITE

1. The site is prepared by running the perimeter (36-foot) colored nylon line around four surveyor stakes spaced so that a square 9 feet on a side is created by the line.

2. The four 9-foot lengths are then laid out like a tic-tac-toe setup, using stakes to attach to the perimeter and at cross-points within the larger square. When completed, this should give a large square, 9 feet on a side, broken into nine equal smaller squares, each 3 feet on a side.

3. The signs are then placed in accordance with the drawing.

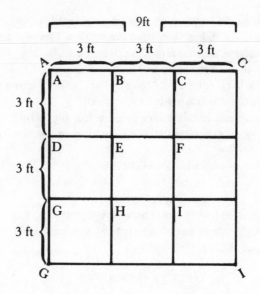

PREPARING THE GRID

1. After it has been laid out, a hole should be dug somewhere in each grid square. These holes should range in depth from 2 to 18 inches, and be at least as big around as a man's fist. Leave the dirt from the hole next to the opening.

2. An individual otherwise completely unconnected with the experiment should be selected to place the target objects. It can be the same person who prepares the grid, but more desirably it is someone else.

3. This person should take the sealed grocery bag with the target objects, the small shovel, and the 3x5 card and envelope. Walking out into the grid, he or she should open the bag and, without looking, reach into it and pull out a target object (which has been earlier sealed into one of the small bags). This person should place one object per hole in four to six holes. Whether an object is placed in a hole or not, before stepping from each square, the hole should be covered by pushing the small mound of earth back into place. In other words, the person placing the objects should go through exactly the same motions at each hole.

4. The actual placement *must* be carried out by this individual *alone*. No one should be with him as he walks over the squares.

5. The placing individual should be asked not to look as he places artifacts. He should simply reach in, take the object, drop it, and then cover it with earth. It is not necessary to make a production over smoothing the earth down. The fact that even the placer is unaware of the object's positioning introduces a triple-blind (no one knows the answer) barrier to telepathic leakage.

6. Before leaving the grid, the 3x5 card should be filled out, showing which holes have been selected. This should be placed in the envelope and sealed. It, in turn, is passed on to the person who selected the individual who placed the objects, in this way freeing that person from any further involvement.

7. The grocery bag should also be resealed and similarly passed on. After the experiment is over, those objects left in the bag should, when added to those selected for burial, account for all artifacts involved in Phase Two.

PREPARING THE QUESTIONNAIRE

1. The questionnaire can be set up by anyone except the person who selected the site and arranged the grid or, if it is someone different, the individual who placed the target objects.

2. The first part of the questionnaire deals with Phase One, the section on psychometry, and Phase Two, which deals with site location, object identification, and location, positioning of object, and psychometrization of objects without having physical contact.

3. The questionnaire should read as described below.

RESPONDENT'S BACKGROUND INFORMATION

Name_____

Age_____ Sex_____ Occupation_____

Date Experiment Conducted _____

Time Experiment Conducted _____

Personal Interests & Hobbies _____

_____

PHASE ONE
PSYCHOMETRY

NOTE: Phase One for each of three objects.

OBJECT # :

1. Identity of object?:
2. How was object used?:
3. In what part of world was object found by archaeologist? (Be as specific as possible, please.):

4. Describe present and/or past scenery of the location at which object was found. (Please state which you are describing.):
5. At what depth was object found?:
6. a. Were any other objects nearby?:
    b. If so, describe them:
7. How did the object come to be located where the archaeologist found it?:
8. Give any details you can about:
    a. the persons who used this object:
    b. the culture they belonged to:
    c. the particular time the object was used:
9. Please give any other information you feel about this object:
10. How confident are you of your answers? Please give a relative evaluation of this with each answer:

PHASE TWO:
LOCATION AND RECONSTRUCTION

NOTE: Answer questions about only one object on this sheet.

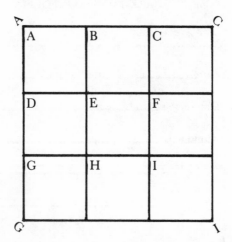

OBJECT # :

1. Letter of grid in which object is located:
2. Size of object:
3. Shape of object:
4. Color of object:
5. Markings, if any, on object:

6. Sketch object on back of this sheet:
7. Identity of object?:
8. How was object used?:
9. In what part of world was object found? (Be as specific as possible, please.):
10. Describe present and/or past scenery of the location at which object was found. (Please state which you are describing.):
11. At what depth was object found?:
12. a. Were any other objects nearby?:
    b. If so, describe them:
13. How did the object come to be located where it was found?:
14. Give any details you can about:
    a. the persons who used this object:
    b. the culture they belonged to:
15. Give any details you can about the present position of the object as it is now buried. (For example, "Lying on its side"; "with the sharp edge pointing straight up"; etc.):
16. Describe, as you see it, the orientation of the object in reference to grid corners A, C, G, & I. (Refer back to the grid drawing on page one of this phase):
17. How confident are you of your answers? Please give a relative evaluation of this with each answer:

CONDUCTING THE EXPERIMENT

1. With the questionnaire completed, and the site and grid prepared, the actual experiment is ready to begin.

2. It is conducted in two phases, with the second phase divided into two segments. The phases are:
    a. Psychometry of objects held in the hand.
    b. Location and reconstruction.

3. The second phase breaks into two segments:
    a. Location at a distance, with no further guidance than topographic maps, and possibly general panoramic photographs.
    b. Location of objects only while walking over the site.

4. Both segments a and b involve questions concerning positioning and psychic reconstruction.

5. Phase One can be conducted in any relaxed surroundings. The target objects are simply handed to the respondent and he or she talks about them. Many people find it easier not to worry about writing down answers and prefer to tape the questions and answers. This should definitely be considered as an option for all portions of the experiment. In this case the questionnaire should be used by the conductor as a guideline only, a basic format to assure that key points get covered. The conductor and respondent

should simply talk about what the psychic perceives, letting the natural flow of conversation direct any additional questions. The key to all psychic work of this type is relaxed human behavior, not restrictive adherence to some rigid format of questions.

6. Phase Two should be run twice. The first time it is run at a distance, with the respondent relying on the map. In this segment, obviously, the first question must be a query as to the site's location. The question here must be phrased with care so that it is clear to the psychic what is being asked. Something on the order of: "Can you locate on this map a grid made up of colored nylon ropes, within whose boundaries artifacts have been placed?" After this has been settled, the conductor can then move on to the questions concerning the grid itself.

7. After *all* respondents have completed this segment, and their answers have been taped or written down, the respondents should then be taken to the site. Ideally this should be done one at a time, to avoid telepathic leakage or body-language cueing. However, if distance or some other factor makes this impossible, the respondents should at least be separated and asked to go over the grid one at a time. A picnic lunch is a good psychological activity, with the respondents leaving, walking some distance to the grid, and with tape or questionnaire tackling the challenge.

8. Respondents should feel free to use any technique that suits them—meditation, dowsing, pendulums, anything that works.

9. At no time during any phase of the experiment should a respondent feel any pressure to answer a question. Nor should he feel that the experiment is some kind of competition.

10. After all answers are in, the sealed envelope should be opened and, following its directions, the targets dug up—and the small bags opened prior to removing from hole, to assure correct judging of position. Respondents may wish to see this, and should be encouraged to do so.

11. The sealed grocery bag should then be opened and a check made to see that targets from the ground plus objects in the bag equals the same twelve objects.

EXPERIMENT CONCLUSION

1. With the answers to the location questions revealed, the experiment is ready for grading.

2. Answers to Phase One should be compared with known facts about artifacts. Attention should be paid to how confident the respondent was.

3. Answers that appear to be wrong, but that the respondent felt strongly about, should receive special attention. It is entirely possible that the psychic reconstruction is the more accurate version of events. A careful researcher will go back with an open mind and check.

4. The answers in Phase Two, Segments a and b, should be compared.

5. As with Phase One, the psychic reconstruction should be evaluated with an open mind.

RECALL OF SEALED PACKETS

Upon completion of the statistical analysis the four sets of documents in the control of the repositors will be recalled, as described above. This concludes the experiment.

BIOGRAPHIES

The biographies of *everyone* involved in the experiment should be included in the papers of the experiment.

FINAL OBSERVATIONS

If all this seems unduly complicated, the reader is advised to review the literature concerning criticism of psychic experiments over the past century. Special attention should be paid to the style and quality as well as the substance of these comments.

OBSERVATIONS

1. If this experiment is carefully run, it will be possible to locate individuals who have paranormal skills useful in psychic archaeology. The researcher can, for instance, learn who is good locating at a distance, who is accurate in doing psychic reconstructions of the past, who can walk over a site and outline things beneath his feet. From such successes a psychic team can be assembled that offers the researcher not only a more efficient method of operation, but insights otherwise unobtainable.

2. The experiment can and should be run several times. Work over the past seventy years has clearly shown that there is a learning curve. The reader is advised to go back over the chapter on complexities.

3. The sites should always be restored as near as possible to their original condition.

II
FIELDWORK PROTOCOL

The principal criticism leveled at the practical application of psychic information is that the sensitive already knew the answer; that in some way—from the story of a grandmother, a book read and seemingly forgotten, a snippet of data from a television news broadcast—he had learned the information later presented as psychically derived.

From a purely archaeological point of view, the source of the information doesn't really matter. Whether the information came from tales learned at the respondent's grandmother's knee or a forgotten high school history lesson, all the archaeologist cares is that the guidance proved to be accurate when subjected to the test of fieldwork, an absolute bar of justice from which there is no appeal. The site and its contents were either there or they were not. Psychic archaeology at its best, however, is far more than just archaeology. The very process of obtaining the information, and its subsequent testing, are a major commentary on the nature of one of philosophy's oldest challenges: What is the mind? Who are we really?

This commentary is valid, though, only if it is protected from attack. From this vantage point it matters very much indeed how the psychic learned about the information he is presenting, and for this reason a very strict protocol must be followed in obtaining the data and assuring their integrity before they are tested by the rite of the spade.

This protocol must have three major functions:

1. Protection of the psychic source from the possibility of telepathic leakage (the psychic is reading the mind of the researcher, who knows the answer to his question, even if he is not consciously aware that he has this information).

2. The establishment of an absolutely sure chronology of events, so that it is clear the information came before any digging was done.

3. An evaluation of the probability that the information provided by the psychic respondent could exist in orthodox nonparanormal sources.

What follows is a simple, straightforward prefieldwork protocol designed to achieve these ends.

PROTOCOL

I. Psychic respondents should each be sent, via registered mail (or closest applicable classification if overseas), a packet of material consisting of three sections. Mailing date and acknowledgment of receipt should be filed in a complete compilation of correspondence. The three sections should be A, B, and C of the following outline.

A. Background profile
B. Resource material
C. Experiment questionnaire

II. The proper procedure should consist of:

A. Background Profile
1. Recent medical history
2. Knowledge of archaeology (self-evaluation in general, and the area in question in particular, including material transmitted to respondents by experiment researchers).
3. General self-perceived sense of physical, mental, and emotional well-being at time respondent fills out answers to the questionnaire, or tapes his or her answers.
4. Expectation of success.
5. Respondent's feelings about researcher attitudes toward himself and experiment.
6. Interest in experiment subject matter.

B. Resource Material
This should be a highly detailed listing of all maps (by number and name), artifacts (by description), photographs, etc. It should be immediately apparent to anyone reading the protocol exactly what the psychics had to work from, if they have worked at a distance, or what artifacts they had to guide them when working on site.

C. Experiment Questionnaire
1. Accompanying each portion of the resource material is a list of questions keyed by designation (questions A to Map A). Questions cover (but not in each instance are all questions applicable):
(a) Location of previously unknown sites.
(b) Surface and subsurface geology.
(c) Identification and location of previously unknown artifacts and ruins.
(d) Culture(s) that produced site.
(e) Culture(s) that produced artifact material to be found at site.
(f) Dating(s) of site.
(g) Dating(s) of ruins.
(h) Dating(s) of artifacts.

(i)   Reconstruction and explanation information.

(j)   Additionally, a first question: "What would you like to tell me about this (site, ruin, artifact)?" should begin each sequence.

(k)   Each answer should be rated by the respondent as to his level of certitude: "How confident of the accuracy of your response are you?" This question should be answered both anecdotally and on a scale of 1 to 10, in which 1 indicates "no confidence" and 10 represents "absolute certainty."

(l)   Also included should be the question: "What is your mental/emotional response to this site (or artifact or ruin)?" Answers such as "I feel very much drawn to this area" could be considered typical.

(m)   A final question should be: "Have you experienced any physical sensations as you focused on this material? Please describe."

2.   Respondents can answer, whether their work is conducted in the presence of the researcher or via mail at a distance, by writing their answers in the space provided on the questionnaire, or by taping their responses, with the questions read to them or by themselves. They should then return *all* material, plus their answers in either form, by registered mail (where applicable), to the project director.

3.   Upon receipt of this material (receipt slips to be included in correspondence compilation), if answers are on tape recordings a transcription should be prepared.

4.   The transcription should be compared with the tape and authenticated individually and separately by two persons, of recognized probity, who are otherwise uninvolved with the project. They will provide sworn depositions attesting to their examination and validation that the tape recordings and the transcriptions agree, and that the latter is a true and valid record of the psychic information.

5.   Handwritten responses will be typed and, as in (4) above, validated.

6.   Four copies of all responses, plus all other pertinent material, will then be made up. These should be compared for completeness and accuracy, and then validated under oath by the same individual as in (4). Each record should contain copies of all notarized or sworn statements. The four packets should then be placed in containers, in the presence of the above validators, and prepared for mailing to four re-

positors. A fifth packet, made up entirely of all originals, should be similarly prepared. This packet, however, will, in the presence of the validators, be sealed in such a way that the seal cannot be removed and opening the container breaks it.

7. One set of Xerox copies, made earlier, should be used as working papers and will be subject to further duplication as need arises. It too, however, should be validated exactly the same as the other sets. This will be done by the same validators as in (4).

8. The four copies should be sent, via registered mail, to four repositors. The repositors should be individuals, geographically separated, of impeccable reputation.

9. The repositors should each respond, via registered mail, stating that they have received the packets and have secured them under their personal control. In notarized statements they should attest that they will hold the material until notified by registered mail that the experiment has been completed and comparisons, based on the working papers, have been made between what was found during the fieldwork and what was psychically predicted. At that time, they will return all material in their possession to the project director, via registered mail.

10. The original of all materials, under seal, should be handed over to the officers of a local bank, who will acknowledge receipt of same through the proper bank form. In a notarized statement they should also state that they will hold the material for one year from the date of receipt in such manner that no one, including all experiment researchers, except the senior bank official will have access to it. (It may be necessary to extend this time period if it takes more then one year to check all psychic data.)

11. In an additional acknowledgment, as in (10) above, they should receive a second packet consisting of all statements received, or papers accumulated, since the five initial compilations were made. This material too should be sealed, as described above. Again one set of working copies will be maintained. This process may be repeated whenever necessary, and in this way all pertinent material should be rendered inviolate from tampering, a complete and verifiable chronology will be established, and all questions of fraud eliminated in the minds of reasonable persons.

D. Fieldwork Goals

After consultation among the members of the research team, with the stipulation that this group can call in others whose contributions they deem important, a list of feasible objectives will be settled on. This need not include every area covered by the questionnaire, but may include other areas that seem promising in light of the psychic information received.

E. Actual Fieldwork

Under the directorship of the most qualified member of the research team, the actual archaeological fieldwork should be carried out. The field director should follow the guidance of the psychic respondents, but will conduct the dig employing the most rigorous methods applicable to the situation. At all times these techniques should meet the highest standards of similar excavations.

F. Evaluation

1. Upon completion of the dig any additional laboratory work deemed necessary by the research team, in consultation with any recognized expert they choose, should be carried out. As in the actual excavation, only the highest standards will be acceptable.

2. When all the best criteria have been met, the results will be compared point for point with the psychic hypotheses provided by the respondents. The evaluating team should consist of the researchers, plus two independent judges, preferably individuals skilled in parapsychological analysis.

3. The evaluation should be made in both anecdotal form, by blind-ranked number on a scale of 1 to 10 (in which 1 is a complete "miss" and 10 a complete "hit") and by linguistic conceptual analysis. An overall subjective evaluation will also be provided by the full evaluating group. In scoring, each member will record his judgments separately.

G. Statistical Analysis

After evaluation is completed, a comprehensive statistical analysis of the entire experiment's record should be conducted by an individual skilled in statistical analysis. This evaluation should attempt to arrive at probability factors as one of its goals.

SOURCES AND NOTES

I. THE GLASTONBURY SCRIPTS

Sources

Interviews with William W. Kenawell, July and August 1975; Correspondence of Stephan Schwartz, 1973-1976.

Notes

1 " 'Wee laid down . . . *nuper Abbas.*' " F. Bligh Bond, *Gate of Remembrance*, 4th ed. p. 53.
"At each session . . . each other." Bond, *Gate of Remembrance*, 2nd ed., p. vi and footnote.

3 " 'JA's hand moved . . . window between them' " *Ibid.*, p. 32.

4 " 'Please give us . . . great church.' " *Ibid.*, p. 35.
" 'Capella St. Edgar . . . *quod vocator.*' " *Ibid.*, p. 35.

5 "It is by . . . St. Mary." Archbishop Ussher, *Britannicarum Ecclesiarum Antiquitates*, "Glastonbury Traditions," H. Kendra Baker, trans., p. 9 *ff.*
"Legend has it . . . easily proved." Geoffrey Ashe, *King Arthur's Avalon*, p. 30 *ff.*

6 "In several of . . . did locate." Bond, *Gate of Remembrance*, 4th ed., pp. 56-65.
"Aside from legend . . . eighth century." Ashe, *King Arthur's Avalon*, pp. 142-145.
"The next one . . . St. Dunstan." *Ibid.*, p. 154.
"In 1184 . . . time on." William W. Kenawell, *The Quest at Glastonbury*, p. 9 *ff.*

7 "It had sufficient . . . and farmhouses." Ashe, *King Arthur's Avalon*, p. 264.
"The family was . . . and schoolmasters." Kenawell, *The Quest at Glastonbury*, p. 20.
"In 1876 . . . business classes." *Ibid.*, p. 21 *ff.*

8 " 'Always ailing . . . into words.' " *Ibid.*, p. 18.

" 'Practically knew . . . or so.' " *Ibid.*, p. 19.
"Bond learned . . . special interests." *Ibid.*, p. 46.

9 "In 1897 . . . a fellow." *Ibid.*, p. 22.
" 'Many of these . . . their own.' " *Ibid.*, p. 33.
"Founded in 1849 . . . Taunton Castle." Schwartz correspondence.
"Soon he was . . . British Architects." Kenawell, *The Quest at Glastonbury*, p. 34.

10 " 'one of the . . . in England.' " *Ibid.*, p. 34.
"In 1894 Bond . . . Louis Mills." *Ibid.*, p. 23.
" 'Your committee notes . . . been secured.' " *Ibid.*, p. 38.
" 'June 6th, 1907 . . . our Calendar.' " *Ibid.*, p. 38.
" 'The stern historian . . . first century.' " *Ibid.*, p. 39.

11 "Essentially, Bond found . . . of Canterbury." Telephone Interview, William Kenawell, 27 July 1975.

12 "Both men, . . . holding hands." Schwartz correspondence.

13 " 'Monks anxious to . . . your minds.' " Bond, *Gate of Remembrance*, 4th ed., p. 37.

14 " 'We worked in . . . in yours.' " *Ibid.*, p. 45.
" 'significant structure at . . . great church,' " Kenawell, *The Quest at Glastonbury*, p. 167.
"Second, if the . . . west end." Bond, *Gate of Remembrance*, 4th ed., p. 46.

15 " 'rewarded by finding . . . thirty-one feet.' " Kenawell, *The Quest at Glastonbury*, p. 175.

16 " 'the theory that . . . the nave.' " *Ibid.*, p. 183.

17 " 'Ye shall find proof . . . west end.' " Bond, *Gate of Remembrance*, 2nd ed., p. 46.

18 "He had been . . . the wall." Kenawell, *The Quest at Glastonbury*, p. 183.
" 'Radulphus Cencellarius, who . . . found it.' " Bond, *Gate of Remembrance*, 2nd ed., p. 105.
" 'Radulphus the Treasurer . . . slew him.' " *Ibid.*, p. 108.

19 "It must have . . . (i.e. Glastonbury)." Bond, *Gate of Remembrance*, 4th ed., p. 106, footnote.
"Abbot Thurstan is . . . historic record." Ashe, *King Arthur's Avalon*, p. 163.
" 'the right forearm . . . fractured.' " Bond, *Gate of Remembrance*, 2nd ed., p. 105, footnote.

21 " 'The respond stands . . . duty here.' " Kenawell, *The Quest at Glastonbury*, p. 221.

22 "Instead, he first . . . the subject." *Ibid.*, p. 263.

24 " 'Digge east beyond . . . great howse.' " Bond, *Gate of Remembrance*, 4th ed., p. 53.
"The special impact . . . Edgar Chapel." Kenawell, *The Quest at Glastonbury*, pp. 292 *ff.*

25 " 'In the summer . . . 1684.' " *Ibid.*, p. 255.

26 " 'Our Abbey was . . . have forgotten.' " Bond, *Gate of Remembrance*, 4th ed., p. 147.
" 'The interim reports . . . advanced technique.' " William Kenawell, correspondence with Radford, June 1962.
"Bond was told . . . same stable.' " Bond, *Gate of Remembrance*, 2nd ed., pp. 66 *ff.*

27 "lay out a . . . 1125?],' " Bond, *The Company of Avalon*, Introduction.

29 " 'Then, when they . . . for beauty,' " Bond, *Gate of Remembrance*, 2nd ed., p. 88.

29 " 'new and faire . . . certain panelling.' " *Ibid.*, p. 88.

30 " 'The gabell was . . . hang there.' " *Ibid.*, p. 88.
"And this information . . . early history." Ashe, *King Arthur's Avalon*, pp. 35-65.
" 'in that . . . save us.' " Bond, *Gate of Remembrance*, 4th ed., p. 90.
"They also knew . . . their wealth." Baskerville, *English Monks and the Suppression of the Monasteries*. pp. 19 *ff.*

31 "In March of . . . the post." Ashe, *King Arthur's Avalon*, p. 252.
"The watchers explained . . . his goodwill." Bond, *Gate of Remembrance*, 2nd ed., p. 90.
" 'Chappells, a many! . . . of them.' " Bond, *Gate of Remembrance*, 4th ed., p. 149.

33 "And what they could . . . they buried." Schwartz correspondence.
"Bond once had . . . ever investigated." Interview with William Kenawell, August 1975.
"A close friend . . . curious figure.' " Ashe, *King Arthur's Avalon*, pp. 247-252.
"Why did Beere . . . buried there." Bond, *Gate of Remembrance*, 4th ed., p. 120.
" 'Wee were borne . . . was well.' " *Ibid.*, p. 120.

34 " 'He ever loved . . . be renewed.' " Bond, *Gate of Remembrance*, 2nd ed., p. 86.

35 " 'We have sat . . . of Heaven.' " *Ibid.*, 4th ed., p. 89.

37 "To begin with . . . and success." Kenawell, *The Quest at Glastonbury*, pp. 55 *ff.*

38 "Through a tiresome . . . diocesan architect." *Ibid.*, pp. 80 *ff.*
"As late as . . . simple wall." *Ibid.*, p. 117.

39 "There were seemingly ... because of it." Interview with William Kenawell, August 1975.
"Bond paid all ... Victorian gentleman." Kenawell, *The Quest at Glastonbury*, p. 25.
" 'Throughout the years ... quiet mind.' " *Ibid.*, p. 25.

40 " 'There is no ... the work.' " Bond, *Gate of Remembrance*, 4th ed., p. 79.

42 " 'the veridical passages ... quire [sic].' " Bond, *Gate of Remembrance*, 2nd ed., p. 70.

43 "By 1921 Bond ... beloved Abbey." Kenawell, *The Quest at Glastonbury*, p. 91.
"He obtained one ... at Glastonbury." *Ibid.*, p. 81.
" 'The Council of ... therefore dissolved.' " *Ibid.*, p. 80.

44 "When he became ... suddenly withdrawn." *Ibid.*, pp. 88 *ff.*
"And when his ... the turnstile." Interview with William Kenawell, July 1976.

45 " 'Sometime during his ... Dunstan's Abbey.' " Kenawell, *The Quest at Glastonbury*, pp. 100 *ff.*
"When Bond realized ... Anglican clergyman." *Ibid.*, p. 116.

47 " 'Intuition must bring ... the method.' " Bond, *Gate of Remembrance*, 2nd ed., p. 157.

48 "He says he ... trance state." *Ibid.*, p. 19.
" 'I cannot find ... monk yet.' " Bond, *Gate of Remembrance*, 4th ed., p. 32.
" 'the material influences ... at fault' " *Ibid.*, p. 37.

49 " 'I think I ... . Latin tongue.' " *Ibid.*, p. 38.
"When either he ... results improved." *Ibid.*, p. 39.
"He discovered ... social sciences." *Ibid.*, p. 25.

50 " 'Beere, Abbot, is ... to perform,' " Bond, *Gate of Remembrance*, 2nd ed., p. 47.

51 " 'Johannes mystified and ... and understand.' " Bond, *Gate of Remembrance*, 4th ed., pp. 92 *ff.*
"Johannes, at the ... 'in pain.' " *Ibid.*, 2nd ed., p. 86.
" 'Why cling I ... seeth yet.' " *Ibid.*, p. 95.

54 " 'Johannes [is] now ... other duties.' " Bond, *Gate of Remembrance*, 4th ed., p. 95.

55 " 'Ye did not ... five feet ... ' " *Ibid.*, p. 173.
" 'there was revealed ... rough stonework.' " *Ibid.*, p. 174.

II. THE EYES WHICH SEE EVERYTHING

Sources

Alice Je Glass affidavit, "Brooch of Mrs. Glass." Warsaw, 15 November 1932; Zenon Koziell affidavit, "A Prophetic Vision." Warsaw, June 1923; F. Pintowski letter to Zofia Ossowiecki Szcecin, 3 June 1952; Jerzy, Olewinski letter to Zofia Ossowiecki, Warsaw, 27 August 1946; Maria Boltus letter to Zofia Ossowiecki, Torun, 24 June 1947; newspaper clipping on Stefan Ossowiecki's honeymoon and screenplay "Eyes Which See Everything," source unknown, 1939; interview with Marian Swida (a series of seven interviews), 8-9 December 1975; interview with M. Grominski, 4 March 1976; Interview with A. Tarnowski, 5 January 1976; interview with T. Jazdzewski, 22 March 1976; interview with J. Grobowski, 23 March 1976; Waldemar Chmielewski correspondence with M. Swida; interview with Stefan Korbonski, 6 January 1976; correspondence between Marian Swida and author, October 1975 to October 1977; interview with Andrew Nowina-Sapinski (a series of five interviews), March to August 1976.

Notes

Where specific page numbers are not given in the notes to this chapter it is because the citation refers to material that was privately translated for the author. In such cases, pagination would be meaningless because of the manner in which the translated work was spaced for ease of research and writing. The situation is further complicated since the Poniatowski manuscript, which obviously is a major source document, has broken pagination, and some portions are still handwritten.

57 " 'You are Mrs. . . . have them.' " Interview with M. Swida, 8 December 1975.

58 "His father had . . . a Pole." *Ibid.*

59 "There is some . . . psychic capabilities." *Ibid.*, 9 December 1975.

"The duly prescribed . . . my ability,' " M. Zawadzka, "Polish Clairvoyant Engineer Stefan Ossowiecki," *Goniec Warszawski*, 1937.

60 "By the time . . . his body." *Ibid.*

"Research being done . . . lightweight objects." Wortz, E.C., Bauer, A.J., Blackwelder, R.F. *et al.*, "Novel Biophysical Information Transfer Mechanisms (NBIT)." Final Report: Document No. 76-13197, Airsearch Manufacturing Company of California, 14 January 1976.

"On one occasion . . . little shaken." Interview with M. Grominski, 4 March 1976.

61 " 'Ah, how many . . . for it.' " Zawadzka, "Polish Clairvoyant."

"In 1898, when . . . his disposal." *Ibid.*

"Ossowiecki took this . . . at home.' " *Ibid.*

"Despite his youth . . . for consultations." *Ibid.*

62 "Gallantly allowed me . . . real teacher." *Ibid.*

"He moved to Dobruz . . . paper factory." Correspondence with M. Swida.

63 " 'I would spend . . . superconscious state.' " Zawadzka, "Polish Clairvoyant."

" 'I conquered time . . . were "superconscious." ' " *Ibid.*

" 'is a state . . . not exist . . . ' " *Ibid.*

"For two months, . . . with Ossowiecki." Interview with M. Swida, April 1976.

64 "He loved good . . . chemical industry." Interview with A. Tarnowski, 5 January 1976.

" 'while I moved . . . among men.' " Zawadzka, "Polish Clairvoyant."

"All that is . . . chemical works." Interview with M. Swida, 8 December 1975.

" 'A tremendous breakthrough . . . help others.' " Zawadzka, "Polish Clairvoyant."

65 "Sometime in 1919 . . . Russian accent." Interview with M. Swida, April 1976.

"Left behind in . . . in time." *Ibid.*

"Always an easy . . . money outright." Interview with T. Jazdzewski, 22 March 1976.

66 " 'I am of . . . spiritual seer.' " Zawadzka, "Polish Clairvoyant."

" 'In 1938, when . . . such experiences.' " Interview with M. Swida, 8 December 1975.

"He had a . . . had used." *Ibid.*

67 "Also as a . . . responsible scientist." Zawadzka, "Polish Clairvoyant."

"What all this . . . of performance." Charles Richet, *Notre Sixième Sens*, p. 161.

68 "He explained that . . . in itself." Ossowiecki, *Świat Mego Ducha*, p. 364.

68 "He wore his . . . be located." Interview with M. Swida, 8 December 1975.
"And when he . . . the piano." *Ibid*, 9 December 1975.
" 'It would seem . . . all psychics.' " Richet, *Notre Sixième Sens*, p. 160.

70 "A wealthy Hungarian . . . sealing wax." Zawadzka, "Polish Clairvoyant.
"The Hungarian had . . . and recorded." *Ibid*.

71 " 'the best of . . . immediately accepted." *Ibid*.
"Fifty people, mostly . . . whole committee.' " *Ibid*.
" 'After ten minutes . . . to speak.' " *Ibid*.
" 'Interesting and convoluted . . . say anything more.' " *Ibid*.

72 " 'All this from . . . 100 per cent success.' " *Ibid*.

73 "Ossowiecki asked that . . . a sister." Interview with T. Jazdzewski, 22 March 1976.

74 "He told his . . . it should." Stanislaw Poniatowski, *Parapsychological Probing of Pre-Historic Cultures: Experiments with Stefan Ossowiecki (1937-1941)*. Unpublished.

74 "all the while . . . in Warsaw." *Ibid*.
"Steiner had spoken . . . supersensible awareness." Witold Balcer, "Report of Experiment with Stefan Ossowiecki," 1935.

75 "Balcer was to . . . the matter." Zawadzka, "Polish Clairvoyant."
"On the appointed . . . a try." Balcer, "Report of Experiment with Stefan Ossowiecki."

76 " 'I see a . . . human foot.' " *Ibid*.
" 'Revealed to our . . . around 1927." *Ibid*.
" 'once more to . . . years old.' " *Ibid*.
" 'Mr. Ossowiecki felt . . . concluded Balcer." *Ibid*.

77 "He had noticed . . . of wine." *Ibid*.

78 "Ossowiecki, although interested . . . the same." Interview with M. Swida, 8 December 1975.
"As if in . . . fifteen years." *Ibid*.

79 "He was considered . . . Polish ethnology." "Stanislaw Poniatowski," *Wielka Encyklopedia Powszechna*, 1967.

79 " 'The possibility of . . . prehistoric cultures.' " Poniatowski, *Pre-Historic Cultures*.
"His plan called . . . crosschecks." *Ibid*.
"In 1937, a . . . smallest fragment.' " Zawadzka, "Polish Clairvoyant."

80 " 'Entrance. A hall . . . small museum.' " *Ibid*.

81 "Included in the . . . be interdisciplinary." Stanislaw Poniatowski, *Clairvoyance*. Unpublished transcript. Poniatowski kept a careful list of all who attended each session, and headed each transcript with the name.
"At about 9:30 . . . leaving the room." Poniatowski, *Pre-Historic Cultures*.

82 " 'Thick, thick forest . . . this moment.' " *Ibid*.
"Fifty-seven minutes . . . my head.' " *Ibid*.

83 "Two weeks later . . . Erazm Majewski." *Ibid*.

84 "On July 24, 1933 . . . the pit!" Herbert Wendt, *In Search of Adam*, pp. 448 *ff*.
"As German writer . . . in appearance." *Ibid., p. 450*.

85 " 'God how far . . . protrude some . . .' " Poniatowski, *Pre-Historic Cultures*.

86 "His eyes were . . . been blocked.' " *Ibid*.
"Consequently, it was . . . same culture." *Ibid*.
"To begin with . . . don't understand.' " *Ibid*.

87 "At various times . . . a weapon." *Ibid.*

" 'Is there water? . . . powerful massive . . . ' " *Ibid.*

88 "Ossowiecki listened carefully . . . La Manche canal,' " *Ibid.*

" 'Prior to World War . . . of life.' " Interview with A. Nowina-Sapinski, 10 May 1976.

"In the fall . . . been lost." Zawadzka, "Polish Clairvoyant."

"It is known . . . spiritual suffering.' " *Ibid.*

89 "The reading, for . . . Italian peninsula." Poniatowski, *Pre-Historic Cultures.*

" 'I cannot get . . . between." *Ibid.*

91 "He also felt . . . my body.' " *Ibid.*

"Poniatowski concluded that . . . dynamics . . . " *Ibid.*

92 "Poniatowski concluded that . . . a concentration." *Ibid.*

"Although she had . . . to error.' " *Ibid.*

93 "There Ossowiecki wrote . . . his life." Newspaper clipping of interview with Ossowiecki, no name given, July 1939.

94 " 'Poland has become . . . unwise leaders.' " V.M. Molotov, cited in *Readings in Twentieth Century European History,* Baltzly, and Salomone, eds., pp. 454 *ff.*

96 "In Warsaw, specifically . . . German citizens." William John Rose, *Poland Old and New,* pp. 256 *ff.*

" 'The Nazis invasion . . . in secret.' " Konrad Jazdzewski, *Poland: Ancient People and Places,* pp. 19 *ff.*

"In one year . . . machinegunned." See Edward Hartwig, *Warsaw.* Introduction by Marck Sadzewkz.

" 'annihilation on the . . . of Poland." *Ibid.*

"One survivor remembers . . . her veil." Interview with J. Grobowski, 23 March 1976.

97 " 'Now this man . . . the ground.' " Poniatowski, *Pre-Historic Cultures.*

98 "She went to . . . in his vision.' " Affidavit of Maria Boltuc, 24 June 1947.

99 "Mrs. Zofia Podkowinska . . . interdicted activities." Poniatowski, *Pre-Historic Cultures.*

" 'intended to carry . . . Mr. Ossowiecki' " *Ibid.*

"From the beginning . . . it appeared." *Ibid.*

"But when Ossowiecki . . . actual excavations." *Ibid.*

100 " 'a Magdalenian engraving . . . funerary customs." *Ibid.*

"It had always . . . later date." *Ibid.*

"Almost three years . . . years before." *Ibid.*

"Years later a . . . now-lost stone." Letter of Prof. Dr. Waldemar Chmielewski to M. Swida.

102 ". . . But Poniatowski presented . . . valid observations." Poniatowski, *Clairvoyance.*

104 "Poniatowski began writing . . . into the fall." *Ibid.*

"Poniatowski was first . . . occupied Czechoslovakia. "Stanislaw Poniatowski," *Wielka Encyklopedia Powszechna,* vol. 9, p. 233.

"His wife was . . . not listen." Interview with M. Swida, 8 December 1975.

105 "Ossowiecki took with . . . had known." *Ibid.*

"Their first stop . . . set up." *Ibid.*

107 "About seven months . . . at Litomierzyce." "Stanislaw Poniatowski" *Wielka Encyklopedia Powszechna,* vol. 9, p. 233.

III. THE SCOTTISH GENERAL AND THE RUSSIAN RODWALKERS

Sources

Interview with Major General James Scott Elliot, F.S.A. (Scot.), 1 March 1976; correspondence with Major General James Scott Elliot, F.S.A. (Scot.), 1976; correspondence between Christopher Bird and Major General James Scott Elliot, F.S.A. (Scot.) 1974-76; correspondence with Christopher Bird, 1973-77; seven interviews with Christopher Bird, 1973-77.

Notes

108 " 'I have no. . . that's all.' " Interview with J. Scott Elliot, March 1976.

"A major general . . . to do." *Ibid.*

" 'I'd never done . . . made sense.' " *Ibid.*

109 " 'Question of efficiency . . . fumbling about.' " *Ibid.*

"He found a . . . practiced it." Interview with C. Bird, 16 April 1976.

111 "At first he . . . very accurate." Interview with J. Scott Elliot, 1 March 1976.

" 'Of course, I . . . very well.' " *Ibid.*

"By 1961 James . . . tunnel flue." Correspondence with C. Bird.

114 "About four years . . . right again." J. Scott Elliot, "An Early Bronze Age Fire Pit at Townfoot Farm by Glencaple," in *The Transactions of the Dumfriesshire and Galloway Natural History and Antiquarian Society*, 3rd series, vol. 49 (1972), p. 20.

"As the workers . . . stone wall." *Ibid.*

"A carbon-14 sample . . . Bronze Age." *Ibid.*

115 " 'The site is . . . known form.' " *Ibid.*, p. 22.

"Two possible explanations . . . spitted meat." *Ibid.*

"No, the General . . . in urns." *Ibid.*

116 "The acidity of . . . remains uncovered." *Ibid.*, p. 23.

"The General had . . . nine years." Interview with J. Scott Elliot, 1 March 1976.

117 "Why not . . . to try." *Ibid.*

"When he was . . . a long time." *Ibid.*

"In 1969, however . . . Swinbrook Cottage." *Ibid.*

"The cottage and . . . any encouragement." Elliot, "Report on the Excavations of Two Sites and Swinbrook Cottage, Swinbrook, Burford." Unpublished site report. p. 1.

"He made no . . . the gardens." Interview with J. Scott Elliot, 1 March 1976.

"Silently in . . . even older." Interview with J. Scott Elliot by C. Bird, n.d.

118 "Less than two . . . General's study." Interview with J. Scott Elliot, 1 March 1976.

" 'provided ample evidence . . . topsoil level.' " Elliot, "Report on the Excavations of Two Sites and Swinbrook Cottage." p. 1.

"A second cut . . . was reached." *Ibid.*, p. 2.

" 'a beautifully laid . . . various levels." *Ibid.*

119 "the site also . . . and flintnaps." *Ibid.*, p. 5.

" 'The pottery group . . . area yet.' " *Ibid.*

"The cottage garden . . . be added.' " *Ibid.*

123 "His 1973 work . . . at Swinbrook." Correspondence be-

tween J. Scott Elliot and C.
Bird, November 1974.

123 " 'main interest . . . me now.' "
Interview with J. Scott Elliot, 1
March 1976.

"Of this Neolithic . . . Salisbury
Plain." J. Scott Elliot, "Field
Notes on Noriston Farm Excavation." Unpublished report, n.d.

126 "Bird was amazed . . . much
longer.' " Interview with C.
Bird, 12 April 1976.

127 " 'here was a man . . . no amateur.' " Interview with C.
Bird, 12 April 1976.

"From the beginning . . . tell you
that.' " Interview with J. Scott
Elliot, 1 March 1976.

128 " 'As far as . . . wholly discredited.' " United States Geologic Survey, *Water Witching*.

"In the face . . . would dowse."
U.S.S.R. Ministry of Com-
munications, *Second Scientific
Technical Seminar on the Biophysical Effect*. (Collected Papers, Moscow, 12 May 1971)

"A vast body . . . were present-
ed." C. Bird, "Dowsing in the
U.S.S.R.," *The American
Dowser*, August 1972, p. 111.

" 'It's perfectly obvious . . .
wouldn't it?' " Interview with
C. Bird, 16 September 1975.

129 " 'at the request . . . of Architecture.' " Aleksandr I.
Pluzhnikov, "Possibilities for
and Results of the Use of the
Biophysical Method in Re-
searching and Restoring Historical and Architectural
Monuments," *The American
Dowser*, August 1974, p. 116.

" 'search for and . . . Biophysical

Method.' " *Ibid.*, p. 117.

"Because of the . . . turning
points." *Ibid.*

130 "Spread out before . . . the
French." "Borodino,"
*Bol'shaia Sovetskaia Entsik-
lopedia*. 2nd ed.

"For the Russians . . . Kurgan-
naya Hill." Armand de
Caulaincourt, *With Napoleon in
Russia*, pp. 94-105.

131 "As many as . . . battle alone."
"Borodino," *Bol'shaia Sovetskaia Entsiklopedia*, 2nd ed.

" 'Of all my . . . called invincible.' " *Ibid.*

"Within hours of . . . was outlined." Pluzhnikov, "Possibilities for and Results of the
Use of the Biophysical Method," (hereafter cited as "The
Biophysical Method"), p. 116.

"On the strength . . . Orthodox
monasteries." *Ibid.*, p. 117.

132 "All had either . . . were lost."
Ibid.

"Pluzhnikov next took . . . the
church." *Ibid.*, p. 118.

"The Godunovs were . . . elected
Czar." "Godunov," *Bol'shaia
Sovetskaia Entsiklopedia*, 2nd ed.

133 "Over the centuries . . . and
memory." "Vyazemy
Bol'shie." *Ibid.*

"Pluzhnikov had the . . . ethno-
historic record." Pluzhnikov,
"The Biophysical Method", p.
117.

" 'within eight hours . . . bell
towers." *Ibid.*

134 " 'the wooden walls . . . the
dowsers." *Ibid.*

"The team had . . . positive response." *Ibid.*, p. 116.

" 'The use of reliable results.' "
Ibid., p. 118.

IV. THE TRANSITIONAL MAN

Sources

Interviews with J. Norman and Ann Emerson (individually and together) between November 1974 and August 1977 (a series of 17); correspondence with
J. Norman and Ann Emerson between January 1975 and August 1977; psychic

reading of George McMullen dealing with argillite carving, 1973; psychic reading of Sheila Conway, 1973; psychic reading of subject known as Sandy, 1973; personal communication to Ann Emerson from George McMullen; personal communication from George McMullen, 1977; correspondence with Sheila Conway, 1977; interview with Allen Tyyska, 11 January 1976; interview with Willem H. C. Tenhaeff, January 1974; Croiset, Gerard. "For Discussion," unpublished paper, n.d.; interview with Gerard Croiset, January 1974.

Notes

136 " 'It is my . . . of reasoning.' " J. Norman Emerson, "Intuitive Archaeology: A Psychic Approach," *New Horizon*, Vol. 1, No. 3 (1974), p. 14

"respondent's accuracy ran . . . eighty percent.' " *Ibid.*

" 'By means of . . . first priority.' " *Ibid.*

138 "Close to ninety . . . their careers." Interview with J. Norman Emerson, 8 January 1976.

" 'Traditional research into . . . psychic persons.' " *Ibid.*

139 "Emerson's trek into . . . fishing together." Interview with J. Norman and Ann Emerson, 9 January 1976.

"In spite of the . . . with me.' " Interview with J. Norman Emerson, 21 November 1974.

"The taciturn, 'totally . . . have guessed.' " Interview with Ann Emerson, 11 January 1976.

" 'George's suggestions made . . . may be.' " Interview with J. Norman Emerson, 9 January 1976.

"Fascinated by the . . . Indian artifacts." Interview with J. Norman and Ann Emerson, 9 January 1976.

140 " 'I was fascinated . . . archaeological means.' " Interview with J. Norman Emerson, 10 January 1976.

"Still Emerson's commitment . . . Jack Harrison Pollack." Interview with J. Norman Emerson, 21 November 1974.

"This work first . . . parapsychological mentor." Jack Harrison Pollack, *Croiset the Clairvoyant*, p. 32.

141 "This time, however, . . . his theory." *Ibid.*

"The experiments began . . . the box.' " *Ibid.*, p. 33.

142 "Although he had . . . psychic discourses." Interview with Willem H.C. Tenhaeff, January 1974.

143 " . . . the small bone . . . this cave." Pollack, *Croiset the Clairvoyant*, p. 32.

" 'A whole troup . . . in height.' " *Ibid.*, p. 33.

" . . . the bone that . . . first session." *Ibid.*

" ' . . . Negroes. But also . . . on their heads.' " *Ibid.*

144 "Dr. van Riet Lowe . . . or turbans.' " *Ibid.*

" 'We now know . . . sight appear . . . . ' " *Ibid.*, p. 35.

145 "There was no doubt . . . arrived ourselves.' " *Ibid.*, p. 193.

" 'I'm not crazy . . . be solved.' " Interview with J. Norman Emerson, 21 November 1974.

146 "One case still . . . for the death." Interview with George McMullen by Ann Emerson. n.d.

"His mother was . . . were wrong." Interview with Ann Emerson, 11 January 1976.

"In his neighborhood . . . as lights." Interview with George McMullen by Ann Emerson, n.d.

147 "Influenced by the . . . information source." Interview with J. Norman Emerson, 10 January 1976.

"In one session . . . Iroquois

times.' " J. Norman Emerson, "Archaeology, Parapsychology and One White Crow." Unpublished paper, April 1975.

148 "During this same . . . County area.' " *Ibid.*

149 "As it turned . . . accompany him." Emerson, "Intuitive Archaeology: A Psychic Approach," p. 16.

" 'First,' said Emerson . . . Indian village.' " Interview with J. Norman Emerson, 11 January 1976.

"Emerson, following with . . . as staples.' " *Ibid.*

"In an effort . . . sunflower seeds." Emerson, "Intuitive Archaeology: A Psychic Approach," p. 16.

"As Emerson sought . . . other Indians." Interview with J. Norman Emerson, 11 January 1976.

" . . . 'felt that he . . . carried out." Emerson, "Intuitive Archaeology: A Psychic Approach," p. 16.

" 'When the study . . . the crops." Interview with J. Norman Emerson, 11 January 1976.

150 " 'To me, it . . . the case.' " *Ibid.*

"Miller came over . . . British Columbia . . . " Emerson, "Intuitive Archaeology: The Argillite Carving." Unpublished paper, March 1974.

151 "Miller told Emerson . . . made it?" Emerson, "Intuitive Archaeology: A Developing Approach." Unpublished paper, November 1974.

"The stone, McMullen . . . as a slave." Emerson, "Intuitive Archaeology: The Argillite Carving."

" 'Here I had . . . the possible.' " Interview with J. Norman Emerson, 10 January 1976.

152 " 'I thought, you . . . would

emerge." *Ibid.*

"The black man . . . black argillite." Psychic readings of George McMullen, 1973.

"As he sat . . . George is right.' " Interview with J. Norman Emerson, January, 1976.

153 "To begin with . . . passed to her." Correspondence with Ann Emerson, n.d.

" 'had been the . . . New World.' " Psychic reading of subject known as Sandy, 1973.

"Ann Emerson, meanwhile . . . showed promise." Interview with J. Norman and Ann Emerson, 9 January 1976.

154 " 'It seemed to . . . listening to.' " Interview with J. Norman Emerson, 10 January 1976.

155 " 'elated to find . . . fundamental disagreements.' " *Ibid.*

" 'He came from . . . miles inland.' " Emerson, "Intuitive Archaeology: The Argillite Carving."

" 'There was a . . . very damp.' " *Ibid.*

" 'He was from . . . Savannah land.' " *Ibid.*

" 'The jungle is . . . down below.' " *Ibid.*

" 'I don't know . . . slave ship . . . ' " *Ibid.*

" ' . . . and he was . . . for them.' " *Ibid.*

" 'They were raided . . . the Americans.' " *Ibid.*

" 'He made the . . . American continent . . . ' " *Ibid.*

" 'Kind of looks . . . Charlotte Islands?' " *Ibid.*

156 "Sheila also reported . . . Majoree." Psychic reading of Sheila Conway, 1973.

" 'no matter how . . . the Indians.' " Interview with J. Norman Emerson, 10 January 1976.

" 'you had to . . . team said.' " *Ibid.*, 21 November 1974.

" 'I had taken . . . to say.' " Interview with Allen Tyyska, 11 January 1976.

157 " 'This little piece . . . that puzzle.' " Emerson, "Intuitive Archaeology: The Argillite Carving."

" 'I would entertain it very seriously.' " Interview with Allen Tyyska, 11 January 1976.

"Almost two years . . . had married!" Correspondence of J. Norman Emerson, n.d.

158 "Ever since then . . . than either.' " Interview with J. Norman Emerson, 11 January 1976.

" 'Much of what . . . be wrong.' " *Ibid,*

159 " 'the information I . . . meant something.' " *Ibid.*, 9 January 1976.

" 'making progress . . . do it all.' " *Ibid.*, 11 January 1976.

"The archaeological record . . . about them." J. V. Wright, *Ontario Prehistory: An Eleven-Thousand Year Archaeological Outline*, pp. 11 *ff.*

" 'We think they . . . about it.' " Interview with J. Norman Emerson, 9 January 1976.

160 " 'the Wind People.' . . . wind chaffing." Extract of Psychic Reading by George McMullen contained in correspondence with Ann Emerson, n.d.

" 'I started to . . . perhaps starvation.' " Interview with J. Norman Emerson, 9 January 1976.

" 'hear him tell . . . terrible winds.' " ' " *Ibid.*

" 'be plugged into . . . be done.' " *Ibid.*

" 'A lot of . . . other questions.' " Interview with Ann Emerson, 11 January 1976.

161 "The only comment . . . a trance.' " *Ibid.*

" 'I project myself . . . asking about.' " Personal communication to Ann Emerson from George McMullen, 1973.

"McMullen maintains he . . . directing him." Personal communication from George McMullen, 1977.

" 'It is as . . . now seeing.' " Interview with J. Norman Emerson, 11 January 1976.

"Having accomplished his . . . site landscape." *Ibid.*, 10 January 1976.

" 'I used to . . . still alive.' " *Ibid.*, 21 November 1974.

162 " 'see their campfires . . . smell them." Personal communication to Ann Emerson from George McMullen, 1974.

" 'he was told . . . listen to.' " Interview with Ann Emerson, 10 January 1976.

"McMullen says he . . . so unaware." ' " *Ibid.*

" 'I don't know . . . to explain.' " Interview with J. Norman and Ann Emerson, 9 January 1976.

163 " 'George is getting . . . is archaeology.' " Interview with J. Norman Emerson, 9 January 1976.

" 'way up in . . . Prince Rupert.' " *Ibid.*

"In some cases . . . more information.' " Personal communication to J. Norman and Ann Emerson from George McMullen, 1974.

"Mrs. Conway 'actually . . . participating in it.' " Interview with Ann Emerson, 11 January 1976.

"Unlike McMullen, she . . . sunrise ceremony." Interview with J. Norman and Ann Emerson, 11 January 1976.

164 "While in this state . . . at once." Interview with Ann Emerson, 11 January 1976.

" 'If I had . . . good start,' " Interview with J. Norman Emerson, 11 January 1976.

V. THE CANADIAN AND THE SEER'S SON

Sources

Interview with J. Norman and Ann Emerson both individually and together between November 1974 and August 1977 (a series of 17); correspondence with J. Norman and Ann Emerson both individually and together between January 1975 and August 1977; interviews with George McMullen, March through September 1977; correspondence with George McMullen, April through November 1977; interview with Hugh Lynn Cayce, May 1974 through September 1977 (a series of 5); correspondence with Hugh Lynn Cayce, February 1973 through November 1977; personal communication with Arch Ogden, May 1975; personal communication with Charles Thomas Cayce, May 1975; interview with Violet Shelley, 16 July 1976; interview with Mark Lehner, May 1975; correspondence with Mark Lehner, May 1975 to June 1976; interview with Mae Gimbert St. Clair, December 1974; interviews with "Richard Roche" (pseudonym), June 1976 to November 1977; correspondence with "Richard Roche," June 1976 through October 1977; personal communication with Linton Satterthwaite, June 1976; interview with Dastur Framroze A. Bode, March 1977 through May 1977 (a series of 3); interview with Jeffrey Furst, May 1976; interview with Frank O. Adams, 12 June 1976; interview with Stephan Goranson, May 1975; correspondence between Violet Shelley and "Richard Roche," March 1974; correspondence between Hugh Lynn Cayce and Professor Ezat O. Negahban, January 1973.

Edgar Cayce Readings: 1) specifically cited in text: 364 series, 378-16, 440-5, 489-1, 1391-1, 5748-6, 5749-8; 2) generally: there are literally hundreds of Edgar Cayce Readings dealing with Egypt, Persia, Yucatan, and the Essenes (far too many to cite here); all are indexed, however, and they are readily available to the public.

George McMullen Readings: 16 June 1975; extracts from several psychic commentaries during Egyptian/Iran trip, October 1975.

Notes

166 "Although many periods . . . Christian era." Based on survey, carried out by the author, of all Edgar Cayce readings.

167 " ' . . . at rare intervals . . . biological evolution.' " Henry Bamford Parkes, *Gods and Men: The Origins of Western Culture,* pp. 77 *ff.*

"Except for a . . . individual facts." Author's survey of Cayce readings.

"Two possible . . . do so." Interview with Hugh Lynn Cayce, May 1974.

168 "Unfortunately, very few . . . their careers." Interview with Mae Gimbert St. Clair, December 1974.

"Edgar Cayce had said . . . now underwater." Edgar Cayce readings, No. 364 Series.

169 "As an example . . . ever asked." Linton Satterthwaite, from a series of reports and bulletins published on his work at Piedras Negras, between 1937 and 1954, published by the University of Pennsylvania Museum and Edgar Cayce Readings; particularly No. 440-5 in which Cayce refers to ''Pennsylvania State Museum" and "Washington" which is the headquarters of the Carnegie Institution, another foundation which sponsored research at this site.

170 "He had, in . . . located there." Edgar Cayce Reading no. 489-1.

"Fortunately, on Tuesday . . . from Jerusalem.' " Edgar Cayce Reading no. 1391-1.

"Nor was this . . . this location. *Ibid.*

171 "the only site . . . for years." Matthew Black, *The Scrolls and Christian Origins*, pp. 3-26.

"These opinion were . . . linen cloth." Millar Burrows, *The Dead Sea Scrolls*, pp. 4 *ff.*

"Its use as . . . A.D. 68." "Dead Sea Scrolls," *Encyclopedia Britannica*, 1973, vol. 7, pp. 116-120.

"Cayce had stated . . . in astrology." Edgar Cayce Reading no. 5749-8.

172 " 'When I showed . . . it up.' " Interview with Hugh Lynn Cayce, 14 May 1975.

"it told the . . . Caucasus Mountains." Recapitulation of Cayce's reconstruction based on author's survey of several dozen readings, and research done by Richard Roche (psuedonym) for monograph *The Ra-Ta Legend*.

174 "Edgar Cayce placed . . . of Giza." Edgar Cayce Reading no. 5748-6.

"By adding up . . . emerging chronology." James Henry Breasted, *A History of Egypt*, pp. 597-601.

175 "Toward the end . . . adobe houses." *Ibid.* pp. 27 *ff.*

"According to this . . . Fourth Dynasty . . . " *Ibid.* pp. 116-120.

"This revelation came . . . other spaces." John David Wortham, *The Genesis of British Egyptology*, pp. 26 *ff.*

176 "Their sole function . . . being crushed." Ahmed Fakhry, *The Pyramids*, p. 120.

"he revealed four . . . three-year reign . . . " *Ibid.* p. 102.

" 'as the sun . . . the river.' " Edgar Cayce Reading no. 378-16. See also E.C. Reading no. 5748-6.

" 'For me it . . . was correct!' "

Interview with Mark Lehner, May 1975.

177 " 'It was the . . . is involved.' " *Ibid.* See also "Egypt: Reflections on a Tour," *ARE Journal*, vol. VIII, no. 2, March 1973, pp. 53-70.

"He began his . . . of religion." Interview with Richard Roche, 14 July 1976.

178 "Soon he realized . . . something there.' " *Ibid.*

" 'without any of . . . Egyptian history.' " *Ibid.*, 19 June 1976.

" 'the general outline . . . story fit.' " *Ibid.*, 26 May 1977.

" 'I looked up . . . a booklet.' " Interview with Violet Shelley, 16 July 1976.

179 " 'He said, 'No, . . . up there.' " *Ibid.*

" 'the very preliminary . . . brief manuscript.' " Correspondence between Violet Shelley and Richard Roche, 1975.

180 " 'There was a . . . avoided altogether.' " See *Egyptian Myths* by Richard Roche for a full treatment of this complex issue.

" 'Osiris does not . . . office's title.' " Richard Roche, *Egyptian Myths*, pp. 15 *ff.*

" 'the case for . . . Jebel Barkal." *Ibid.* pp. 49-53.

181 " 'As the cultural . . . the job,' " Interview with Richard Roche, 14 July 1976.

"a temple 'of . . . greater size.' " A.J. Arkell, *A History of the Sudan from the Earliest Times to 1821*, 2nd ed., p. 280.

" 'evidence that Cayce's . . . major discoveries.' " Interview with Richard Roche, 14 July 1976.

182 " 'not with the . . . out there.' " Interview with Hugh Lynn Cayce, June 1976.

"The time is . . . is unclear." This brief presentation of Cayce's Persian reconstruction is a composite based on dozens of individual readings, each containing some portion of the whole.

183 "He states that . . . about 6600 B.C. " Jeffrey Furst, *Edgar Cayce's Story of Jesus,* pp. 96-98. "This disagrees with . . . century date." Henry Bamford Parkes, *Gods and Men: The Origins of Western Culture,* p. 135. "An ancient Iranian . . . 330 B.C." George Rawlinson, *Five Great Monarchies of the Ancient Eastern World,* vol. II, p. 320 *ff.*

184 " 'In Palestine the . . . in Athens.' " Parkes, *Gods and Men,* p. 77. " 'When Zoroaster lived . . . be possible.' " Interview with Frank O. Adams, 12 June 1976.

185 " 'Zoroaster was *not* . . . original sources.' " Interview with Framroze Bode, 22 May 1977. See also Framroze Bode, *The Zoroastrian Doctrine of the Soul,* unpublished doctoral dissertation, The Philosophical Research Society, August, 1957, ch. 2 for a complete discussion of the Zoroastrian Chronology from the point of view of a scholar who is also an authentic and native Zoroastrian priest.

187 " 'How could we . . . prove it?' " Interview with Hugh Lynn Cayce, May 1975.

188 " 'to the material . . . first place.' " Interview with J. Norman Emerson, 9 January 1976. "On Sunday, October . . . than willing." *Ibid.* " 'I took a . . . you get?'·' " Interview with Hugh Lynn Cayce, June 1976. " 'and the city . . . inland water.' " Psychic reading of George McMullen, 16 June 1975.

"He also correctly . . . be correct." Psychic reading of George McMullen, 16 June 1975, and subsequent evaluation by J. Norman Emerson. See "Intuitive Archaeology: Egypt and Iran," *ARE Journal* vol. 9 no. 2 (March 1976), pp. 55-65.

189 "Back in Canada . . . many misgivings.' " Interview with George McMullen by Ann Emerson, June 1975.

190 " 'no way I . . . those words.' " Interview with J. Norman Emerson, 10 January 1976. " 'like a spare wheel . . . know it.' " *Ibid.* "Finally, able to . . . lies beyond.' " Interview with Ann Emerson, 10 January 1976.

191 " 'It was as . . . psychic abilities.' " Interview with George McMullen, 16 November 1977. "Hugh Lynn, for . . . been intended." Interview with Hugh Lynn Cayce, June 1976. "There he outlined . . . the Sphinx . . ." Psychic reading of George McMullen, October 1975.

192 " 'hall of records' . . . the Nile.' " Psychic reading of George McMullen, October 1975. See also "Intuitive Archaeology: Egypt and Iran," *ARE Journal,* vol. 9, no. 2 (March 1976), pp. 55-65.

193 "her information, unknown . . . every particular." Interview with J. Norman Emerson, 10 January 1976. " 'It is all . . . to see.' " *Ibid.* See also, Emerson, "Intuitive Archaeology: Egypt and Iran." " 'sense of control . . . to do.' " Interview with J. Norman Emerson, 12 January 1976.

194 "It was agreed . . . with records." Emerson, "Intuitive Archaeology: Egypt and

Iran," pp. 62-65.
"There, he said . . . center here.'" Psychic reading of George McMullen, October 1975.
"'Although I have . . . support them.'" Interview with J. Norman Emerson, 12 January 1976.
195 "No sooner did . . . be discovered." Emerson, "Intuitive Archaeology: Egypt and Iran," pp. 63-65.
"When Emerson mentioned . . . found intact." Interview with

J. Norman Emerson, 13 January 1976.
196 "He asked if . . . to read." Emerson, "Intuitive Archaeology: Egypt and Iran," p. 65.
"At the appointed . . . same period.'" *Ibid.*
"To Emerson, while . . . great relief.'" Interview with J. Norman Emerson, 13 January 1976.
"'I had always . . . the details.'" *Ibid.*, 12 January 1976.
197 "'We are just . . . any find.'" *Ibid.*, 10 January 1976.

VI. CHILDREN OF THE CHANGE

Sources
Interview with Charles Garrad, 16 January 1976; correspondence with Charles Garrad, 1976; interview with J. Norman and Ann Emerson, January 10, 1976; psychic reading of Sheila Conway, 4 August 1973; interview with J. Norman Emerson, 9 January 1976; interview with Rolling Thunder, December 1973; interview with C. W. Weiant, 20 March 1976; psychic reading of George McMullen, 19 May 1973.

Notes
198 "'By four o'clock . . . cannot say.'" Interview with Charles Garrad, 16 January 1976.
"'Look here . . . I'll quit!'" *Ibid.*
199 "'But while the dust . . . from view.'" Correspondence with Charles Garrad, 1976.
200 "Pipes that had . . . them out." Interview with Garrad.
201 "'I didn't know . . . a photograph.'" *Ibid.*
202 "He was sure . . . been doing." *Ibid.*
"'I too have . . . been hallucinating.'" *Ibid.*
"Emerson assured him . . . was working." Interview with J. Norman and Ann Emerson, 10 January 1976.
203 "'Ehwae was significant . . . clear about.'" Interview with Garrad.
"'It was windy. . . . deep trance.'" *Ibid.*
204 "'There is a. . . .made new.'"

Psychic reading of Sheila Conway, 4 August 1973.
"'The babies died. . . . no more.'" *Ibid.*
205 "The French priests . . . no resistance." Interview with J. Norman Emerson, 9 January 1976.
"Her statements about . . . powerful rituals." Joseph K. Long, "Shamanism: Trance, Hallucinogens, and Psychical Events." Unpublished paper, 1973.
"Rolling Thunder has . . . Virginia Beach." Boyd, *Rolling Thunder*, pp. 189 *ff.*
"He is an . . . implied perceptions." Douglas Boyd, *Rolling Thunder*, pp. 48 *ff.*
"He has even . . . flowing water." Interview with Rolling Thunder, December 1973.
206 "The account of . . . Indian behavior." Interview with J. Nor-

man Emerson, 9 January 1976.

206 " 'Many people came . . . wore it.' " Psychic reading of Conway.·

" 'looking at . . . sort of thing.' " Interview with Garrad.

" 'So what we're . . . friendly 'cousins.' " *Ibid.*

" 'everything this psychic. . . . ethno-historic sources.' " *Ibid.*

207 "he did not . . . the literature." Interview with C. W. Weiant, 20 March 1976.

" 'certain members . . . has died.' " C. W. Weiant, "Parapsychology and Anthropology," *Manas*, vol. 13 no. 15, (13 April 1961).

" 'the process is . . . shelter shakes.' " *Ibid.*

"It might be . . . pagan ritual." *Ibid.*

"it is known . . . was lost." Conrad Heidenreich, *Huronia: A History and Geography of the Huron Indians 1600-1650*, p. 53.

" 'any evidence of . . . Jesuit priest.' " Interview with Garrad.

208 " 'A kindly man. . . . they say.' " Psychic reading of Sheila Conway, 9 September 1973.

"As early as . . . de Trevisa." *Oxford English Dictionary*, compact ed., 1971, p. 480.

209 " 'the leather thong . . . his neck.' " Psychic reading of Conway, 9 September 1973.

"Years before a . . . most venerated saints." Interview with Garrad.

"In fact . . . earlier experimentation." *Ibid.*

210 " 'I see tall . . . rough candles.' " Psychic reading of Sheila Conway, July 1973.

210 " 'The owner of . . . the left.' " *Ibid.*

211 " 'a straight archaeologist . . . little experience.' " Interview with Garrad.

"One, the oldest . . . +120 years." C. S. Reid, "Psychometrics and Settlement Patterns: Field Tests on Two Iroquoian Sites." n.d. Unpublished paper.

"The other site . . . A.D. 1300." *Ibid.*

"At the Boys site . . . be traced." *Ibid.*

213 "On May 19 . . . warm Saturday." " J. Norman Emerson, "Archaeology, Parapsychology, and One White Crow." 1975. Unpublished paper.

" 'a nondescript . . . can tell you.' " Interview with J. Norman Emerson, 10 January 1976.

"The palisade, he said . . . some distance." Psychic reading of George McMullen, 19 May 1973.

214 "For another . . . the village." *Ibid.*

"He said the . . . two people." *Ibid.*

"A week later . . . been excavated." Reid, "Psychometrics and Settlement Patterns."

"The Boys site . . . twelve inches." *Ibid.*

218 "No such palisade . . . ever existed." *Ibid.*

" 'a palisade line . . . its construction.' " *Ibid.*

" 'within eighteen inches . . . predicted location.' " *Ibid.*

"His outline of . . . were exact." *Ibid.*

" 'On the basis . . . ceremonial structure.' " *Ibid.*

220 "Sewell looked exactly . . . north edge." Interview with J. Norman Emerson, 10 January 1976.

"It was . . . he indicated." Reid, "Psychometrics and Settlement Patterns."

"McMullen then told ... the site." *Ibid.*
221 "Reid had his ... test trench." *Ibid.*

"Then the diggers ... post holes." *Ibid.*
"As the trench ... have been." *Ibid.*

VII. INTO THE AMERICAN MAINSTREAM

Sources

Interviews with Clarence W. Weiant between March and April 1976 (a series of 3); correspondence with Clarence W. Weiant between March 1976 and September 1976; interviews with Joseph K. Long between November 1974 and September 1977 (a series of 6); correspondence with Joseph K. Long between November 1974 and November 1977; interview with James Officer, 25 March 1976; interview with Robert Van de Castle, July 1975; interviews with Philip Staniford, September 1977 through December 1977 (a series of 3); interview with David E. Jones, December 1977.

Notes

222 "At the end ... our necks,'" Matthew W. Stirling, "Discovering the New World's Oldest Dated Work of Man," *National Geographic*, vol. 76, no. 2 (August 1939), p. 201. "the oldest recorded date ... 291 B.C. " *Ibid.*, pp. 211-213.

223 "He had begun ... photographic film." Interview with Clarence W. Weiant, 20 March 1976. " 'She had to ... their portraits.'" *Ibid.*

224 "A full study ... early 1960s." See *The World of Ted Serios* by Jule Eisenbud, M.D. New York: Morrow, 1967. "Weiant had already ... New York." Interview with Clarence W. Weiant, 20 March 1976.

225 " 'persuading Columbia University ... school subject.'" *Ibid.*, 26 March 1976.

226 " 'I had started ... across it.'" *Ibid.*, 20 March 1976. "There was about fifty ... jungled-covered." Clarence W. Weiant, *An Introduction to the Ceramics of Tres Zapotes Veracruz, Mexico,* Smithsonian Institution Bureau of American Ethnology Bulletin 139, 1943, pp. 1-3.

" 'had not been ... that work.'" Interview with Clarence W. Weiant, 20 March 1976. " 'He was an old ... looking for.'" *Ibid.*, 26 March 1976.

228 "He had gone ... flood plain." *Ibid.*, 20 March 1976. " 'Within twenty minutes ... the ground.'" *Ibid.*

230 " 'ninety percent of ... that winter." Weiant, *Ceramics of Tres Zapotes*, p. 9.

232 "The Ranchito group ... of cultures." *Ibid.*, pp. 7-9.

233 " 'a series of ... this cemetery.'" *Ibid.*, p. 7. "(even Stela C ... of stone.)" Stirling, "Oldest Dated Work," pp. 211 *ff.* " 'since there was ... first suggested.'" Interview with Clarence W. Weiant, 20 March 1976. " 'had made a very ... scientific test.'" *Ibid.*

234 " 'No one would ... like that.'" *Ibid.*, 26 March 1976. " 'the tremendous implications ... Tres Zapotes,'" Correspondence with Clarence W. Weiant, June 1976.

236 " 'A significant revolution ... scientific manner.'" An open letter of John R. Swanton to

all listed members of the American Anthropological Association, 1952.

236 " 'the thunderbolt has . . . science consists.' " *Ibid.*
" 'a set of . . . be damned.' " *Ibid.*

237 " 'I am delighted . . . December meeting.' " Correspondence between Ignacio Bernac and Clarence W. Weiant, 1961.
" 'I had no . . . discussion unnecessary.' " Interview with Clarence W. Weiant, 20 March 1976.
" 'She said she . . . and why.' " *Ibid.*

238 " 'It was the . . . the voice.' " Interview with Joseph K. Long, 20 November 1974.
" 'It was incredible. . . . on them.' " *Ibid.*

239 " 'Apparently, it just . . . understand why.' " *Ibid.*
" 'was a case . . . and all.' " *Ibid.*
" 'had read not . . . don't happen.' " *Ibid,* 16 September 1975.
" 'I didn't have a . . . myself committed.' " *Ibid.*
" 'not a word . . . even now.' " *Ibid.*

240 " 'Going over my . . . been taught.' " *Ibid,* 20 November 1974.
" 'Parapsychology and anthropology . . . psi research.' " *Ibid,* 9 November 1975.
" 'Some effort had . . . 1959 effort.' " *Ibid,* 20 November 1974.
"First he got . . . psychic research.' " *Ibid.*

241 " 'only in this . . . to anthropology.' " *Ibid,* 21 March 1976.

" 'We couldn't have . . . other decision,' " Interview with James Officer, 25 March 1976.

242 " 'You're either lying . . . you've got.' " Transcript of Session 703, Rhine-Swanton Symposium on Parapsychology and Anthropology. For a finished transcript of this entire symposium, see *Extrasensory Ecology: Parapsychology and Anthropology,* Joseph K. Long, ed. However, a number of papers *not* actually presented are included and all have been edited; some are very severely at variance with what was actually said that day.

243 "Bharati, a former . . . all of it!' " *Ibid.*

244 " 'How much more . . . believe it?' " Anonymous aside, Rhine-Swanton Symposium.
" 'I take special umbrage . . . of science.' "Transcript, Rhine-Swanton Symposium.

245 " 'Excuse this contraption . . . in India.' " *Ibid.*
"At San Diego . . . cultural anthropology . . . " See "Inside Out: Anthropological Communication of Alternate Realities," *Phoenix,* vol. 1, no. 1 (summer 1977), pp. 36-46.
"at Florida Technological . . . psychic respondents." See "An Experiment with Non-Scientific Discovery Procedures in Archaeology," *Phoenix,* vol. 1, no. 2 (winter 1977), and "Folsum Ethnography," *Phoenix,* vol. 1, no. 3 (summer 1978).

VIII. KUHN, CONTEXT, AND REVOLUTION

Notes

All notes are taken from: Thomas S. Kuhn, *The Structure of Scientific Revolutions.* Chicago: University of Chicago Press, 1962.

IX. PREJUDICE, PAIN, and PARADIGM

Sources

Interview with an anonymous archaeologist, 17 May 1976; open letter to members of the Society of American Archaeology from its president, Charles McGimsey III, 1974.

Notes

263 " 'I'm trying to . . . I wouldn't.' " Interview with archaeologist, (Interview 9-27), May 17 1976.

266 "They studied everything . . . Age sequence." Lucretius, *The Nature of Things*, J.S. Watson, trans., vol. 5, p. 1288.

"In spite of . . . great museums," W.H. Van Loon, *The Arts*, pp. 19 *ff.*

267 "James Ussher . . . of creation . . . " Walter W. Taylor, *A Study of Archaeology*, p. 15.

"Dr. John Lightfoot . . . of October!" *Ibid.*, p. 15.

"Despite the surprisingly . . . Marco Canal," Herbert Wendt, *In Search of Adam: The story about Man's Quest for Truth about His Earliest Ancestors*, p. 10.

"Some researchers held . . . were found." To be found in: *An Universal History from the Earliest Account of Time*, Allen, *et al*, (published 1754) vol. 24, p. 117.

269 "For years he . . . had mutated." Walter Ralegh, *The Historie of the World*, book 1, chapters 7 and 8, pp. 83-112.

270 "He coolly and . . . such theories." Wendt, *In Search of Adam*, pp. 163-166.

271 . . ."in the dead of an eighteenth . . . go back to sleep." *Ibid.*, p. 149.

"but also in the task . . . under observation." Panchanan Mitra, *A History of American Anthropology*, Section 9.

273 "He surveyed a . . . Monticello." Thomas Jefferson, *The Works of Thomas Jefferson*, coll. and ed. by Paul Leicester Ford, vol. 3, pp. 499-517.

"As he dug . . . the positioning." *Ibid.*

"Jefferson saw digging : . . . cultural interpretation." *Ibid.*

"his comparison of . . . on individuals . . . " Gordon R. Willey and Jeremy A. Sabloff, *A History of American Anthropology*, pp. 36-40.

275 "He had worked . . . written record." See William Flinders Petrie, "Sequences in Prehistoric Remains," *Journal of the Royal Anthropological Institute of Great Britain and Ireland*, vol. 29, pp. 295-301.

"Petrie had theorized . . . technical sophistication." *Ibid*, pp. 295-301.

"He began by . . . known world " V. Gordon Childe, *The Dawn of European Civilization*, 1st ed., (1925) pp. 27-28. (This proposition is not so much formally stated in Childe's works as it is the unspoken premise.)

"Childe made another . . . to Europe." *Ibid.*, p. xiii and pp. 27-30.

276 " 'To begin with . . . their ideas.' " Clyde Kluckhohn, "The Conceptual Structure in Middle American Studies," *The Maya and Their Neighbors*, pp. 43-44.

276 " 'the task will . . . fact compare.' " Walter W. Taylor, *A Study of Archaeology*, p. 43. " 'It is the blunt . . . of man)' " *Ibid*, p. 93.

277 "Leonardo da Vinci . . . the observation.'' ''Dendrocronology,'' *Encyclopedia Britannica*, (1973) vol. 7, p. 234.

278 "Soon he concluded . . . technically sophisticated." Bryant Bannister, "Dendrocronology," *Science in Archaeology*, pp. 191-205.

"Ferguson's dates agreed . . . hieroglyphic dates!" *Ibid*.

"Childe had based . . . as dolmens." Childe, *The Dawn of European Civilization*, pp. 131-134.

"The issue was . . . Ban Chiang." Peter Gwynne with Stephen G. Michaud, "The Roots of Man," *Newsweek*, May 31, 1976, p. 75.

279 " 'These people had . . . that time.' " *Ibid*.

"dolmens in general . . . mathematical abilities." See Gerald S. Hawkins, *Stonehenge Decoded*.

"Even ancient man . . . record them.'' See Alexander Marshack, *The Roots of Civilization: The Cognitive Beginnings of Man's First Art, Symbol and Notation*. (This fascinating book traces man's cognitive beginnings and demonstrates previously unsuspected levels of pre-historic analytical sophistication through the study of Paleolithic and later artifacts.)

"Heinrich Schliemann's disputed . . . hardly arose." Schliemann is generally thought to have been unique in his beliefs; he was not. Almost a century earlier the Englishman John Bacon Sawney Morritt entertained similar views. See: *Sir Walter Scott: The Great Unknown*, vol. 1, p. 295.

281 "Thus few, if . . . were possible." Willey and Sabloff, *A History of American Archaeology*, p. 194.

282 " 'Archaeology is anthropology or it is nothing.' " See Gordon R. Willey and Philip Phillips, *Method and Theory in American Archaeology*.

"By the 1950s . . . certain laws." Willey and Sabloff, *A History of American Anthropology*, pp. 193-211.

"While not going . . . did exist." *Ibid*.

283 " 'We do not . . . explanations demand.' " *Ibid.*, p. 195.

284 " 'keep your heads . . . innovative approaches.' " Open letter to members of the Society of American Archaeology from its president, Charles McGimsey III, 1974.

X. COMPLEXITIES, PROMISE, and IMPLICATIONS

Sources

Interview with Margaret Mead, 22 November 1974; correspondence with Margaret Mead, December 1974 to January 1975; interviews with Ingo, Swann between May 1975 and September 1977 (a series of 5); correspondence with Ingo Swann between April 1975 and January 1978; interviews with Aron Abrahamsen between November 1973 and July 1977 (a series of 6); correspondence with Aron Abrahamsen between December 1971 and November 1977; Edgar Cayce readings, a number of which contain references to a treasure at Kelly's Ford, Virginia; interviews with Hugh Lynn Cayce between May 1968 and June 1976 (a series of 7); correspondence with Hugh Lynn Cayce between August 1971 and October 1977; interviews with J. Norman and Ann Emerson, both individually and together between November 1974 and August 1977 (a series of 17); correspondence with J. Norman and Ann Emerson between February 1975 and October 1977; interview with Gladys Davis Turner, 15 May 1975.

Notes

290 " '*Capella St. Edgar . . . St. Edgar . . .*' " F. Bligh Bond, *Gate of Remembrance*, 2d ed., p. 35.

" 'A new scientific . . . familiar with it.' " Max K.E.L. Planck, *Scientific Autobiography and Other Papers*, pp. 33 *ff.*

295 "Werner Heisenberg, a successor . . . experiment's outcome;" Werner Heisenberg, *Philosophic Problems of Nuclear Science*, pp. 77-94.

" 'it . . . only succeeded . . . the sparkle.' " Ingo Swann, *To Kiss Earth Goodbye*, p. 26.

"This apparatus, encased . . . been requested." *Ibid.*, pp. 57-61.

296 " 'I simply grew . . . do it.' " Interview with Ingo Swann, 15 May 1976.

" 'The trouble with . . . be destroyed.' " Interview with Margaret Mead, 22 November 1974.

298 " 'sometimes it was . . . such matters.' " Interview with Ingo Swann, 15 May 1976.

299 " 'Such conversation distracts . . . for conversation.' " Stanislaw Poniatowski, *Clairvoyance and the Method of Its Application for Scientific Research.* "Two psychologists had . . . bad experiment.' " Poniastowski, *Clairvoyance.*

300 " 'The attitudes of . . . psychic realms' " Interview with Aron Abrahamsen, 15 May 1976.

302 "At the height . . . turned earth." Edgar Reading Series and Associated Papers.

"With everything but . . . treasure hunt." Interview with Hugh Lynn Cayce, September 1969.

" 'At the time, I . . . it appears.' " *Ibid.*

304 " 'Every instrument of . . . observational errors.' " Poniatowski, *Clairvoyance.*

" . . . when Ossowiecki spoke . . . like 'iron.' " *Ibid.*

305 " 'I needed to . . . was seeing." Interview with J. Norman Emerson, 10 January 1976.

" 'was quite literally . . . strange skull.' " *Ibid.*

" 'fang teeth' were . . . further thought." J. Norman Emerson, "Intuitive Archaeology and Related Matter: Summary and Potential." Unpublished paper presented at meeting of the Canadian Archaeological Association, March 1975.

" 'The impressions pass . . . other impressions.' " Jack Harrison Pollack, *Croiset, The Clairvoyant,* p. 191.

306 "But there is . . . could, remember." Interview with Hugh Lynn Cayce, 15 May 1975.

307 "Indeed, personal waking . . . all episodes." *Ibid.*, November 1972.

"On more than one . . . but English." Interview with Gladys Davis Turner, 15 May 1975.

308 " 'because of experiences . . . of interest?' " Interview with J.

Norman Emerson, 10 January 1976.

309 " 'they were people . . . any different.' " *Ibid.*

" 'It is very . . . to tell.' " Interview with Aron Abrahamsen, 22 September 1975.

311 " " 'he appeared . . . knew before.' " *Ibid.*, 11 January 1976.

" 'It is as if . . . you know.' " *Ibid.*, 10 January 1976.

315 " 'intuition, but also . . . mathematical formalism.' " Fritjof Capra, *The Tao of Physics.*

BIBLIOGRAPHY

I. THE GLASTONBURY SCRIPTS

Alcock, Leslie. *Arthur's Britain: History and Archaeology.* London: Allen Lane, 1971.

Ashe, Geoffrey. *King Arthur's Avalon: The Story of Glastonbury.* London and Glasgow: Fontana Books, 1973.

Baskerville, Geoffrey. *English Monks and the Suppression of the Monasteries.* New Haven: Yale University Press, 1937.

Bede, the Venerable. *A History of the English Church and People.* Edited and translated by Leo Sherley-Price. Revised by R. E. Latham. Harmondsworth, Middlesex, England: Penguin, 1968.

Bond, Frederick Bligh, and Lea, Thomas S. *A Preliminary Investigation of the Cabala Contained in the Coptic Gnostic Books and of a Similar Gematria in the Greek Text of the New Testament.* Oxford: Basil Blackwell, 1917.

———— *An Architectural Handbook of Glastonbury Abbey.* Bristol: Edward Eveard, 1910.

————, with Camm, Dom B. *Roodscreens and Roodlofts.* 2 vols. London: Pitman & Sons, 1909.

———— *The Gospel of Philip the Deacon (Through the Hand of Hester Dowden.)* New York: Macoy, 1932.

———— *The Secret of Immortality.* Boston: Marshall Jones, 1934.

———— *The Gate of Remembrance: The Story of the Psychological Experiment which Resulted in the Discovery of the Edgar Chapel at Glastonbury.* 2nd and 4th eds. rev. Oxford: Basil Blackwell, 1918, 1921.

———— *The Hill of Vision: A Forecast of the Great War and of Social Revolution with the Coming of the New Race.* London: Constable, 1919.

———— *The Company of Avalon: A Study of the Script of Brother Symon, Sub-Prior of Winchester Abbey in the Time of King Stephen.* Oxford: Basil Blackwell, 1924.

Brock, E.J., and Brock, J.S. *The Legend of the Holy Thorn of Glastonbury.* Glastonbury: Brock, 1843.

Chadwick, Nora K. *Early Brittany.* Cardiff: University of Wales Press, 1969.

Geoffrey of Monmouth, Bishop of Asaph. *The History of the Kings of Britain.* (c. 1150). Translated with an introduction by Lewis Thorpe. London: Folio Society, 1969.

Kenawell, William W. *The Quest at Glastonbury: A Biographical Study of Frederick Bligh Bond.* New York: Helix/Garrett, 1965.

Knowles, David, and Hadcock, R. Neville. *Medieval Religious Houses, England and Wales.* London: Longman, 1971.

Knowles, Brooke, C.L.N, and Vera, C.M.; editors. *The Heads of Religious Houses, England and Wales 940-1216.* Cambridge: Cambridge University Press, 1972.

Lewis, Lionel S. *St. Joseph of Arimathea at Glastonbury, or, The Apostolic Church of Britain.* 4th ed. enlarged. London and Oxford: A. R. Mowbray, 1927.

353

Maltwood, K.E. *A Guide to Glastonbury's Temple of the Stars.* London: James Clark, 1964.

—— *Airview Supplement to A Guide to Glastonbury's Temple of the Stars.* London: Watkins, 1937.

Robinson, Joseph Armitage. *The Times of Saint Dunstan.* Ford Lectures delivered at University of Oxford, 1922. Oxford: Clarendon Press, 1969.

Somerset Archaeological and Natural History Society. *Proceedings.* Vols. from 1907-1919 (except 1918).

Treharne, Reginald Francis. *Essays on Thirteenth-Century England.* London: the Historical Association, 1971.

Ussher, James, Archbishop of Armagh. "Glastonbury Traditions, Concerning Joseph of Arimathea," *Britannicarum Ecclesiarum Antiquitates.* (1639). Translated by H. Kendra Baker. London: Covenant, 1928.

Waite, Arthur Edward. *The Hidden Church of the Holy Grail: Its Legends and Symbolism.* London: Rebman, 1909.

Warner, Richard. *An History of the Abbey of Glaston and of the Town of Glastonbury.* Bath: R. Cruttwell, 1826.

II. THE EYES WHICH SEE EVERYTHING

Balcer, Witold. "Report of Experiment with Stefan Ossowiecki." 1935. Unpublished report.

Baltzly, Alexander, and Salomone, A. William. *Readings in Twentieth-Century European History.* New York: Appleton-Century-Crofts, 1950.

Besterman, Theodore. "An Experiment in 'Clairvoyance' with M. Stefan Ossowiecki." *Proceedings,* Society for Psychical Research, vol. 41. pp. 345-51.

Bordes, François. *The Old Stone Age.* Translated by J. E. Anderson. New York: McGraw-Hill, 1968.

Borzmowski, Andrzey. "Experiments with Ossowiecki." *International Journal of Parapsychology,* vol. 7, no. 3 (1965), pp. 259-84.

—— "Parapsychology in Poland." *International Journal of Parapsychology,* vol. 4, no. 4 (1962), pp. 59-74.

Buell, Raymond Leslie. *Europe: A History of Ten Years.* (With the Aid of the Staff of the Foreign Policy Association.) New York: Macmillan, 1928.

Dingwall, E.J. "An Experiment with the Polish Medium Stefan Ossowiecki." *Proceedings,* Society for Psychical Research, vol. 21 (1924), pp. 259-63.

Hartwig, Edward. *Warsaw.* (Poland): Sport I Turystyka, n.d.

Jazdzewski, Konrad. *Poland.* Translated by Maria Abramowicz and Robin Place. Ancient Peoples and Places Series. General editor, Glyn Daniel. New York and Washington: Praeger, 1965.

Karski, Jan. *Story of a Secret State.* Boston: Houghton Mifflin, 1944.

Ossowiecki, Stefan. *Swiat Mego Ducha: I Wizje Przyszlosci (My Spiritual World and Visions of the Future).* Rev. ed. by Marian Swida. Privately printed, 1976.

Ostrander, Sheila, and Schroeder, Lynn. *Psychic Discoveries Behind the Iron Curtain.* Englewood Cliffs, N.J.: Prentice-Hall, 1970.

——, editors. *The ESP Papers: Scientists Speak Out from Behind the Iron Curtain.* New York: Bantam, 1976.

Poniatowski, Stanislaw. *Parapsychological Probing of Prehistorical Cultures: Experiments with Stefan Ossowiecki (1939-1941).* Unpublished.

Richet, Charles. *Notre Sixième Sens.* Paris: Editions Montaigne n.d.

Rose, William John. *Poland Old and New.* London: G. Bell & Sons, 1948.

Stanislaw Poniatowski. *Wielka Encyklopedia Powszechna.* Warsaw: P.W.N. Panst Owe Wydawnictwo, 1967.

Stefan Ossowiecki. In *International Journal of Parapsychology,* vol. 7, no. 3 (1965), p. 275.

Wendt, Herbert. *In Search of Adam.* Translated by James Cleugh. Cambridge, Mass.: Riverside, 1956.

Wortz, E.C.; Bauer, A.J.; Blackwelder, R.F.; et al. "Novel Biophysical Information Transfer Mechanisms (NBIT)." Final Report: Document No. 76-13197. Airsearch Manufacturing Co. of California. 20 January 1975. Unpublished.

————"Biophysical Aspects of Parapsychology." Document No. 75-11096A. Airsearch Manufacturing Co. of California. 20 January 1975. Unpublished.

Zawadzka, M. Polish Clairvoyant Engineer Stefan Ossowiecki. *Goniec Warszawski.* Warsaw: n.p. 1937.

III. THE SCOTTISH GENERAL AND THE RUSSIAN RODWALKERS

Bird, Christopher. "Dowsing in the U.S.S.R.," *The American Dowser,* August 1972, pp. 110-120.

Bol'shaia Sovestskaia Entsyklopedia. 2nd ed. 51 vols. Moscow: Goseud Nauchnoe Izd-vo, 1949-1958.

Elliot, James Scott. "An Early Bronze Age Fire Pit at Townfoot Farm by Glencaple. *The Transactions of the Dumfriesshire and Galloway Natural History and Antiquarian Society,* 3rd ser., vol. 49 (1972), pp. 20-23.

———— "Report on the Excavation of Two Sites at Swinbrook Cottage, Swinbrook, Burford." Unpublished site report.

———— Chieveley Manor, Berkshire, Orchard Site. Unpublished field notes.

————Untitled field notes referring to site MR 74872160 at Nuriston Farm near Petersfield, Hampshire. With dowsing maps and excavation notations. Unpublished.

Harrison, W. "Detection of Graves and Underground Objects by Dowsing." *New Horizons* (Journal of the New Horizons Research Foundation incorporating transactions of the Toronto Society for Phsychical Research.) vol. 1, no. 4 (July 1974), pp. 155-159.

Hawkins, Gerald S., in collab. with John B. White. *Stonehenge Decoded.* New York: Dell Publishing Co., 1965.

Paulenkov, Florentii Fedorovich. *Entsiklopedicheskii Slovar.* St. Petersburg: Ehlikh, 1899.

Pluzhnikov, Aleksandr Ivanovich. "Possibilities for and Results of the Use of the Biophysical Method in Researching and Restoring Historical and Architectural Monuments." Translated by Cyril Muromcew and Christopher Bird. *The American Dowser,* August 1974, pp. 116-18.

Second International Congress on Psychotronic Research. Monte Carlo: The International Association for Psychotronic Research, 1975.

Second Scientific Technical Seminar on the Biophysical Effect (Document L61607). Moscow: KHOZU Typographers, U.S.S.R. Ministry of Communications, . 1971.

Tromp, S.W. "Review of the Possible Physiological Causes of Dowsing. *International Journal of Parapsychology*. vol. 10, no. 4 (1968), pp. 363-391.

United States Geologic Survey brochure. *Water Witching*. Washington, D.C.: Government Printing Office, 1966.

Vogt, E. von Zartman and Hyman, Ray. *Water Witching U.S.A.* Chicago: University of Chicago Press, 1959.

IV. THE TRANSITIONAL MAN

Emerson, J. Norman. "Intuitive Archaeology: A Psychic Approach," *New Horizons*. vol. 1, no. 3 (January 1974), p. 18.

———— "Intuitive Archaeology: A Developing Approach." November 1974. Unpublished report.

————"Intuitive Archaeology: The Argillite Carving." March 1974. Unpublished report.

————"Intuitive Archaeology and Related Matters." (Presented as introductory remarks at the 8th Annual Meeting, Canadian Archaeological Association) March 1975. Unpublished.

————"Intuitive Archaeology and Related Matters: Summary and Potential." (Presented as summary remarks at a special seminar on Psychic Archaeology, 8th Annual Meeting, Canadian Archaeological Association.) March 1975. Unpublished.

————"Archaeology, Parapsychology and One White Crow." April 1975. Unpublished report.

————"Psychic Archaeology." *Psychic Magazine*, September/October 1975, pp. 23-25.

————and Russell, Fr. William. "The Cahiague Village Palisade." A report submitted to the Archaeological and Historic Sites Board of the Province of Ontario. Toronto: 1965.

Pollack, Jack Harrison. *Croiset the Clairvoyant*, New York: Bantam, 1965 (orig. pub. Doubleday, 1964).

Proceedings of the Parapsychological Institute of the State University of Utrecht (Netherlands), no. 1 (December 1960).

Wright, J.V. *Ontario Prehistory: An Eleven-Thousand-Year Archaeological Outline*. Ottawa: National Museum of Man, 1972.

V. THE CANADIAN AND THE SEER'S SON

Arkell, A.J. *A History of the Sudan from the Earliest Times to 1821*. 2nd ed. London: University of London, 1961.

Belzoni, G. *Narrative of the Operations and Recent Discoveries within the Pyramids, Temples, Tombs, and Excavations in Egypt and Nubia; and of a Journey to the Coast of the Red Sea, in Search of the Ancient Berenice; and Another to the Oasis of Jupiter Ammon*. 2 vols. London: John Murray, 1821.

Black, Matthew. *The Scrolls and Christian Origins: Studies in the Jewish Background of the New Testament.* New York: Scribner's Sons 1961.

Bode, Framroze A. *Man Soul Immortality in Zoroastrianism.* Bombay: K.R. Cama Oriental Institute, 1960.

———*The Zoroastrian Doctrine of the Soul.* Unpublished doctoral dissertation, Los Angeles: The Philosophical Research Society, 1957.

Boylan, Patrick. *Thoth the Hermes of Egypt: A Study of Some Aspects of Theological Thought in Ancient Egypt.* London: Humphrey Milford, Oxford University Press, 1922.

Breasted, James Henry. *A History of Egypt: From the Earliest Times to the Persian Conquest.* London: Hodder & Stoughton, 1905.

———*The Dawn of Conscience.* New York: Scribner's Sons, 1939.

Budge, E.A. Wallis. *Osiris and the Egyptian Resurrection.* 2 vols. London: Philip Lee Warner, 1811.

———*The Gods of the Egyptians or Studies in Egyptian Mythology.* 2 vols. Chicago: Open Court, 1904.

———*The Papyrus of Ani* (The Book of the Dead). 2 vols. London: The Medici Society, 1913.

Burrows, Millar. *The Dead Sea Scrolls.* New York: Viking, 1955.

Ceram, C.W. *Hands on the Past: Pioneer Archaeologists Tell Their Own Story.* New York: Knopf (Borzoi), 1966.

———*Gods, Graves and Scholars: The Story of Archaeology.* Translated by E.B. Garside and Sophie Wilkins. Toronto: Bantam, 1972.

Dawson, Miles M. *The Ethical Religion of Zoroaster.* New York: Macmillan, 1931.

Denon, Vivant. *Voyage Dans La Basse et La Haute Egypt: pendant les Campagnes du General Bonaparte.* (My Journeys to Upper and Lower Egypt during the Campaigns of General Bonaparte.) 2 vols. London: Aux Frais et pour Compte De M. Peltier, 1802.

Dhalla, Manackji Nusservanji. *History of Zoroastrianism.* Bombay: K.R. Cama Oriental Institute, 1963.

———*Zoroastrian Civilization: from Earliest Times to the Downfall of the Last Zoroastrian Empire 651 A.D.* New York: Oxford University Press, 1922.

Dinshaw, Viccaji. *The Date and Country of Zarathustra: A Contribution to the Controversy.* Hyderabad-Deccan: A Venoogopaul Pillai & Sons (Printers), 1912.

Ebers, G. *Egypt: Descriptive, Historical and Picturesque.* Translated by Clara Bell. 2 vols. London: Cassell, Petter, Galpin, 1877-79.

Edwards, Amelia B. *Pharaohs, Fellahs, and Explorers.* New York: Harper & Brothers, 1891.

Emerson, J. Norman. "Intuitive Archaeology: Egypt and Iran." *The ARE Journal.* Vol XI, No. 2, March 1976.

Emery, Walter B. *Archaic Egypt.* Harmondsworth, Middlesex, England: Penguin, 1961.

Fakhry, Ahmed. *The Pyramids.* Chicago: The University of Chicago Press, 1961.

Fraser, J.G. *Adonis, Attis, Osiris.* 2 vols. New Hyde Park, N.Y.: University Books, 1961.

Furst, Jeffrey. *Edgar Cayce's Story of Jesus.* London: Spearman, 1968.

Goranson, Stephan. Unpublished report dealing with correlations between

Edgar Reading 5749-8 (Essene involvement with astrology, numerology, and phrenology) and Dead Sea Scrolls, 26 February 1975.

Jackson, A.V. Williams. *Persia Past and Present*. New York: Macmillan, 1909.

———*Zoroaster: Prophet of Ancient Iran.* New York: Columbia University Press, 1926.

Karaka, Dosabhai Framji. *History of the Parsis: Including Their Manners, Customs, Religion, and Present Position*. 2 vols. London: Macmillan, 1884.

Kittler, Glen D. *Edgar Cayce on the Dead Sea Scrolls*. Hugh Lynn Cayce, editor. New York: Paperback Library, 1970.

Lamberg-Karlovsky, C.C. & Martha. "An Early City in Iran." *Scientific American*. June 1971.

Lehner, Mark. "Reflections on a Tour." *The ARE Journal*. vol. 8, no. 2 (March 1973).

———*The Egyptian Heritage*. Virginia Beach, Virginia: ARE Press, 1974.

Maspero, Gaston, *Manual of Egyptian Archaeology and Guide to the Study of Antiquities in Egypt*. London: H. Grevel, 1895.

———*History of Egypt, Chaldea, Syria, Babylonia and Assyria*. Edited by A.H. Sayce. Translated by M.L. McClure. 13 vols. London: The Grolier Society, n.d.

Parkes, Henry Bamford. *Gods and Men: The Origins of Western Culture*. New York: Afred A. Knopf, 1959.

Perkins, Justin. *A Residence of Eight Years in Persia, Among the Nestorian Christians with Notices of the Muhammedans*. Andover: Allen, Morrill & Wardwell, 1843.

Petrie, William Flinders. *The Pyramids and Temples of Giza*. London: Field & Tuer, 1883.

———*Religion and Conscience in Ancient Egypt*. New York: Scribner's Sons, 1898.

———*A History of Egypt*. 6 vols. New York: Scribner's Sons, 1924.

Rawlinson, George. *Five Great Monarchies of the Ancient Eastern World*. 3 vols. New York: Scribner, Welford and Co., 1887.

Robinson, Lytle W. *The Story of Creation* (Monograph). Virginia Beach, Virginia: Edgar Cayce Publishing Co., 1959.

Roche, Richard (pseudo.). *Egyptian Myths and the Rafa Story*. Virginia Beach, Virginia: ARE Press, 1975.

Satterthwaite, Linton. "Thrones at Piedras Negras." *Bulletin, University of Pennsylvania Museum*, vol 7, no. 1 (1937).

———"Some Central Peten Maya Architectural Traits at Piedras Negras." *Los Mayas Antiguos*. Mexico: El Collegio de Mexico, 1941.

———"Piedras Negras Arch: Architecture." *Bulletin of the University of Pennsylvania Museum*. 1943-54. (A continuing series of reports on Satterthwaite's work at this site.)

———"Concepts and Structures of Maya Calendrical Arithmetics." Joint publication of University of Pennsylvania Museum and Philadelphia Anthropological Society, no. 3 (1947).

———"Radiocarbon Dates and the Maya Correlation Problem." *American Antiquity*, vol 21, 1956.

———"Maya Long Count." *El Mexico Antiguo*, vol. 9 (1961).

————and Ralph, E.K. "Radiocarbon Dates and the Maya Correlation Problem." *American Antiquity,* vol. 26 (1960).

Schance, Don A. "Woman 'Recalls' Life with Pharaohs." *Los Angeles Times.* 2 November 1977, pt. 1, p. 16.

Smyth, C. Piazzi. *Life and Work at the Great Pyramid.* 3 vols. Edinburgh: Edmonston and Douglas, 1867.

Talmon, Shemaryaho. "The New Covenanters of Qumran." *Scientific American,* vol 225, no. 5 (November 1971).

Vyse, Howard. *Operations Carried on at the Pyramids of Gizeh in 1837: with an Account of a Voyage into Upper Egypt.* 3 vols. London: James Fraser, 1840.

Wauchope, Robert. *Lost Tribes and Sunken Continents: Myth and Method in the Study of American Indians.* Chicago: The University of Chicago Press, 1962.

————and Wiley, Gordon R., *Handbook of Middle American Indians.* vol 3, pt. 2. (Archaeology of Southern Mesoamerica.) Austin: University of Texas Press, 1965.

West, E.W., Ed., *Avesta Pahlavi and Ancient Persian Studies.* (Memorial Volume to Shams-Ul-Ulama Dastur Peshotanji Behramji Sanjana by European Academics) Bombay, 1899.

Wilkinson, J.G. *Manners and Customs of the Ancient Egyptians.* 3 vols. London: John Murray, 1827.

————*Manners and Customs of the Ancient Egyptians.* (Second Series) 2 vols and suppl. London: John Murray, 1841.

Wortham, John David. *The Genesis of British Egyptology 1549-1906.* Norman, Oklahoma: University of Oklahoma Press, 1971.

VI. CHILDREN OF THE CHANGE

Barnouw, V. "Siberian Shamanism and Western Spiritualism." *Journal of the American Society for Psychical Research,* vol. 36 (1942).

Biggar, H.P., editor. *The Works of Samuel de Champlain.* A series of 6 vols. by various translators. Toronto: Champlain Society of Toronto, 1922—. Vol. 1, 1922; vol 2, 1925; vol. 3, 1929; vol. 4, 1932; vol. 5, 1933; vol. 6, 1936; and a vol. of plates and maps, n.d.

Boyd, Douglas *Rolling Thunder.* New York: Delta, 1974.

The Compact Edition of the Oxford English Dictionary. Oxford: Oxford University Press, 1971.

Emerson, J. Norman. "Archaeology, Parapsychology and One White Crow." Unpublished paper, April 1975.

Heidenreich, Conrad. *Huronia: A History and Geography of the Huron Indians 1600-1650.* Ontario: McClelland & Stewart Ltd., 1971.

Morison, Samuel Eliot. *Samuel de Champlain: Father of New France.* Boston: Atlantic Monthly Press, 1972.

Reid, C.S. "Psychometrics and Settlement Patterns: Field Tests on Two Iroquoian Sites." Unpublished paper, n.d.

Tuck, James A. "The Iroquois Confederacy." *Scientifc American,* vol. 224, no. 2 (February 1971).

Weiant, Clarence W. "Parapsychology and Anthropology." *Manas,* vol. 13, no. 15 (1960).

Wright, J.V. *The Ontario Iroquois Tradition.* Bulletin 210. Ottawa: National Museum of Canada, 1966.

———*Ontario Prehistory: An Eleven-Thousand-Year Archaeological Outline.* Ottawa: National Museum of Man, 1972.

VII. INTO THE AMERICAN MAINSTREAM

Eisenbud, Jule. *The World of Ted Serios: "Thoughtographic" Studies of an Extraordinary Mind.* New York: Morrow, 1967.

Essays in Historical Anthropology of North America. (Published in honor of John R. Swanton). Publication 3588. Washington: Smithsonian Institution, 25 May 1940.

Jones, David E. "An Experiment with Non-Scientific Discovery Procedures in Archaeology." *Phoenix,* vol. 1, no. 2 (Winter 1977).

———"Folsom Ethnography." *Phoenix,* vol. 1, no. 3 (Summer 1978).

Long, Joseph K. "Paranormal Events: Toward a Comprehensive Theory." Unpublished paper delivered at 72nd annual meeting of AAA, 1973.

———"Introduction to and Resume of Ethnographic Literature on Parapsychology." Unpublished paper delivered at 73rd annual meeting of AAA, November 1974.

———"Psi: The Scientific Debate." 1974. Unpublished paper.

———"Shamanism: Trance, Hallucinogens, and Psychical Events." In, *Realm of the Extra-Human,* A. Bharati, editor. The Hague: Mouton, 1976. (First presented as a paper at IXth International Congress of Anthropological and Ethnological Sciences, 27 August–8 September 1973).

———*Extrasensory Ecology: Parapsychology and Anthropology.* Metuchen, N.J.: Scarecrow, 1977. (This book presents the papers of the Rhine-Swanton Symposium on Parapsychology and Anthropology at the 1974 annual meeting of the AAA. However, papers are included which were not actually presented at that time, and many papers have been subsequently altered, some almost beyond recognition, for publication in this book.)

Rhine-Swanton Symposium on Parapsychology and Anthropology Transcript, November 1974 annual meeting of AAA. (For a published version of these papers see Long, Joseph K. *Extrasensory Ecology: Parapsychology and Anthropology* above.)

Staniford, Philip. "Inside Out: Anthropological Communication of Alternative Realities." *Phoenix,* vol. 1, no. 1 (Summer 1977).

Stirling, Matthew W. "Discovering the New World's Oldest Dated Work of Man." The *National Geographic* magazine, vol. 76, no. 2, August 1939.

Swanton, John Reed. "Superstition—But Whose?" Open letter to all listed fellows of the American Anthropological Association. n.d. but circa 1953.

Weiant, Clarence W. *An Introduction to the Ceramics of Tres Zapotes, Veracruz, Mexico.* Bulletin 139. Washington: Smithsonian, Bureau of American Ethnology, 1943.

———"Parapsychology and Anthropology." *Manas,* vol. 13, no. 15, (1960).

VIII. KUHN, CONTEXT, AND REVOLUTION

Burtt, Edwin Arthur. *The Metaphysical Foundation of Modern Physical Science.* New York: Harcourt, Brace, 1925.

Dreistadt, Roy. "The Prophetic Achievements of Geniuses and Types of Extrasensory Perception." *Psychology,* vol. 8, no. 2 (1971).

Jeans, James. *The New Background of Science.* Cambridge: Cambridge University Press, 1933.

Koestler, Arthur. *The Roots of Coincidence.* New York: Random House, 1972.

Kuhn, Thomas S. *The Copernican Revolution: Planetary Astronomy in the Development of Western Thought.* Cambridge: Harvard University Press, 1957.

———*Structure of Scientific Revolutions.* Chicago: Phoenix-The University of Chicago Press, 1962.

Mihalasky, John. "Extrasensory Perception in Management." *Advanced Management Journal,* July 1967.

———"ESP: Can It Play a Role in Idea-Generation?" *Mechanical Engineering,* December 1972.

Stent, Gunther S. "Prematurity and Uniqueness in Scientific Discovery." *Scientific American,* December 1972.

IX. PREJUDICE, PAIN, AND PARADIGM

Adams, Robert McC. "Some Hypotheses on the Development of Early Civilizations." *American Antiquity,* vol. 21, no. 3 (1956).

Allen, William L. and Richardson, James B. III. "The Reconstruction of Kinship from Archaeological Data: The Concepts, Methods, and the Feasibility." *American Antiquity,* vol. 36, no. 1 (1971).

———*An Universal History from the Earliest Account of Time.* Compiled from original authors. London: T. Osborne (Printer), 1754.

Anderson, Keith M. "Ethnographic Analogy and Archaeological Interpretation." *Science,* vol. 163, no. 3863 (1969).

Ascher, Robert. "Analogy in Archaeological Interpretation." *Southwestern Journal of Anthropology,* vol. 17, no. 4 (1961).

Atwater, Caleb. "Descriptions of the Antiquities Discovered in the State of Ohio and Other Western States." *Transactions and Collections of the American Antiquarian Society,* vol. 1 (1820).

Bacon, Edward. *Digging for History: Archaeological Discoveries throughout the World 1945-1959.* New York: John Day, 1960.

Bannister, Bryant. "Dendrochronology." *Science in Archaeology.* D. Brothwell and E. Higgs, editors. London: Thames, 1963. (Revised and published by Praegar, 1969).

Binford, Lewis R. "Archaeology as Anthropology." *American Antiquity,* vol. 28, no. 2 (1962).

———"Some Comments on Historical Versus Processual Archaeology." *Southwestern Journal of Anthropology,* vol. 24, no. 3 (1968).

Caldwell, Joseph R. "The New American Archaeology." *Science,* vol. 129, no. 3345 (1959).

Ceram, C.W., (pseudonym for Kurt W. Marek) editor, *Hands on the Past: Pioneer Archaeologists Tell Their Own Story.* New York: Knopf, 1966.

Chamberlain, Alexander F. "Thomas Jefferson's Ethnological Opinions." *American Anthropologist,* (new series) vol. 9 (1907).

Childe, V. Gordon. "A Prehistorian's Interpretation of Diffusion." *Independence, Convergence, and Borrowing in Institutions, Thought, and Art.* Cambridge: Harvard University Press, 1937.

————*The Dawn of European Civilization.* 2nd ed. New York: Knopf, 1939.

Cook, Sherburne F. "Physical Analysis as a Method for Investigating Prehistoric Habitation Sites." *University of California Archaeological Survey Report No. 7.* Berkeley, 1950.

————"Dating Prehistoric Bone by Chemical Analysis." *Viking Fund Publications in Anthropology,* no. 28 (1960).

Daniel, Glyn E. *A Hundred Years of Archaeology.* London: Gerald Duckworth, 1950.

Darwin, Charles. *The Descent of Man and Selection in Relation to Sex.* 2 vols. New York: D. Appleton, 1872.

————*On the Origin of the Species: by Means of Natural Selection, or the Preservation of Favored Races in the Struggle for Life.* New York: D. Appleton, 1873.

"Dendrochronology." *Enclyclopedia Britannica.* vol. 7. 1973 ed.

Dunnell, Robert C. "Seriation Method and its Evaluation." *American Antiquity,* vol. 35, no. 3 (1970).

Fagan, Brian M., editor. *Introductory Readings in Archaeology.* Boston: Little, Brown, 1970.

Figgins, Jesse D. "The Antiquity of Man in America." *Natural History,* vol. 27, no. 3 (1927).

Fritz, John M. and Plog, Fred T. "The Nature of Archaeological Explanation." *American Antiquity,* vol. 35, no. 4 (1970).

Gwynne, Peter with Michaud, Stephen G., "The Roots of Man," *Newsweek,* 31 May 1976.

Hawkes, C.F.C. "Archaeological Theory and Method: Some Suggestions from the Old World." *American Anthropologist,* vol. 56, no. 1 (1954).

Hawkins, Gerald S. *Stonehenge Decoded.* New York: Dell Publishing Co., 1965.

Jaspers, Karl. *The Origin and Goal of History.* New Haven: Yale University Press, 1953.

Jefferson, Thomas. *The Works of Thomas Jefferson.* Paul Leicester Ford, editor. 12 vols. New York: Putnam's (Federal edition), 1904.

Johnson, Edgar. *Sir Walter Scott: The Great Unknown.* 2 vols. New York: Macmillan, 1970.

Kluckhohn, Clyde. "The Conceptual Structure in Middle American Studies." *The Maya and Their Neighbors.* Clarence L. Hay, editor. New York: D. Appleton-Century (Limited Editon), 1940.

Kuhn, Thomas S. *The Structure of Scientific Revolutions.* Chicago: Phoenix-The University of Chicago Press, 1962.

Longacre, William A. "Archaeology as Anthropology: A Case Study." *Science,* vol. 144, no. 3625 (1964).

Lowie, Robert H. *The History of Ethnological Theory.* New York: Holt, Rinehart and Winston, 1937.

————"Reminiscences of Anthropological Currents in America Half a Century Ago." *American Anthropologist,* vol. 58, no. 6 (1956).

Lucretius, *The Nature of Things*. Translated by John Selby Watson. London: Bohn, 1851.

MacNeish, Richard S. "Early Man in the Andes." *Scientific American*, vol. 224, no. 4 (April 1971).

Marshack, Alexander. *The Roots of Civilization: The Cognitive Beginnings of Man's First Art, Symbol and Notation*. New York: McGraw-Hill, 1972.

Meggers, Betty and Evans, Clifford. "Review of 'Method and Theory in American Archaeology.' " *American Antiquity*, vol. 24, no. 2 (1958).

Mitra, Panchanan. *A History of American Anthropology*. Calcutta: Calcutta University Press, 1933.

Petrie, William M. "Sequences in Prehistoric Remains," *Journal of the Royal Anthropological Institute of Great Britain and Ireland*, vol. 29 (1899).

Penniman, T.K. *A Hundred Years of Anthropology*. London: Gerald Duckworth, 1935.

Ralegh, Walter. *The Historie of the World*. In 5 books. London: Walter Burre, 1614.

Rapport, Samuel and Wright, Helen, Eds. *Archaeology*. New York: Washington Square Press, 1964.

Rouse, Irving. "The Classification of Artifacts in Archaeology." *American Antiquity*, vol. 25, no. 3 (1960).

Rowe, John H. "Stratigraphy and Seriation." *American Antiquity*, vol. 26, no. 3 (1961).

——"Stages and Periods in Archaeological Interpretation." *Southwest Journal of Anthropology*, vol. 18, no. 1 (1962).

———"The Renaissance Foundations of Anthropology." *American Anthropologist*, vol. 67, no. 1 (1965).

——"Diffusionism and Archaeology." *American Antiquity*, vol. 31, no. 3 (1966).

Swanson, Earl H. Jr. "Theory and History in American Archaeology." *Southwestern Journal of Anthropology;* vol. 15 (1959).

Taylor, Walter W. *A Study of Archeology*. Carbondace and Edwardsville, Ill.: Southern Illinois University, 1967.

Toynbee, Arnold. *An Historian's Approach to Religion*. London: Oxford University Press, 1956.

Van Loon, Hendrik W. *The Arts*. New York: Simon and Schuster, 1937.

Wauchope, Robert. *Lost Tribes and Sunken Continents: Myth and Method in the Study of American Indians*. Chicago: The University of Chicago Press, 1962.

Wendt, Herbert. *In Search of Adam: The Story of Man's Quest for the Truth about His Earliest Ancestors*. Translated by James Cleugh. Boston: Houghton Mifflin, 1956.

Willey, Gordon R. and Phillips, Philip. *Method and Theory in American Archaeology*. Chicago: The University of Chicago Press, 1958.

——and Sabloff, Jeremy A. *A History of American Archaeology*. San Francisco: W.H. Freeman, 1973.

X. COMPLEXITIES, PROMISE, AND IMPLICATIONS

Beauregard, O. de. "Time Symmetry and Interpretation of Quantum Mechanics." *Foundations of Physics*. Lecture delivered at Boston Colloquium for the Philosophy of Science, February 1974.

Bohm, David. "Quantum Theory as an Indication of a New Order in Physics: Part A, the Development of New Orders as Shown Through the History of Physics." *Foundations of Physics*, vol. 1, no. 4, 1971. "Part B: Implicate and Explicate Order in Physical Law," vol. 3, no. 2, 1973.

Bond, Frederick Bligh. *The Gate of Remembrance: The Story of the Psychological Experiment which Resulted in the Discovery of the Edgar Chapel at Glastonbury.* 2nd ed., Oxford: Basil Blackwell, 1918.

Broglie, Louis de. *The Revolution in Physics: A Non-mathematical Survey of Quanta.* Translated by Ralph W. Niemeyer. New York: Noonday, 1953.

Burtt, Edwin A. *The Metaphysical Foundations of Modern Physical Science.* New York: Harcourt, Brace, 1925.

Capra, Fritjof. *The Tao of Physics: An Exploration of the Parallels between Modern Physics and Eastern Mysticism.* Berkeley: Shambhala, 1975.

Chauvin, Rémy. "To Reconcile Psi and Physics." *The Journal of Parapsychology,* vol. 34, no. 3 (1970).

Clark, Ronald W. *Einstein: The Life and Time.* New York: World, 1971.

Darwin, C.G. *The New Conceptions of Matter.* New York: Macmillan, 1931.

Emerson, J. Norman. "Intuitive Archaeology and Related Matters: Summary and Potential." Unpublished paper presented at a meeting of the Canadian Archaeological Association, March 1975.

Forwald, Haakon. *Mind, Matter and Gravitation: A Theoretical and Experimental Study.* New York: Parapsychology Foundation, 1969.

Hall, Manly P. *Man: Grand Symbol of Mysteries.* Los Angeles: Philosophical Research Society, 1972.

Harman, Willis. *An Incomplete Guide to the Future.* San Francisco: San Francisco, 1976.

Heisenberg, Werner. *The Physical Principles of Quantum Theory.* Chicago: University of Chicago Press, 1930.

——*Philosophic Problems of Nuclear Science.* London: Faber and Faber, 1952.

Jeans, James. *The Mysterious Universe.* Cambridge: Cambridge University Press, 1930.

——*The New Background of Science.* Cambridge: Cambridge University Press, 1933.

——*Physics and Philosophy.* Cambridge: Cambridge University Press, 1943.

Karagulla, Shafica. *Breakthrough to Creativity.* Los Angeles: De Vorss, 1967.

Koestler, Arthur. *The Roots of Coincidence,* New York: Random House, 1972.

Krippner, Stanly and Rubin, Daniel, editors, *Galaxies of Life: The Human Aura in Acupuncture and Kirilian Photography.* New York: Interface, 1973.

Kuhn, Thomas S. *The Structure of Scientific Revolutions.* Chicago: Phoenix-University of Chicago Press, 1962.

Kuhns, Richard. *Structures of Experience: Essays on the Affinity between Philosophy and Literature.* New York: Harper Torchbooks, 1974.

Le Shan, Lawrence. *Toward a General Theory of the Paranormal.* New York: Parapsychology Foundation, 1969.

Long, Joseph K. *Extrasensory Ecology: Parapsychology and Anthropology.* Metuchen, N.J.: Scarecrow, 1977.

Love, Jeff. *The Quantum Gods: The Origin and Nature of Matter and Consciousness.* Tisbury, Wiltshire, England: Compton Russell Element, 1976.

McConnell, R.A. "ESP and Creditability in Science." *American Psychologist,* vol. 24, no. 5 (May 1969).

Mishlove, Jeffrey. *The Roots of Consciousness: Psychic Liberation through History, Science and Experience.* New York: Random-Bookworks, 1975.

Mitchell, Edgar D. *Psychic Exploration: A Challenge for Science.* John White, editor, New York: Putnam's, 1974.

Oyle, Irving. *Time Space and Mind.* Millbrae, California: Celestial Arts, 1976.

Penfield, Wilder. *The Mystery of the Mind: A Critical Study of Consciousness and the Human Brain.* Princeton: Princeton University Press, 1975.

Persinger, M.A.; Ludwig, H.W., and Ossenkopp, K-P. "Psychophysical Effects of Extremely Low Frequency Electromagnetic Fields: A Review. *Perceptual and Motor Skills.* Monograph Supplement 3-V36, 1973.

Planck, Max. K.E.L. *The Universe in the Light of Modern Physics.* London: Allen and Unwin, 1937.

———*Scientific Autobiography and Other Papers.* Translated by F. Gaynor. New York: Philosophical Library, 1949.

Pollack, Jack Harrison. *Croiset the Clairvoyant.* New York: Bantam, 1965.

Poniatowski, Stanislaw. "Clairvoyance and the Method of its Application for Scientific Research." Unpublished manuscript apparently part of a greater work by the same author, *Parapsychological Probing of Pre-Historical Cultures: Experiments with Stefan Ossowiecki (1937-1941).*

Puthoff, Harold E. and Targ, Russell E. "A Perceptual Channel for Information Transfer over Kilometer Distances: Historical Perspective and Recent Research." *Proceedings of the Institute for Electrical and Electronic Engineering,* vol. 64, no. 3 (1976).

Rapoport, Anatol. *Science and the Goals of Man; A Study in Semantic Orientation.* New York: Harper & Brothers, 1950.

Rhine, Joseph Banks. *New World of the Mind.* London: William Sloane, 1953.

Rose, Steven. *The Conscious Brain.* New York: Knopf, 1973.

Scientific American Editors. *Scientific American Resource Library: Readings on Altered States of Awareness.* San Francisco: W.H. Freeman, 1971.

Smith, C.V.M. *The Brain: Towards an Understanding.* New York: Putnam's, 1970.

Swann, Ingo. *To Kiss Earth Goodbye.* New York: Hawthorn, 1975.

Targ, Russell and Puthoff, Harold E. "Information Transfer under Conditions of Sensory Shielding." *Nature,* vol. 252, (October 1974).

———*Mind Reach: Scientists Look at Psychic Ability.* New York: Delacorte/ Eleanor Friede, 1977.

Tart, Charles. "Card Guessing Tests: Learning Paradigm or Extinction Paradigm?" *Journal of the American Society for Psychical Research;* vol. 60, no. 1 (January 1966).

Tart, Charles, Ed., *Altered States of Consciousness.* New York: John Wiley, 1969.

Toynbee, Arnold. *An Historian's Approach to Religion.* London: Oxford University Press, 1956.

Walker, Evan Harris. "Application of the Quantum Theory of Consciousness to the Problem of Psi Phenomena." *Proceedings,* Fifteenth Annual Convention of Parapsychological Association, 1972.

———"Consciousness in the Quantum Theory of Measurement" (pt. 1 of series), *Journal for the Study of Consciousness*, vol. 5 (1972).

———"The Nature of Consciousness: The Relation to the Foundations of Quantum Mechanics and its Philosophical Implication." Unpublished paper, presented at SIMS Conference, University of Massachusetts, 1971.

Weil, Andrew. *The Natural Mind: A New Way of Looking at Drugs and the Higher Consciousness*. Boston: Houghton Mifflin, 1972.

INDEX

A

Abrahamsen, Aron, 300-301, 309
Adams, Frank O., 184
Aerial surveying, 280
Arthur, King, 6
Asafetida bags, 208-209
Astral projection, 67
Atkinson, R. J. C., 279-80
Atlantis, 168-69, 173, 279
Auras, 59, 62
Automatic writing, 1-5, 12-14, 48-55
Avalon, Isle of, 6

B

Balcer, Witold, 74-76, 78, 81-82, 96
Bartlett, Capt. John Allen, 1-5, 12-14, 27, 42
Beere, Abbot Richard, 5, 31, 33-34
Berckhemer, Dr., 84
Bernal, Dr. Ignacio, 237
Besterman, Theodore, 67
Bharati, Agehananda, 241, 243-45
Binford, Lewis R., 282
Biophysical effect, 128-29, 133, see also Dowsing
Bird, Christopher, 126-29
Bishop of Bath and Wells, 10, 11-12
Blake, William, 31
Blomfield, Sir Arthur, 8
Bode, Framroze, 185
Boltuc, Maria, 98
Bond, Frederick Bligh, 1-56, 284
  automatic writing and, 2-5, 12-14, 28-34
  childhood of, 7-8
  Glastonbury excavation and, 15-20, 24, 26
  marriage of, 39-40, 49
Bond, Rev. Frederick Hookey, 7
Bryant, Johannes, 14, 15, 34-36, 50-52

C

Cannon, John, 25, 26
Camelot, 6

Canterbury, Archbishopric of, 6, 30
Carbon-14 dating, 26, 175, 277-78
Caroe, W.D., 12, 37-40
Carrington, Hereward, 223
Catastrophe Theory, 268
Cayce, Edgar, 139, 165-75, 312-13
  Egyptian history and, 172-81, 191-93
  Iranian history and, 182-86, 193-96
Cayce, Hugh Lynn, 165-68, 171, 175, 182, 187-97
Celts, 5-6
Cencellarius, Radulphus, 18-20, 29
Childe, V. Gordon, 274-76, 277-78
Children, sensitive, 8
Christianity
  Celtic church and, 5-6
  Henry VIII and, 7, 30-31
  historical archaeology and, 265-70
Clairvoyance, 60-61, 62, 65, 70-72, 99
Clairaudience, 207
Communication, Scientific, 251-54
Company of Avalon, The, 27-28
Conway, Sheila, 154-56, 163-64, 203-11
Copernicus, Nicholas, 72-73
Cremation, Bronze Age, 115-16
Croiset, Gerard, 140, 142-45
Cromwell, Thomas, 31
Crowe, Catherine, 8
Cultural evolution, 282-83
Cuvier, Georges, 268, 270-71

D

Darwin, Charles Robert, 268, 269
Darwin, Erasmus, 269
Dead Sea Scrolls, 170-71
Decline effect, 296-97
De la Carrière, Alietta, 92-93
Dendrochronology, 277-78
Dingwall, E. J., 67
Douglas, A. E., 278
Dowsing, 33, 91, 116
  map, 108-27
  Russian science and, 128
Doyle, Sir Arthur Conan, 44
Dunstan, Saint, 6

E

Eady, Dorothy Louise, 193
Eawulf of Edgarley, 18-20, 29
Edgar, King, 5, 6
Edgar Chapel, 5, 13, 14, 15, 24-25, 33-34, 42
  Gematria and, 38-39
Edmund, King, 6
Edmund Ironside, King, 6
Efimovich, Grigori, 64
Eggstones, 22-23
Egypt, 172-81, 191-93
Eisenbud, Dr. Jule, 224, 242, 244
Elizabeth I, 7
Elliot, James Scott, 108-27
Emic, 42, 243-44
Emerson, Ann, 139, 153, 160-63, 190
Emerson, J. Norman, 136-40, 145-68
  H.L. Cayce and, 187-97
  C. Garrad and, 202-203
  C.S. Reid and, 211, 213, 220
  J. Long and, 241, 243
Essenes, 170-71
Etic, 243-44
Evolution, 268-71
Eyes Which See Everything, The, 93

F

Farelly, Frances, 134
Ferguson, Charles W., 277-78
Fielding, Everard, 40-41, 42
Fossilization, 267
Francis, Rev. William H., 45

G

Garrad, Charles, 198-211
  S. Conway and, 203-11
  J.N. Emerson and, 202-203
Gate of Remembrance, The, 26, 28, 40-41, 47, 55
Geley, Gustave, 67, 68
Gematria, 25-26, 38
Glastonbury Abbey
  automatic writing and, 1-5, 12-14, 28-34
  excavation of, 15-20, 24, 26
  history of, 5-7
Glastonbury Scripts, The, 27-28
Godunov, Boris Fedorovich, 132-33
Gravier, Dr. M., 67
Gulielmus Monachus, 3-5, 34

H

Hanson, Charles, F., 8
Hawkins, Gerald S., 279
Henry VII, King, 33
Henry VIII, King, 7, 30-31
Henser, Witold, 81
Herlewin, Abbot, 28
Historical archaeology, 265-71
Hope, Lord Charles, 67
Hugh, Bishop of Lincoln, 27

I

Ine of Wessex, King, 6
Iran, history of, 182-86, 193-96

J

Jefferson, Thomas, 273-74
Jones, David E., 245
Jonky, Dionizy, 70-72
Joseph of Arimathea, 5-6, 27

K

Kamienski, Michael, 81, 82
Kluckhohn, Clyde, 276
Kosieradzki, 82
Kuhn, Thomas S.
  philosophy of science of, 247-67

L

Latin, monastery, 52
Lawicki, Ludivik, 99
Leclerc, Louis, 268
Lehner, Mark, 176-77, 190, 193
Libby, Willard Frank, 277
Lightfoot, Peter, 29-30
Limitation of standard knowledge, 311-12
Lloyd, Philip, 27
Location, problem of, 279-81, 285
Long, Joe, 238-41, 245-46, 284
Long, Col. William, 25, 26, 41
Loretto Chapel, 24-25, 33-34, 55
Lukasiewicz, Jan, 81

M

McCullough, Jean, 167
McGimsey, Charles, 283-84, 285

McMullen, George, 139, 146-52, 159-60, 161-63
  H.L. Cayce and, 181, 187-97
  C.S. Reid and, 213-21
McMullen, Lottie, 139
Manczarski, Stefan, 81
Map dowsing, 116-18, 123-27
Marshack, Alexander, 279
Marston, Dr. Alvan Theophilus, 84
Mead, Margaret, 237, 248, 296-97
Medicine men, Indian, 205
Miller, Jack, 150-52
Mills, Mary Louis, 10, 39-40

N

Negahban, Ezat O., 182, 187, 195-96
Norwina-Sapinski, Andrew, 88

O

Observer effect, 297-99
Ogden, Arch, 188
Ossowiecki, Stephen, 57-107, 284
  childhood of, 58-61
  marriage of, 92-93
  Poniatowski and, 79-92, 96-104
Ossowiecki, Zofia Swida, 57, 93, 104-107

P

Paradigm, scientific role of, 249-51, 253-60
Petrie, William M. Flinders, 274-75
Pitt-Rivers, Augustus Henry Lane-Fox, 9
Pluzhnikov, Aleksandr Ivanovich, 129-35
Podkiwinska, Zofia, 99
Poland, occupation of, 93-96, 99-100, 104-105
Pollen analysis, 25-27, 149-50, 175
Poniatowski, Stanislaw, 79-92, 96-107, 284
Post molds, 118-19
Pottery dating, 275
Psychic archaeology
  F.B. Bond and, 5, 7, 12-14, 17-19, 27-28, 40-42, 47-55
  J.N. Emerson and, 136-40, 145-68
  C. Garrad and, 198-203
  need for, 284-89, 313-15

  orthodox science and, 271-72, 290, 314-15
  S. Ossowiecki and, 72-73, 75-78, 81-92, 96-104
  A. I. Pluzhnikov and, 129-35
  S. Poniatowski and, 79-92, 96-104
  qualifications of, 291-313
Psychic Science, 43
Psychometry, 63, 69-70, 139-40,
  see also Psychic archaeology *and* Time travel

Q

Qumran, 170-71

R

Racial memory, 75
Radiation, perception of, 129
Radlinski, Stefan, 82, 96
Raleigh, Sir Walter, 268-69
Rasputin, 64
Rayevskii Redoubt, 130-31
Reconstruction, problem of, 281-82, 285-86
Reid, C S., 158, 211-21
  and G. McMullen, 213-21
Reutiner, A., 67
Rhine, Dr. J. B., 81, 241-42
Rhine, Dr. Louisa, 81
Richet, Dr. Charles R., 67, 68, 78
Reformation, 30, 31
Robinson, Dean Joseph Armitage, 37-40, 43
Roche, Richard (pseudonym), 177-81
Rodwalking, 129
Rolling Thunder, 205
Roodscreens and Roodlofts, 9-10
Rossi, Josephine, 223-24
Royal Institute of British Architects, 9

S

Saint Dunstan's Chapel, 24-25
St. Michael Chapel, 41
Schliemann, Heinrich, 180, 279
Science, philosphy of, 247-67
Séance, 48
Sensitives, 48, 70, 97
  condition of, 294, 295
  use of, 291-313

Seriation, 175
Sethe, K., 180
Shaw, Thomas, 191
Shelly, Violet, 178
Somerset Archaeological and Natural
   History Society, 9, 10-11, 38
Smith, Hester Travers, 27
Staniford, Philip, 245
Starocherkassk, 134
Steiner, Dr. Rudolf, 74-75, 169
Steinheim man, 84
Stirling, Matthew W., 225-26, 233
Stonehenge, 123, 279
Stratigraphy, 175
Strong, William Duncan, 225, 233
Structure of Scientific Revolutions, The,
   248
Stulginski, Antoni, 61, 62
Subconscious, 63
Superconsciousness, 63, 74, 92
Survivalism, 70
Swann, Ingo, 295-96, 298
Swanton, John Reed, 235-37
Swida, Marian, 66

T

Taylor, Walter W., 276-77
Tegoma, Emilio, 226-28, 233
Telekinesis, 60-61, 64, 65, 97
Tenhaeff, Dr. W. H., 140-43
Telepathy, 69, 70, 72, 206-207
Thermoluminescence, 26
Thurstan, Abbot, 18, 19
Time Travel, 72-75, 82-92, 96-104
Trance, 48
Tres Zapotes, 225-35
Turner, Gladys Davis, 167

Tyyska, Allen, 156-57

U

Ussher, James, 267
Ussher, W.A.E., 23

V

Valkhoff, Dr. Marius, 140-45
Vasiliev, Leonid, 129
Van Riet Lowe, C. 144-45
Von Schrenck-Notzing, Dr., 67
Vyazemy, Bol'shie, 132-33
Vyse, Colonel Richard W.H., 175-76

W

Walker, Dr. Evan Harris, 243, 295
Wallace, Alfred Russel, 268
Wallace, Ronald L., 245
Westminster Abbey, 30
Weiant, Clarence W., 207, 223, 237,
   246
   Mexican excavation and, 224-35
Whiting, Abbot, 5
Whiting, Richard, 31
Willey, Gordon R., 282, 283
Wolsey, Cardinal Thomas, 31
World of My Spirit and Visions of the
   Future, The, 88
Wrobel, 61-63, 65, 80, 95
Wyandot Indians, 198-206

Z

Zoroastrianism, 183-86, 193

Hampton Roads Publishing Company

. . . for the evolving human spirit

HAMPTON ROADS PUBLISHING COMPANY publishes books on a variety of subjects, including metaphysics, spirituality, health, visionary fiction, and other related topics.

We also create on-line courses and sponsor an *Applied Learning Series* of author workshops. For a current list of what is available, go to www.hrpub.com, or request the ALS workshop catalog at our toll-free number.

For a copy of our latest trade catalog, call toll-free, 800-766-8009, or send your name and address to:

HAMPTON ROADS PUBLISHING COMPANY, INC.
1125 STONEY RIDGE ROAD • CHARLOTTESVILLE, VA 22902
e-mail: hrpc@hrpub.com • www.hrpub.com